Helicopter Flying Handbook

First published in 2019
First Skyhorse Publishing edition 2021

Skyhorse Publishing books may be purchased in bulk at special discounts for sales promotion, corporate gifts, fund-raising, or educational purposes. Special editions can also be created to specifications. For details, contact the Special Sales Department, Skyhorse Publishing, 307 West 36th Street, 11th Floor, New York, NY 10018 or info@skyhorsepublishing.com.

Skyhorse® and Skyhorse Publishing® are registered trademarks of Skyhorse Publishing, Inc.®, a Delaware corporation.

Visit our website at www.skyhorsepublishing.com.

10 9 8 7 6 5 4 3 2 1

Library of Congress Cataloging-in-Publication Data is available on file.

Cover design by Federal Aviation Administration

Print ISBN: 978-1-5107-6720-1
eBook ISBN: 978-1-5107-6721-8

Printed in China

Helicopter Flying Handbook

FAA-H-8083-21B

2019

Skyhorse Publishing

Preface

The Helicopter Flying Handbook is designed as a technical manual for applicants who are preparing for their private, commercial, or flight instructor pilot certificates with a helicopter class rating. Certificated flight instructors may find this handbook a valuable training aid, since detailed coverage of aerodynamics, flight controls, systems, performance, flight maneuvers, emergencies, and aeronautical decision-making is included. Topics such as weather, navigation, radio navigation and communications, use of flight information publications, and regulations are available in other Federal Aviation Administration (FAA) publications.

This handbook conforms to pilot training and certification concepts established by the FAA. There are different ways of teaching, as well as performing, flight procedures and maneuvers, and many variations in the explanations of aerodynamic theories and principles. This handbook adopts a selective method and concept to flying helicopters. The discussion and explanations reflect the most commonly used practices and principles. Occasionally the word "must" or similar language is used where the desired action is deemed critical. The use of such language is not intended to add to, interpret, or relieve a duty imposed by Title 14 of the Code of Federal Regulations (14 CFR). Persons working towards a helicopter rating are advised to review the references from the applicable practical test standards (FAA-S-8081-3 for recreational applicants, FAA-S-8081-15 for private applicants, and FAA-S-8081-16 for commercial applicants). Resources for study include FAA-H-8083-25, Pilot's Handbook of Aeronautical Knowledge, and FAA-H-8083-1, Weight and Balance Handbook, as these documents contain basic material not duplicated herein. All beginning applicants should refer to FAA-H-8083-25, Pilot's Handbook of Aeronautical Knowledge, for study and basic library reference.

It is essential for persons using this handbook to become familiar with and apply the pertinent parts of 14 CFR and the Aeronautical Information Manual (AIM). The AIM is available online at www.faa.gov. The current Flight Standards Service airman training and testing material and learning statements for all airman certificates and ratings can be obtained from www.faa.gov.

This handbook supersedes FAA-H-8083-21A, Helicopter Flying Handbook, dated 2012. Gyroplane information can be found in the FAA-H-8083-20, Gyroplane Flying Handbook.

This handbook is available for download, in PDF format, from www.faa.gov.

This handbook is published by the United States Department of Transportation, Federal Aviation Administration, Airman Testing Branch, P.O. Box 25082, Oklahoma City, OK 73125.

Comments regarding this publication should be emailed to AFS630comments@faa.gov.

Acknowledgments

The Helicopter Flying Handbook was produced by the Federal Aviation Administration (FAA) with the assistance of Safety Research Corporation of America (SRCA). The FAA wishes to acknowledge the following contributors:

Federation of American Scientists (www.fas.org) for rotor system content used in Chapter 5

Kaman Aerospace, Helicopters Division for image of Kaman used in Chapter 5

Burkhard Domke (www.b-domke.de) for images of rotor systems (Chapters 1 and 4)

New Zealand Civil Aviation Authority for image of safety procedures for approaching a helicopter (Chapter 9)

Shawn Coyle of Eagle Eye Solutions, LLC for images and content used in Chapter 10

Dr. Pat Veillette for information used on decision-making (Chapter 13)

Additional appreciation is extended to the Helicopter Association International (HAI), United States Helicopter Safety Team (USHST), Leonardo Helicopters, Aircraft Owners and Pilots Association (AOPA), and the AOPA Air Safety Foundation for their technical support and input.

Table of Contents

Chapter 10

Advanced Flight Maneuvers 10-1

Chapter 11

Helicopter Emergencies and Hazards 11-1

Chapter 12

Night Operations 12-1

Chapter 13

Effective Aeronautical Decision-Making 13-1

Glossary G-1

Index I-1

Introduction to the Helicopter

Introduction

A helicopter is an aircraft that is lifted and propelled by one or more horizontal rotors, each rotor consisting of two or more rotor blades. Helicopters are classified as rotorcraft or rotary-wing aircraft to distinguish them from fixed-wing aircraft, because the helicopter derives its source of lift from the rotor blades rotating around a mast. The word "helicopter" is adapted from the French hélicoptère, coined by Gustave de Ponton d'Amécourt in 1861. It is linked to the Greek words helix/helikos ("spiral" or "turning") and pteron ("wing").

As an aircraft, the primary advantages of the helicopter are due to the rotor blades that revolve through the air, providing lift without requiring the aircraft to move forward. This lift allows the helicopter to hover in one area and to take off and land vertically without the need for runways. For this reason, helicopters are often used in congested or isolated areas where fixed-wing aircraft are not able to take off or land. *[Figures 1-1* and *1-2]*

Piloting a helicopter requires adequate, focused and safety-orientated training. It also requires continuous attention to the machine and the operating environment. The pilot must work in three dimensions and use both arms and both legs constantly to keep the helicopter in a desired state. Coordination, timing and control touch are all used simultaneously when flying a helicopter.

Although helicopters were developed and built during the first half-century of flight, some even reaching limited production; it was not until 1942 that a helicopter designed by Igor Sikorsky reached full-scale production, with 131 aircraft built. Even though most previous designs used more than one main rotor, it was the single main rotor with an antitorque tail rotor configuration that would come to be recognized worldwide as the helicopter.

Figure 1-1. *Search and rescue helicopter conducting a pinnacle approach.*

Figure 1-2. *Search and rescue helicopter landing in a confined area.*

Turbine Age

In 1951, at the urging of his contacts at the Department of the Navy, Charles H. Kaman modified his K-225 helicopter with a new kind of engine, the turbo-shaft engine. This adaptation of the turbine engine provided a large amount of horsepower to the helicopter with a lower weight penalty than piston engines, heavy engine blocks, and auxiliary components. On December 11, 1951, the K-225 became the first turbine-powered helicopter in the world. Two years later, on March 26, 1954, a modified Navy HTK-1, another Kaman helicopter, became the first twin-turbine helicopter to fly. However, it was the Sud Aviation Alouette II that would become the first helicopter to be produced with a turbine engine.

Reliable helicopters capable of stable hover flight were developed decades after fixed-wing aircraft. This is largely due to higher engine power density requirements than fixed-wing aircraft. Improvements in fuels and engines during the first half of the 20th century were critical factors in helicopter development. The availability of lightweight turbo-shaft engines in the second half of the 20th century led to the development of larger, faster, and higher-performance helicopters. While smaller and less expensive helicopters still use piston engines, turboshaft engines are the preferred powerplant for helicopters today.

The turbine engine has the following advantages over a reciprocating engine:

- Less vibration
- Increased aircraft performance
- Reliability
- Ease of operation

Uses

Due to the unique operating characteristics of the helicopter—its ability to take off and land vertically, to hover for extended periods of time, and the aircraft's handling properties under low airspeed conditions—it has been chosen to conduct tasks that were previously not possible with other aircraft or were too time- or work-intensive to accomplish on the ground. Today, helicopters are used for transportation, construction, firefighting, search and rescue, and a variety of other jobs that require its special capabilities. *[Figure 1-3]*

Figure 1-3. *The many uses for a helicopter include search and rescue (top), firefighting (middle), and construction (bottom).*

Rotor System

The helicopter rotor system is the rotating part of a helicopter that generates lift. A rotor system may be mounted horizontally, as main rotors are, providing lift vertically; and it may be mounted vertically, such as a tail rotor, to provide lift horizontally as thrust to counteract torque effect. In the case of tilt rotors, the rotor is mounted on a nacelle that rotates at the edge of the wing to transition the rotor from a horizontal mounted position, providing lift horizontally as thrust, to a vertical mounted position providing lift exactly as a helicopter.

The rotor consists of a mast, hub, and rotor blades. *[Figure 1-4]* The mast is a hollow cylindrical metal shaft which extends upwards from and is driven by the transmission. At the top of the mast is the attachment point for the rotor blades called the hub. The rotor blades are then attached to the hub by several different methods. Main rotor systems are classified according to how the main rotor blades are attached and move relative to the main rotor hub. There are three basic classifications: semirigid, rigid, or fully articulated, although some modern rotor systems use an engineered combination of these types. All three rotor systems are discussed with greater detail in Chapter 4, Helicopter Components, Sections, and Systems.

With a single main rotor helicopter, a torque effect is created as the engine turns the rotor. This torque causes the body of the helicopter to turn in the opposite direction of the rotor (Newton's Third Law: Every action has an equal and opposite reaction, as explained in Chapter 2, Aerodynamics of Flight). To eliminate this effect, some sort of antitorque control must be used with a sufficient margin of power available to allow the helicopter to maintain its heading and prevent the aircraft from moving unsteadily. The three most common controls used today are the traditional tail rotor, Fenestron (also called a fantail), and the NOTAR®. All three antitorque designs will be discussed in Chapter 4, Helicopter Components, Sections, and Systems.

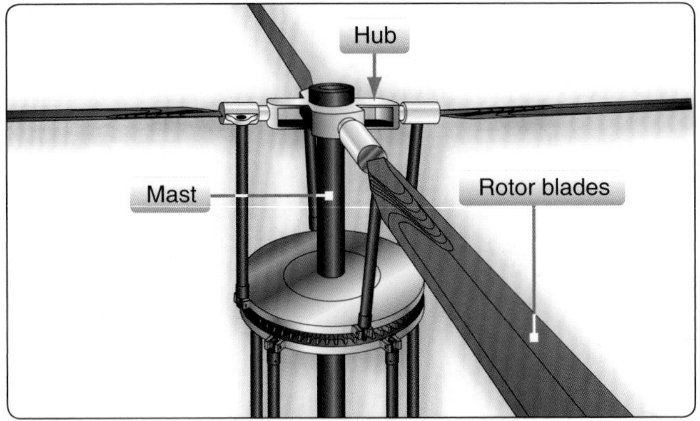

Figure 1-4. *Basic components of the rotor system.*

Rotor Configurations

Most helicopters have a single, main rotor but require a separate rotor to overcome torque which is a turning or twisting force. This is accomplished through a variable pitch, antitorque rotor or tail rotor. This is the design that Igor Sikorsky settled on for his VS-300 helicopter shown in *Figure 1-5.* It has become the recognized convention for helicopter design, although designs do vary. Helicopter main rotor designs from different manufacturers rotate in one of two different directions (clockwise or counter-clockwise when viewed from above). This can make it confusing when discussing aerodynamic effects on the main rotor between different designs, since the effects may manifest on opposite sides of each aircraft. For clarity, throughout this handbook, all examples use a counter-clockwise rotating main rotor system when viewed from above.

For clarity, throughout this handbook, all examples use a counter-clockwise rotating main rotor system when viewed from above.

Tandem Rotor

Tandem rotor (sometimes referred to as dual rotor) helicopters have two large horizontal rotor assemblies, instead of one main assembly and a smaller tail rotor. *[Figure 1-6]* Single rotor helicopters need a tail rotor to neutralize the twisting momentum produced by the single large rotor. Tandem rotor helicopters, however, use counter-rotating rotors, each canceling out the other's torque. Counter-rotating rotor blades will not collide with and destroy each other if they flex into the other rotor's pathway. This configuration has the advantage of being able to hold more weight with shorter blades, since there are two blade sets. Also, all the power from the engines can be used for lift, whereas a single rotor helicopter must use some power to counter main rotor torque. Because of this, tandem helicopters make up some of the most powerful and fastest rotor system aircraft.

Figure 1-5. *Igor Sikorsky designed the VS-300 helicopter incorporating the tail rotor into the design.*

Figure 1-6. *Tandem rotor helicopters.*

Coaxial Rotors

Coaxial rotors are a pair of rotors turning in opposite directions, but mounted on a mast, with the same axis of rotation, one above the other. This configuration is a noted feature of helicopters produced by the Russian Kamov helicopter design bureau. *[Figure 1-7]*

Intermeshing Rotors

Intermeshing rotors on a helicopter are a set of two rotors turning in opposite directions, with each rotor mast mounted on the helicopter with a slight angle to the other so that the blades intermesh without colliding. *[Figure 1-8]* This arrangement allows the helicopter to function without the need for a tail rotor. It has high stability and powerful lifting capability. This configuration is sometimes referred to as a synchropter. The arrangement was developed in Germany

Figure 1-7. *Coaxial rotors.*

Figure 1-8. *HH-43 Huskie with intermeshing rotors.*

for a small anti-submarine warfare helicopter, the Flettner Fl 282 Kolibri. During the Cold War the American Kaman Aircraft company produced the HH-43 Huskie, for USAF firefighting purposes. The latest Kaman K-MAX model is a dedicated sky crane design used for construction work.

Tail Rotor

The tail rotor is a smaller rotor mounted vertically or near-vertically on the tail of a traditional single-rotor helicopter. The tail rotor either pushes or pulls against the tail to counter the torque. The tail rotor drive system consists of a drive shaft powered from the main transmission and a gearbox mounted at the end of the tail boom. *[Figure 1-9]* The drive shaft may consist of one long shaft or a series of shorter shafts connected at both ends with flexible couplings. The flexible couplings allow the drive shaft to flex with the tail boom.

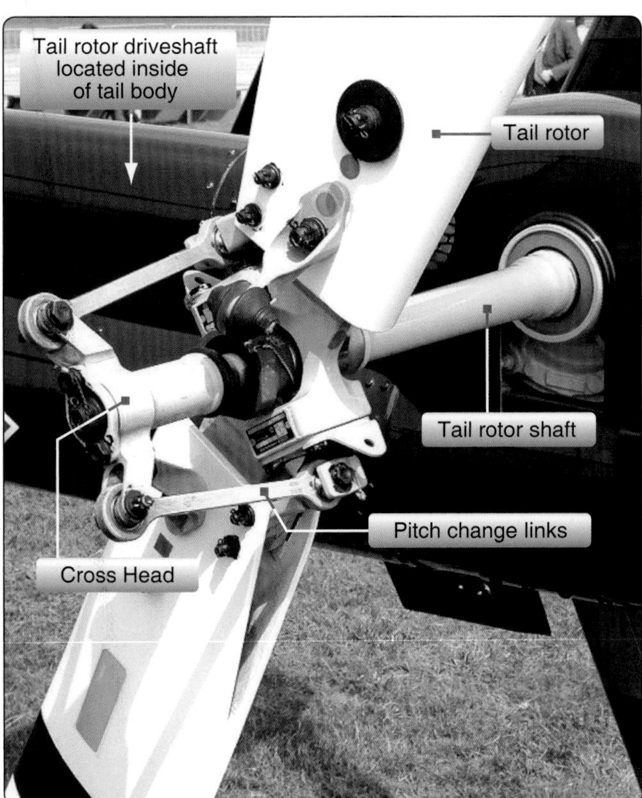

Figure 1-9. *Basic tail rotor components.*

The gearbox at the end of the tail boom provides an angled drive for the tail rotor and may also include gearing to adjust the output to the optimum rotational speed typically measured in revolutions per minute (rpm) for the tail rotor. On some larger helicopters, intermediate gearboxes are used to angle the tail rotor drive shaft from along the tail boom or tailcone to the top of the tail rotor pylon, which also serves as a vertical stabilizing airfoil to alleviate the power requirement for the tail rotor in forward flight. The pylon (or vertical fin) may also provide limited antitorque within certain airspeed ranges if the tail rotor or the tail rotor flight controls fail.

Controlling Flight

A helicopter has four primary flight controls:

- Cyclic
- Collective
- Antitorque pedals
- Throttle

Cyclic

The cyclic control is usually located between the pilot's legs and is commonly called the "cyclic stick" or simply "cyclic." On most helicopters, the cyclic is similar to a joystick; however, Robinson helicopters have unique T-bar cyclic control systems. A few helicopters have cyclic controls that descend into the cockpit from overhead while others use side cyclic controls.

The control is called the cyclic because it can vary the pitch of the rotor blades throughout each revolution of the main rotor system (i.e., through each cycle of rotation) to develop unequal lift (thrust). The result is to tilt the rotor disk in a particular direction, resulting in the helicopter moving in that direction. If the pilot pushes the cyclic forward, the rotor disk tilts forward, and the rotor produces a thrust in the forward direction. If the pilot pushes the cyclic to the side, the rotor disk tilts to that side and produces thrust in that direction, causing the helicopter to hover sideways. *[Figure 1-10]*

Collective

The collective pitch control, or collective, is located on the left side of the pilot's seat with a pilot-selected variable friction control to prevent inadvertent movement. The collective changes the pitch angle of all the main rotor blades

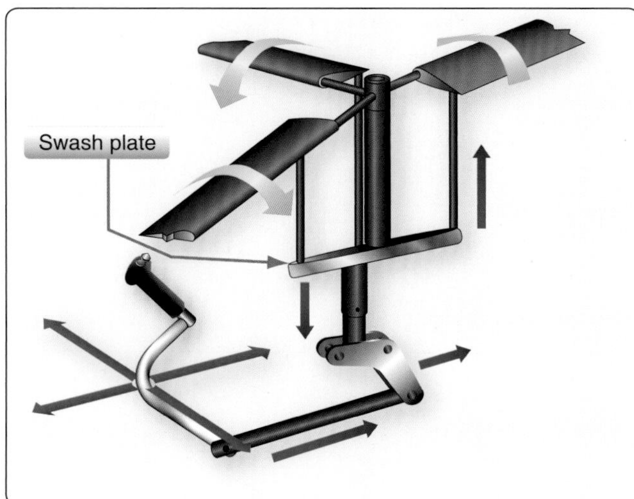

Figure 1-10. *Cyclic controls changing the pitch of the rotor blades.*

collectively (i.e., all at the same time) and independently of their positions. Therefore, if a collective input is made, all the blades change equally, increasing or decreasing total lift or thrust, with the result of the helicopter increasing or decreasing in altitude or airspeed.

Antitorque Pedals

The antitorque pedals are located in the same position as the rudder pedals in a fixed-wing aircraft and serve a similar purpose, namely to control the direction in which the nose of the aircraft is pointed. Application of the pedal in a given direction changes the pitch of the tail rotor blades, increasing or reducing the thrust produced by the tail rotor, causing the nose to yaw in the direction of the applied pedal. The pedals mechanically change the pitch of the tail rotor, altering the amount of thrust produced.

Throttle

Helicopter rotors are designed to operate at a specific rpm. The throttle controls the power produced by the engine, which is connected to the rotor by a transmission. The purpose of the throttle is to maintain enough engine power to keep the rotor rpm within allowable limits to produce enough lift for flight. In single-engine helicopters, if so equipped, the throttle control is typically a twist grip mounted on the collective control, but it can also be a lever mechanism in fully governed systems. Multi-engine helicopters generally have a power lever or mode switch for each engine. *[Figure 1-11]* Helicopter flight controls are discussed in greater detail throughout Chapter 4, Helicopter Components, Sections and Systems.

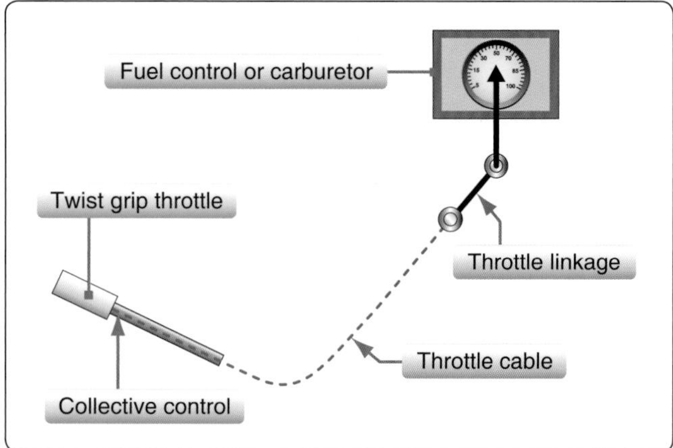

Figure 1-11. *The throttle control mounted at the end of the collective control.*

Flight Conditions

There are two basic flight conditions for a helicopter: hover and forward flight. Hovering is the most challenging part of flying a helicopter. This is because a helicopter generates its own gusty air while in a hover, which acts against the fuselage and flight control surfaces. The end result is the need for constant control inputs and corrections by the pilot to keep the helicopter where it is required to be. Despite the complexity of the task, the control inputs in a hover are simple. The cyclic is used to eliminate drift in the horizontal direction that is to control forward and back, right and left. The collective is used to maintain altitude. The pedals are used to control nose direction or heading. It is the interaction of these controls that makes hovering so difficult, since an adjustment in any one control requires an adjustment of the other two, creating a cycle of constant correction.

Displacing the cyclic forward initially causes the nose to pitch down, with a resultant increase in airspeed and loss of altitude. Aft cyclic initially causes the nose to pitch up, slowing the helicopter and causing it to climb; however, as the helicopter reaches a state of equilibrium, the horizontal stabilizer helps level the helicopter to minimize drag, unlike

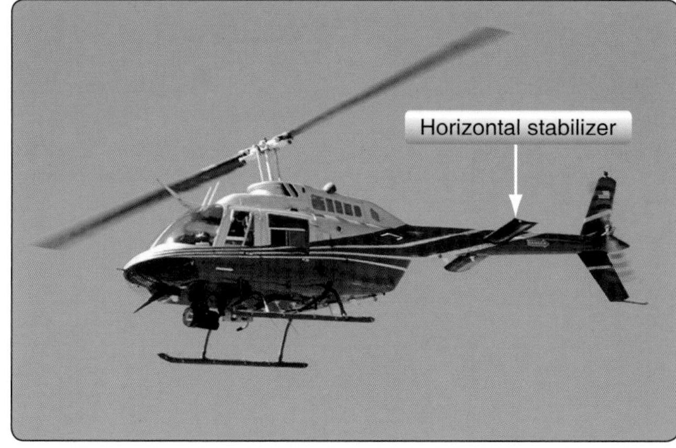

Figure 1-12. *The horizontal stabilizer helps level the helicopter to minimize drag during flight.*

an airplane. *[Figure 1-12]* Therefore, the helicopter has very little pitch deflection up or down when the helicopter is stable in a flight mode. The variation from absolutely level depends on the particular helicopter and the horizontal stabilizer function.

Increasing collective (power) while maintaining a constant airspeed induces a climb while decreasing collective causes a descent. Coordinating these two inputs, down collective plus aft cyclic or up collective plus forward cyclic, results in airspeed changes while maintaining a constant altitude.

The pedals serve the same function in both a helicopter and a fixed-wing aircraft, to maintain balanced flight. This is done by applying pedal input in whichever direction is necessary to center the ball in the turn and bank indicator. Flight maneuvers are discussed in greater detail throughout Chapter 9, Basic Flight Maneuvers.

Chapter Summary

This chapter gives the reader an overview of the history of the helicopter, its many uses, and how it has developed throughout the years. The chapter also introduces basic terms and explanations of the helicopter components, sections, and the theory behind how the helicopter flies.

Chapter 2

Aerodynamics of Flight

Introduction

This chapter presents aerodynamic fundamentals and principles as they apply to helicopters. The content relates to flight operations and performance of normal flight tasks. It covers theory and application of aerodynamics for the pilot, whether in flight training or general flight operations.

Gravity acting on the mass (the amount of matter) of an object creates a force called weight. The rotor blade below weighs 100 lbs. It is 20 feet long (span) and is 1 foot wide (chord). Accordingly, its surface area is 20 square feet. *[Figure 2-1]*

The blade is perfectly balanced on a pinpoint stand, as you can see in *Figure 2-2* from looking at it from the end (the airfoil view). The goal is for the blade to defy gravity and stay exactly where it is when we remove the stand. If we do nothing before removing the stand, the blade will simply fall to the ground. Can we exert a force (a push or pull) opposite gravity that equals the 100 lb. weight of the blade? Yes, for example, electromagnetic force could be used. In helicopters, however, we use aerodynamic force to oppose weight and to maneuver.

Every object in the atmosphere is surrounded by a gas that exerts a static force of 2,116 lb per square foot (a force times a unit area, called pressure) at sea level. However, that pressure is exerted equally all over the blade (top and bottom) and therefore does not create any useful force on the blade. We need only create a difference of a single pound of static pressure differential per square foot of blade surface to have a force equal to the blade's weight (100 lb of upward pressure opposite 100 lb downward weight).

Total pressure consists of static pressure and, if the air is moving, dynamic pressure (a pressure in the direction of the air movement). As shown in *Figure 2-3*, if dynamic pressure is increased the static pressure will decrease. Due to the design of the airfoil, the velocity of the air passing over the upper surface will be greater than that of the lower surface, leading to higher dynamic pressure on the upper surface than on the lower surface. The higher dynamic pressure on the upper surface lowers the static pressure on the upper surface. The static pressure on the bottom will now be greater than the static pressure on the top. The blade will experience an upward force. With just the right amount of air passing over the blade the upward force will equal one pound per square foot. This upward force is equal to, and acts opposite the blade's weight of 100 lb. So, if we now remove the stand, the blade will defy gravity and remain in its position (ignoring rearward drag for the moment).

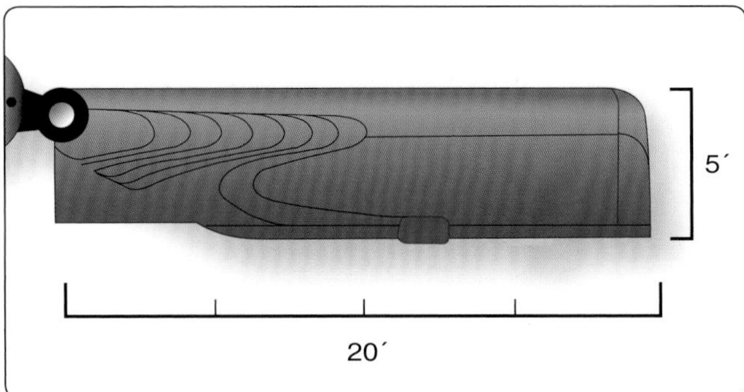

Figure 2-1. *Area of a blade.*

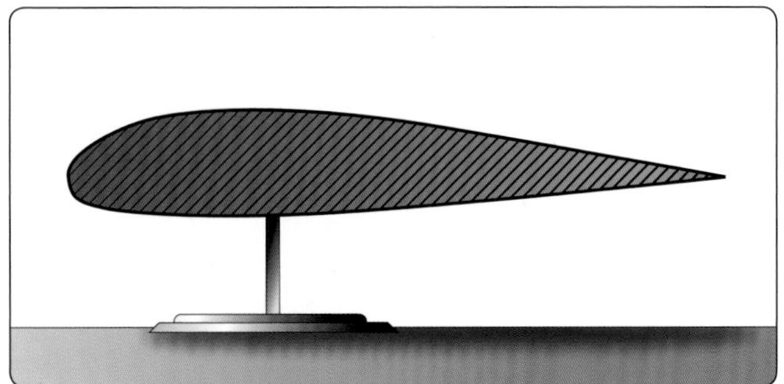

Figure 2-2. *Profile of an airfoil.*

The force created by air moving over an object (or moving an object through the air) is called aerodynamic force. Aero means air. Dynamic means moving or motion. Accordingly, by moving the air over an airfoil we can change the static pressures on the top and bottom thereby generating a useful force (an aerodynamic force). The portion of the aerodynamic force that is usually measured perpendicular to the air flowing around the airfoil is called lift and is used to oppose weight. Drag is the portion of aerodynamic force that is measured as the resistance created by an object passing through the air (or having the air passed over it). Drag acts in a streamwise direction with the wind passing over the airfoil and retards forward movement.

Forces Acting on the Aircraft

Once a helicopter leaves the ground, it is acted upon by four aerodynamic forces; thrust, drag, lift, and weight. Understanding how these forces work and knowing how to control them with the use of power and flight controls are essential to flight. *[Figure 2-3]* They are defined as follows:

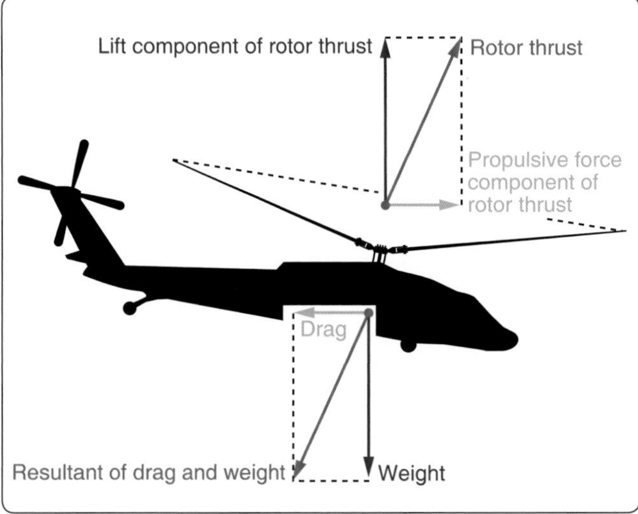

Figure 2-3. *Four forces acting on a helicopter in forward flight.*

- Lift—opposes the downward force of weight, is produced by the dynamic effect of the air acting on the airfoil and acts perpendicular to the flightpath through the center of lift.
- Weight—the combined load of the aircraft itself, the crew, the fuel, and the cargo or baggage. Weight pulls the aircraft downward because of the force of gravity. It opposes lift and acts vertically downward through the aircraft's center of gravity (CG).
- Thrust—the force produced by the power plant/ propeller or rotor. It opposes or overcomes the force of drag. As a general rule, it acts parallel to the longitudinal axis. However, this is not always the case, as explained later.
- Drag—a rearward, retarding force caused by disruption of airflow by the wing, rotor, fuselage, and other protruding objects. Drag opposes thrust and acts rearward parallel to the relative wind.

For a more in-depth explanation of general aerodynamics, refer to the Pilot's Handbook of Aeronautical Knowledge.

Lift

Lift is generated when an object changes the direction of flow of a fluid or when the fluid is forced to move by the object passing through it. When the object and fluid move relative to each other and the object turns the fluid flow in a direction perpendicular to that flow, the force required to do this work creates an equal and opposite force that is lift. The object may be moving through a stationary fluid, or the fluid may be flowing past a stationary object—these two are effectively identical as, in principle, it is only the frame of reference of the viewer which differs. The lift generated by an airfoil depends on such factors as:

- Speed of the airflow
- Density of the air
- Total area of the segment or airfoil
- Angle of attack (AOA) between the air and the airfoil

The AOA is the angle at which the airfoil meets the oncoming airflow (or vice versa). In the case of a helicopter, the object is the rotor blade (airfoil) and the fluid is the air. Lift is produced when a mass of air is deflected, and it always acts perpendicular to the resultant relative wind. A symmetric airfoil must have a positive AOA to generate positive lift. At a zero AOA, no lift is generated. At a negative AOA, negative lift is generated. A cambered or nonsymmetrical airfoil may produce positive lift at zero, or even small negative AOA.

The basic concept of lift is simple. However, the details of how the relative movement of air and airfoil interact to produce the turning action that generates lift are complex. In any case causing lift, an angled flat plate, revolving cylinder, airfoil, etc., the flow meeting the leading edge of the object is forced to split over and under the object. The sudden change in direction over the object causes an area of low pressure to form behind the leading edge on the upper surface of the object. In turn, due to this pressure gradient and the viscosity of the fluid, the flow over the object is accelerated down along the upper surface of the object. At the same time, the flow forced under the object is rapidly slowed or stagnated causing an area of high pressure. This also causes the flow to accelerate along the upper surface of the object. The two sections of the fluid each leave the trailing edge of the object with a downward component of momentum, producing lift. *[Figure 2-4]*

Bernoulli's Principle

Bernoulli's principle describes the relationship between internal fluid pressure and fluid velocity. It is a statement of the law of conservation of energy and helps explain why an airfoil develops an aerodynamic force. The concept of conservation of energy states energy cannot be created or destroyed and the amount of energy entering a system must also exit. Specifically, in this case the "energy" referred to is the dynamic pressure (the kinetic energy of the air—more velocity, more kinetic energy) and static air pressure (potential energy). These will change among themselves, but the total pressure energy remains constant inside the tube.

A simple tube with a constricted portion near the center of its length illustrates this principle. An example is running water through a garden hose. The mass of flow per unit area (cross-sectional area of tube) is the mass flow rate. In *Figure 2-5,* the flow into the tube is constant, neither accelerating nor decelerating; thus, the mass flow rate through the tube must be the same at stations 1, 2, and 3. If the cross-sectional area at any one of these stations—or any given point—in the tube is reduced, the fluid velocity must increase to maintain a constant mass flow rate to move the same amount of fluid through a smaller area. The continuity of mass flow causes the air to move faster through the venturi. In other words, fluid speeds up in direct proportion to the reduction in area.

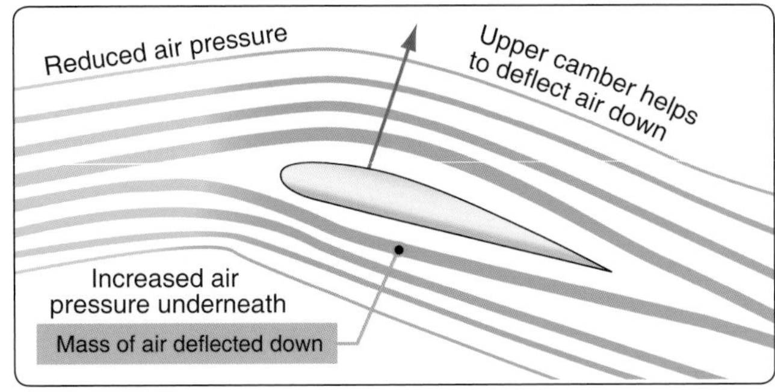

Figure 2-4. *Production of lift.*

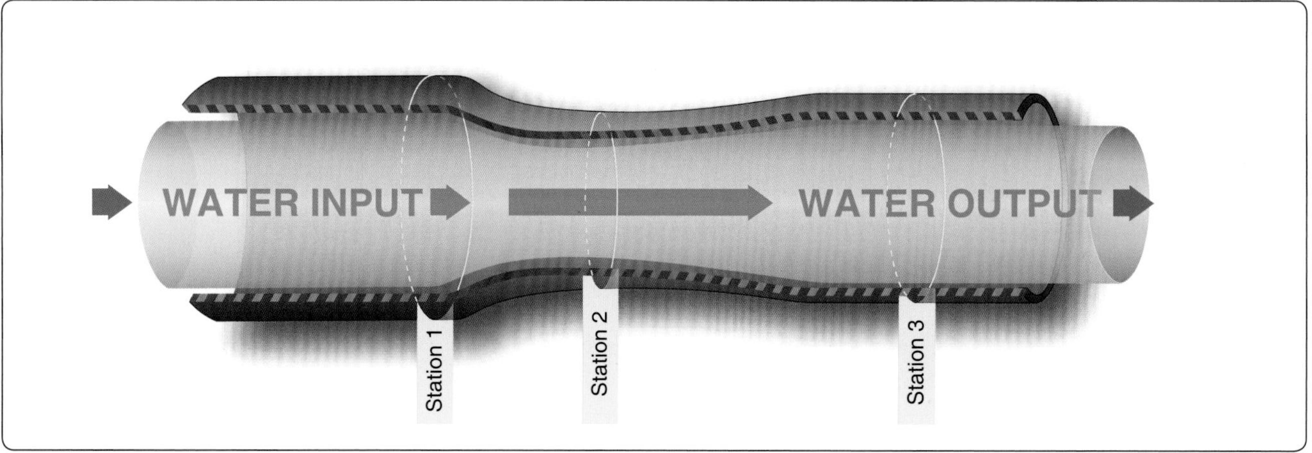

Figure 2-5. *Water flow through a tube.*

Bernoulli ($P_{total} = P_{dynamic} + P_{static}$) states that the increase in velocity will increase the streamwise dynamic pressure. Since the total pressure in the tube must remain constant, the static pressure on the sides of the venturi will decrease. Venturi effect is the term used to describe this phenomenon.

Figure 2-6 illustrates plates of one square foot in the dynamic flow and on the sides of the tube indicating static pressure, with corresponding pressure. At point 2, it is easier to visualize the static pressure reduction on the top of the airfoil as compared to the bottom of the airfoil, which is depicted as outside of the tube and therefore at ambient static pressure. Keep in mind with actual blades it is not a simple as this example because the bottom static pressure is influenced by blade design and blade angle, among other things. However, the basic idea is that it is the static pressure differential between the top and bottom multiplied by the surface area of the blade that generates the aerodynamic force.

Venturi Flow

While the amount of total energy within a closed system (the tube) does not change, the form of the energy may be altered. Pressure of flowing air may be compared to energy in that the total pressure of flowing air always remains constant unless energy is added or removed. Fluid flow pressure has two components—static and dynamic pressure. Static pressure is the pressure component measured in the flow but not moving with the flow as pressure is measured. Static pressure is also known as the force per unit area acting on a surface. Dynamic pressure of flow is that component existing as a result of movement of the air. The sum of these two pressures is total pressure. As air flows through the constriction, static pressure decreases as velocity increases. This increases dynamic pressure. *Figure 2-7* depicts the bottom half of the constricted area of the tube, which resembles the top half of an airfoil. Even with the top half of the tube removed, the air

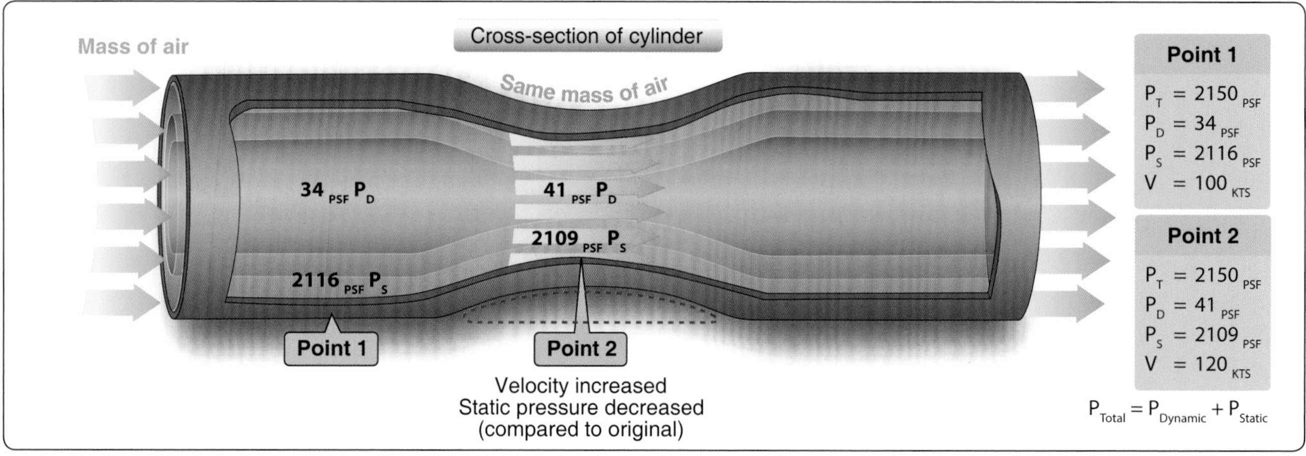

Figure 2-6. *Venturi effect.*

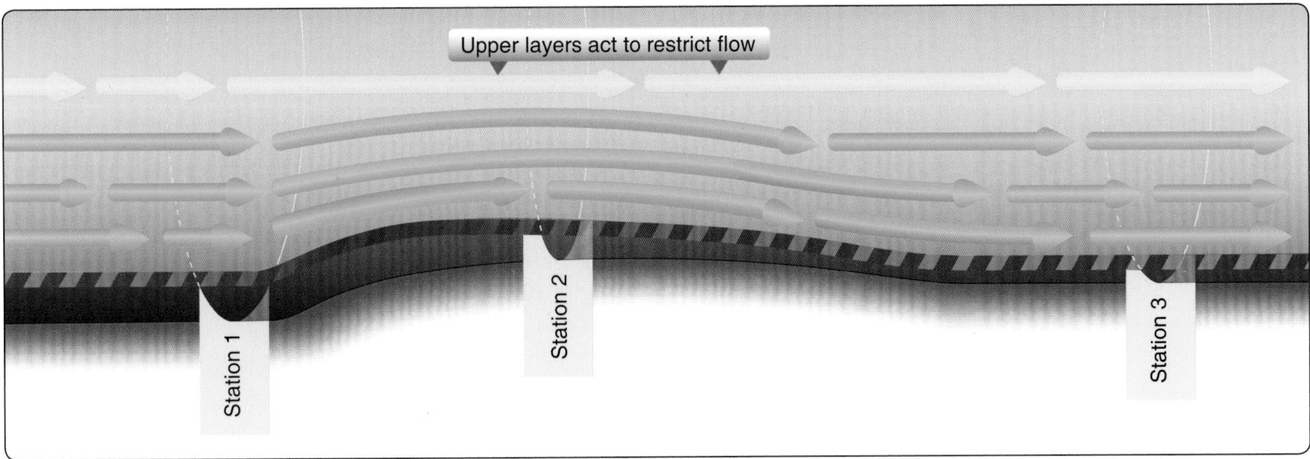

Figure 2-7. *Venturi flow.*

still accelerates over the curved area because the upper air layers restrict the flow—just as the top half of the constricted tube did. This acceleration causes decreased static pressure above the curved portion and creates a pressure differential caused by the variation of static and dynamic pressures.

Newton's Third Law of Motion

Additional lift is provided by the rotor blade's lower surface as air striking the underside is deflected downward. According to Newton's Third Law of Motion, "for every action there is an equal and opposite reaction," the air that is deflected downward also produces an upward (lifting) reaction.

Since air is much like water, the explanation for this source of lift may be compared to the planing effect of skis on water. The lift that supports the water skis (and the skier) is the force caused by the impact pressure and the deflection of water from the lower surfaces of the skis.

Under most flying conditions, the impact pressure and the deflection of air from the lower surface of the rotor blade provides a comparatively small percentage of the total lift. The majority of lift is the result of decreased pressure above the blade, rather than the increased pressure below it.

Weight

Normally, weight is thought of as being a known, fixed value, such as the weight of the helicopter, fuel, and occupants. To lift the helicopter off the ground vertically, the rotor disk must generate enough lift to overcome or offset the total weight of the helicopter and its occupants. Newton's First Law states: "Every object in a state of uniform motion tends to remain in that state of motion unless an external force is applied to it." In this case, the object is the helicopter whether at a hover or on the ground and the external force applied to it is lift, which is accomplished by increasing the pitch angle of the main rotor blades. This action forces the helicopter into a state of motion, without it the helicopter would either remain on the ground or at a hover.

The weight of the helicopter can also be influenced by aerodynamic loads. When you bank a helicopter while maintaining a constant altitude, the "G" load or load factor increases. The load factor is the actual load on the rotor blades at any time, divided by the normal load or gross weight (weight of the helicopter and its contents). Any time a helicopter flies in a constant altitude curved flightpath, the load supported by the rotor blades is greater than the total weight of the helicopter. The tighter the curved flightpath is, the steeper the bank is; the more rapid the flare or pullout from a dive is, the greater the load supported by the rotor. Therefore, the greater the load factor must be. *[Figure 2-8]*

To overcome this additional load factor, the helicopter must be able to produce more lift. If excess engine power is not available, the helicopter either descends or has to decelerate in order to maintain the same altitude. The load factor and, hence,

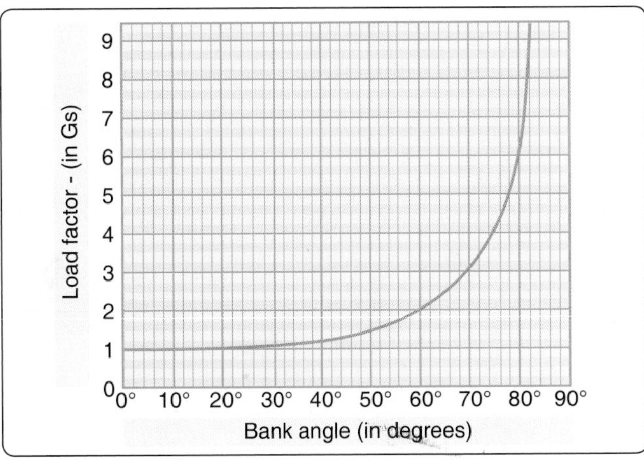

Figure 2-8. *The load factor diagram allows a pilot to calculate the amount of "G" loading exerted with various angles of bank.*

apparent gross weight increase is relatively small in banks up to 30°. Even so, under the right set of adverse circumstances, such as high-density altitude, turbulent air, high gross weight, and poor pilot technique, sufficient or excess power may not be available to maintain altitude and airspeed. Pilots must take all of these factors into consideration throughout the entire flight from the point of ascending to a hover to landing. Above 30° of bank, the apparent increase in gross weight soars. At 30° of bank, or pitch, the apparent increase is only 16 percent, but at 60°, it is twice the load on the wings and rotor disk. For example, if the weight of the helicopter is 1,600 pounds, the weight supported by the rotor disk in a 30° bank at a constant altitude would be 1,856 pounds (1,600 + 16 percent (or 256)). In a 60° bank, it would be 3,200 pounds; in an 80° bank, it would be almost six times as much, or 8,000 pounds. It is important to note that each rotor blade must support a percentage of the gross weight. In a two-bladed system, each blade of the 1,600-pound helicopter as stated above would have to lift 50 percent or 800 pounds. If this same helicopter had three rotor blades, each blade would have to lift only 33 percent, or 533 pounds. One additional cause of large load factors is rough or turbulent air. The severe vertical gusts produced by turbulence can cause a sudden increase in AOA, resulting in increased rotor blade loads that are resisted by the inertia of the helicopter.

Each type of helicopter has its own limitations that are based on the aircraft structure, size, and capabilities. Regardless of how much weight one can carry or the engine power that it may have, they are all susceptible to aerodynamic overloading. Unfortunately, if the pilot attempts to push the performance envelope the consequence can be fatal. Aerodynamic forces effect every movement in a helicopter, whether it is increasing the collective or a steep bank angle. Anticipating results from a particular maneuver or adjustment of a flight control is not good piloting technique. Instead pilots need to truly understand the capabilities of the helicopter under any and all circumstances and plan never to exceed the flight envelope for any situation.

Thrust

Thrust, like lift, is generated by the rotation of the main rotor disk. In a helicopter, thrust can be forward, rearward, sideward, or vertical. The resultant lift and thrust determines the direction of movement of the helicopter.

The solidity ratio is the ratio of the total rotor blade area, which is the combined area of all the main rotor blades, to the total rotor disk area. This ratio provides a means to measure the potential for a rotor disk to provide thrust and lift. The mathematical calculations needed to calculate the solidity ratio for each helicopter may not be of importance to most pilots but what should be are the capabilities of the rotor disk to produce and maintain lift. Many helicopter accidents are caused from the rotor disk being overloaded. Simply put, pilots attempt maneuvers that require more lift than the rotor disk can produce or more power than the helicopter's powerplant can provide. Trying to land with a nose high attitude along with any other unfavorable condition (i.e., high gross weight or wind gusts) is most likely to end in disaster.

The tail rotor also produces thrust. The amount of thrust is variable through the use of the antitorque pedals and is used to control the helicopter's yaw.

Drag

The force that resists the movement of a helicopter through the air and is produced when lift is developed is called drag. Drag must be overcome by the engine to turn the rotor. Drag always acts parallel to the relative wind. Total drag is composed of three types of drag: profile, induced, and parasite.

Profile Drag

Profile drag develops from the frictional resistance of the blades passing through the air. It does not change significantly with the airfoil's AOA but increases moderately when airspeed increases. Profile drag is composed of form drag and skin friction. Form drag results from the turbulent wake caused by the separation of airflow from the surface of a structure. The amount of drag is related to both the size and shape of the structure that protrudes into the relative wind. *[Figure 2-9]*

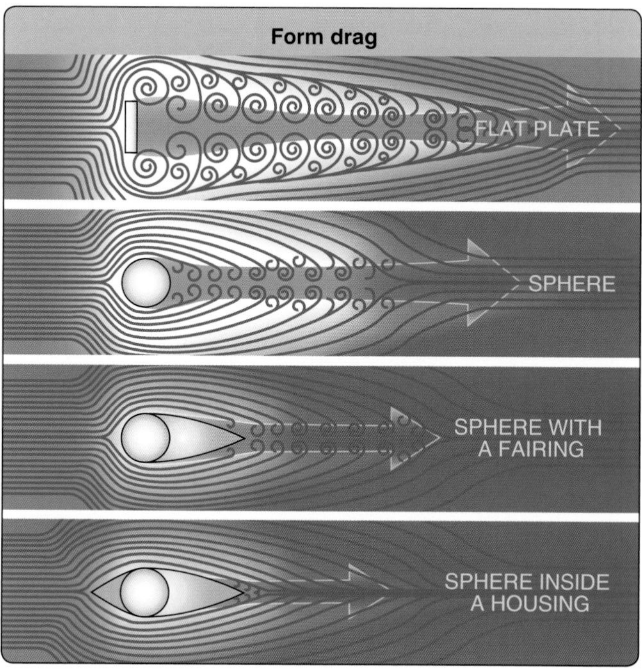

Figure 2-9. *It is easy to visualize the creation of form drag by examining the airflow around a flat plate. Streamlining decreases form drag by reducing the airflow separation.*

Skin friction is caused by surface roughness. Even though the surface appears smooth, it may be quite rough when viewed under a microscope. A thin layer of air clings to the rough surface and creates small eddies that contribute to drag.

Induced Drag

Induced drag is generated by the airflow circulation around the rotor blade as it creates lift. The high-pressure area beneath the blade joins the low-pressure area above the blade at the trailing edge and at the rotor tips. This causes a spiral, or vortex, which trails behind each blade whenever lift is being produced. These vortices deflect the airstream downward in the vicinity of the blade, creating an increase in downwash. Therefore, the blade operates in an average relative wind that is inclined downward and rearward near the blade. Because the lift produced by the blade is perpendicular to the relative wind, the lift is inclined aft by the same amount. The component of lift that is acting in a rearward direction is induced drag. *[Figure 2-10]*

As the air pressure differential increases with an increase in AOA, stronger vortices form, and induced drag increases. Since the blade's AOA is usually lower at higher airspeeds, and higher at low speeds, induced drag decreases as airspeed increases and increases as airspeed decreases. Induced drag is the major cause of drag at lower airspeeds.

Parasite Drag

Parasite drag is present any time the helicopter is moving through the air. This type of drag increases with airspeed. Non-lifting components of the helicopter, such as the cabin, rotor mast, tail, and landing gear, contribute to parasite drag. Any loss of momentum by the airstream, due to such things as openings for engine cooling, creates additional parasite drag. Because of its rapid increase with increasing airspeed, parasite drag is the major cause of drag at higher airspeeds. Parasite drag varies with the square of the velocity; therefore, doubling the airspeed increases the parasite drag four times.

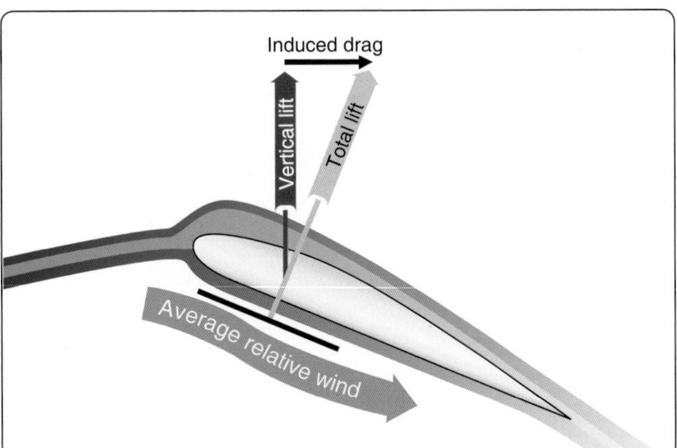

Figure 2-10. *The formation of induced drag is associated with the downward deflection of the airstream near the rotor blade.*

Total Drag

Total drag for a helicopter is the sum of all three drag forces. *[Figure 2-11]* As airspeed increases, parasite drag increases, while induced drag decreases. Profile drag remains relatively constant throughout the speed range with some increase at higher airspeeds. Combining all drag forces results in a total drag curve. The low point on the total drag curve shows the airspeed at which drag is minimized. This is the point where the lift-to-drag ratio is greatest and is referred to as L/D_{MAX}. At this speed, the total lift capacity of the helicopter, when compared to the total drag of the helicopter, is most favorable. This is an important factor in helicopter performance.

Airfoil

Helicopters are able to fly due to aerodynamic forces produced when air passes around the airfoil. An airfoil is any surface producing more lift than drag when passing through the air at a suitable angle. Airfoils are most often associated with production of lift. Airfoils are also used for stability (fin), control (elevator), and thrust or propulsion (propeller or rotor). Certain airfoils, such as rotor blades, combine some of these functions. The main and tail rotor blades of the helicopter are airfoils, and air is forced to pass around the blades by mechanically powered rotation. In some conditions, parts of the fuselage, such as the vertical and horizontal stabilizers, can become airfoils. Airfoils are carefully structured to accommodate a specific set of flight characteristics.

Airfoil Terminology and Definitions

- Blade span—the length of the rotor blade from center of rotation to tip of the blade.

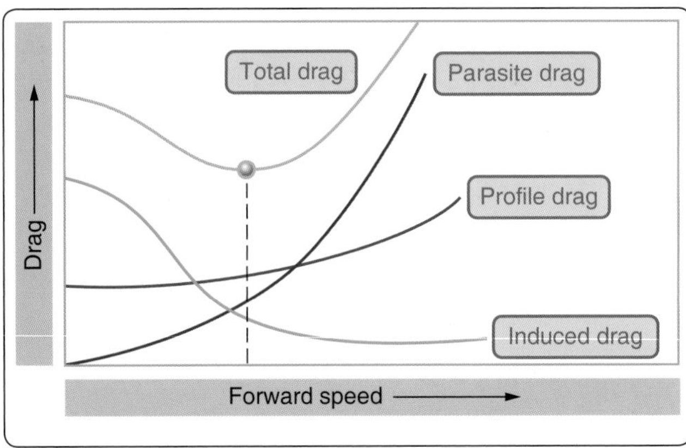

Figure 2-11. *The total drag curve represents the combined forces of parasite, profile, and induced drag and is plotted against airspeed.*

- Chord line—a straight line intersecting leading and trailing edges of the airfoil. *[Figure 2-12]*
- Chord—the length of the chord line from leading edge to trailing edge; it is the characteristic longitudinal dimension of the airfoil section.
- Mean camber line—a line drawn halfway between the upper and lower surfaces of the airfoil. *[Figure 2-12]*

  The chord line connects the ends of the mean camber line. Camber refers to curvature of the airfoil and may be considered as curvature of the mean camber line. The shape of the mean camber is important for determining aerodynamic characteristics of an airfoil section. Maximum camber (displacement of the mean camber line from the chord line) and its location help to define the shape of the mean camber line. The location of maximum camber and its displacement from the chord line are expressed as fractions or percentages of the basic chord length. By varying the point of maximum camber, the manufacturer can tailor an airfoil for a specific purpose. The profile thickness and thickness distribution are important properties of an airfoil section.
- Leading edge—the front edge of an airfoil. *[Figure 2-12]*
- Flightpath velocity—the speed and direction of the airfoil passing through the air. For airfoils on an airplane, the flightpath velocity is equal to true airspeed (TAS). For helicopter rotor blades, flightpath velocity is equal to rotational velocity, plus or minus a component of directional airspeed. The rotational velocity of the rotor blade is lowest closer to the hub and increases outward towards the tip of the blade during rotation.
- Relative wind—defined as the airflow relative to an airfoil and is created by movement of an airfoil through the air. This is rotational relative wind for rotary-wing aircraft and is covered in detail later. As an induced airflow may modify flightpath velocity, relative wind experienced by the airfoil may not be exactly opposite its direction of travel.
- Trailing edge—the rearmost edge of an airfoil.
- Induced flow—the downward flow of air through the rotor disk.
- Resultant relative wind—relative wind modified by induced flow.
- AOA—the angle measured between the resultant relative wind and chord line.
- Angle of incidence (AOI)—the angle between the chord line of a blade and rotor hub. It is usually referred to as blade pitch angle. For fixed airfoils, such as vertical fins or elevators, angle of incidence is the angle between the chord line of the airfoil and a selected reference plane of the helicopter.
- Center of pressure—the point along the chord line of an airfoil through which all aerodynamic forces are considered to act. Since pressures vary on the surface of an airfoil, an average location of pressure variation is needed. As the AOA changes, these pressures change, and the center of pressure moves along the chord line.

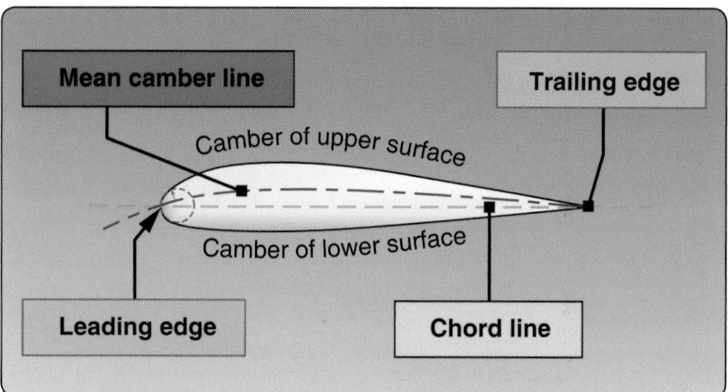

Figure 2-12. *Aerodynamic terms of an airfoil.*

Airfoil Types

Symmetrical Airfoil

The symmetrical airfoil is distinguished by having identical upper and lower surfaces. *[Figure 2-13]* The mean camber line and chord line are the same on a symmetrical airfoil, and it produces no lift at zero AOA. Most light helicopters incorporate symmetrical airfoils in the main rotor blades.

Nonsymmetrical Airfoil (Cambered)

The nonsymmetrical airfoil has different upper and lower surfaces, with a greater curvature of the airfoil above the chord line than below. *[Figure 2-13]* The mean camber line and chord line are different. The nonsymmetrical airfoil design can produce useful lift at zero AOA. A nonsymmetrical design

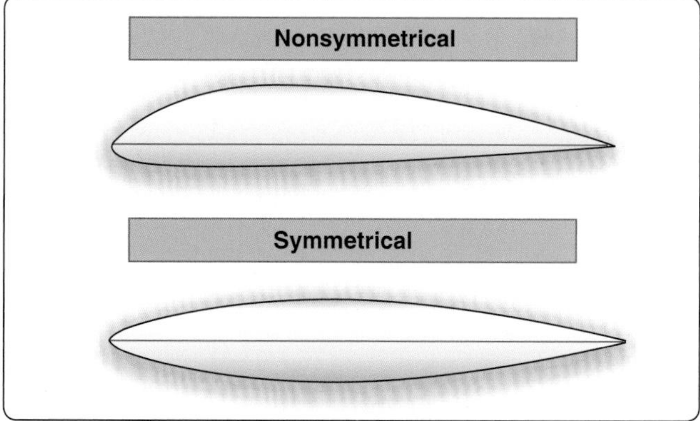

Figure 2-13. *The upper and lower curvatures are the same on a symmetrical airfoil and vary on a nonsymmetrical airfoil.*

has advantages and disadvantages. The advantages are more lift production at a given AOA than a symmetrical design, an improved lift-to-drag ratio, and better stall characteristics. The disadvantages are center of pressure travel of up to 20 percent of the chord line (creating undesirable torque on the airfoil structure) and greater production costs.

Blade Twist

Because of lift differential due to differing rotational relative wind values along the blade, the blade should be designed with a twist to alleviate internal blade stress and distribute the lifting force more evenly along the blade. Blade twist provides higher pitch angles at the root where velocity is low and lower pitch angles nearer the tip where velocity is higher. This increases the induced air velocity and blade loading near the inboard section of the blade. *[Figure 2-14]*

Rotor Blade and Hub Definitions

- Hub—on the mast, the attaching point for the root of the blade, and the axis about which the blades rotate. *[See Figure 1-7]*
- Tip—the farthest outboard section of the rotor blade
- Root—the inner end of the blade and is the point that attaches to the hub
- Twist—the change in blade incidence from the root to the outer blade

The angular position of the main rotor blades (as viewed from above, as they rotate about the vertical axis of the mast) is measured from the helicopter's longitudinal axis, and usually from its nose. The radial position of a segment of the blade is the distance from the hub as a fraction of the total distance.

Airflow and Reactions in the Rotor Disk

Relative Wind

Knowledge of relative wind is essential for an understanding of aerodynamics and its practical flight application for the pilot. Relative wind is airflow relative to an airfoil. Movement of an airfoil through the air creates relative wind. Relative wind moves in a direction parallel to but opposite of the movement of the airfoil. *[Figure 2-15]*

There are two parts to wind passing a rotor blade:

- Horizontal part—caused by the blades turning plus movement of the helicopter through the air *[Figure 2-16]*
- Vertical part—caused by the air being forced down through the rotor blades plus any movement of the air relative to the blades caused by the helicopter climbing or descending *[Figures 2-17* and *2-18]*

Rotational Relative Wind (Tip-Path Plane)

The rotation of rotor blades as they turn about the mast produces rotational relative wind (tip-path plane). The term rotational refers to the method of producing relative wind. Rotational relative wind flows opposite the physical flightpath of the airfoil, striking the blade at 90° to the leading edge and parallel to the plane of rotation; and it is constantly changing in direction during rotation. Rotational relative wind velocity is highest at blade tips, decreasing uniformly to zero at the axis of rotation (center of the mast). *[Figure 2-19]*

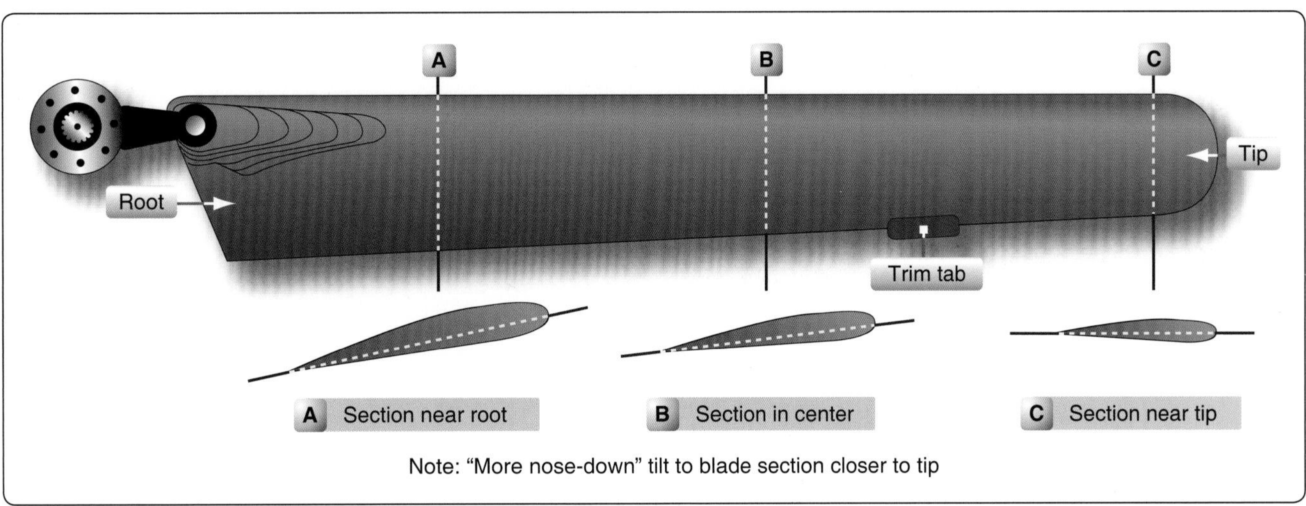

Figure 2-14. *Blade twist.*

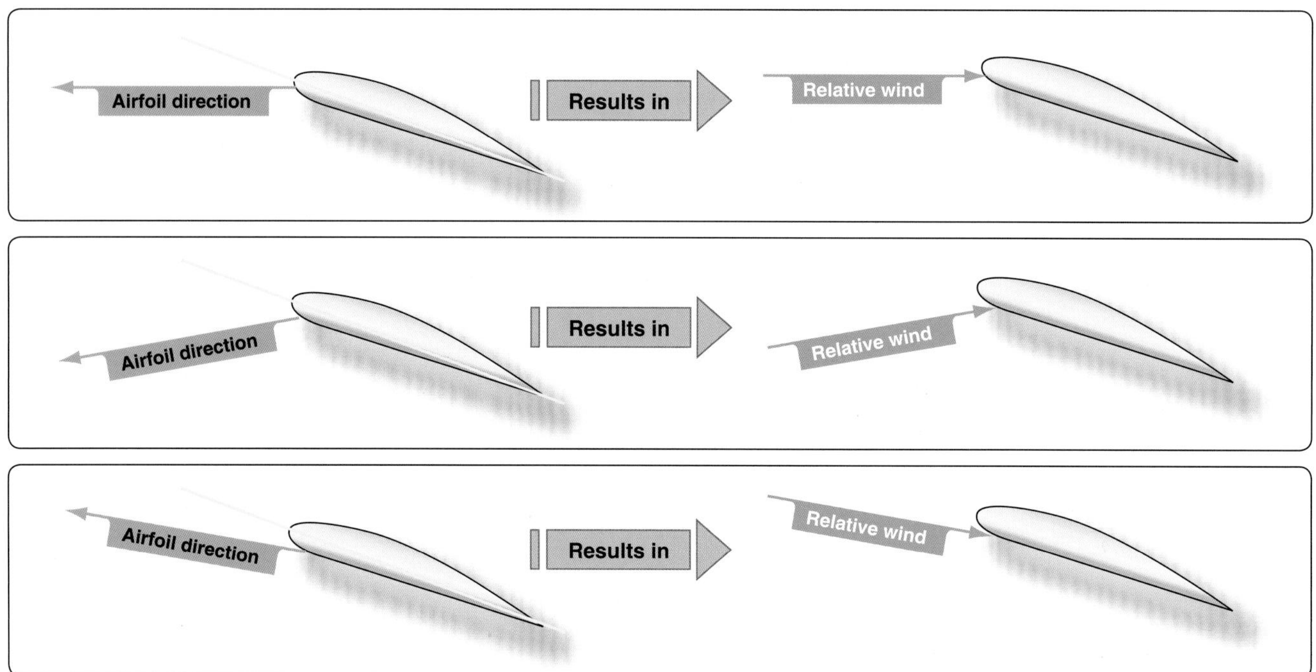

Figure 2-15. *Relative wind.*

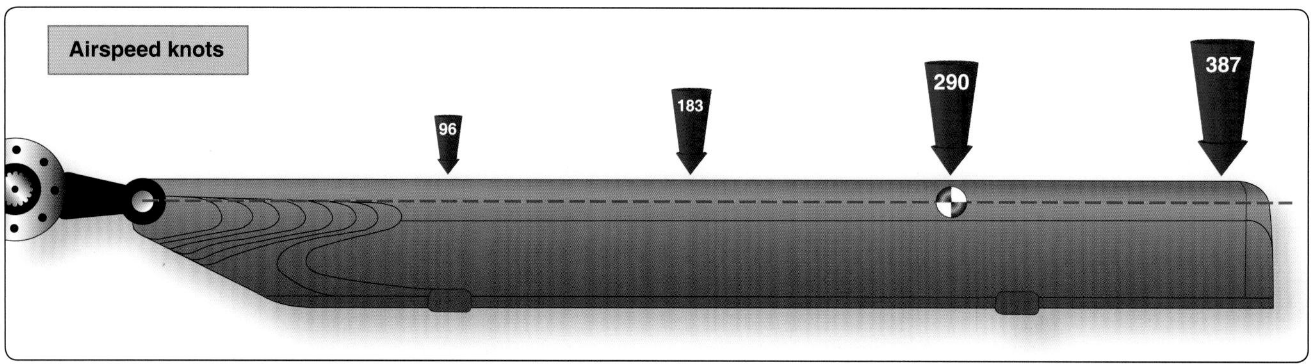

Figure 2-16. *Horizontal component of relative wind.*

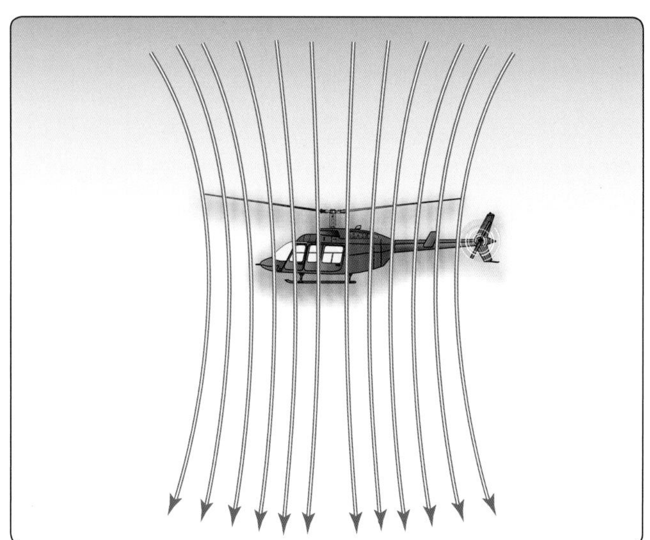

Figure 2-17. *Induced flow.*

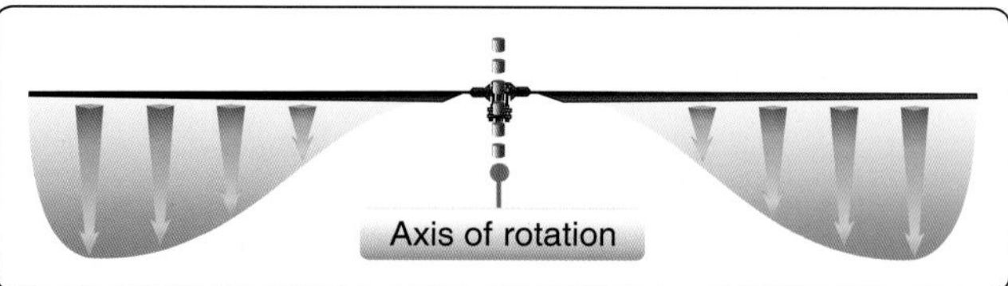

Figure 2-18. *Normal induced flow velocities along the blade span during hovering flight. Downward velocity is highest at the blade tip where blade speed is highest. As blade speed decreases nearer the center of the disk, downward velocity is less.*

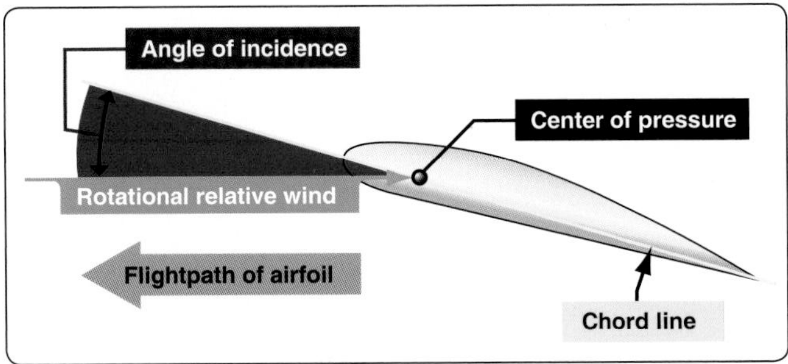

Figure 2-19. *Rotational relative wind.*

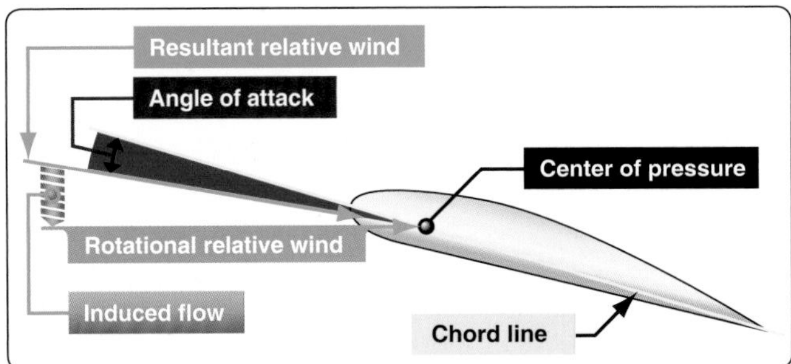

Figure 2-20. *Resultant relative wind.*

Resultant Relative Wind

The resultant relative wind at a hover is rotational relative wind modified by induced flow. This is inclined downward at some angle and opposite the effective flightpath of the airfoil, rather than the physical flightpath (rotational relative wind). The resultant relative wind also serves as the reference plane for development of lift, drag, and total aerodynamic force (TAF) vectors on the airfoil. *[Figure 2-20]* When the helicopter has horizontal motion, airspeed further modifies the resultant relative wind. The airspeed component of relative wind results from the helicopter moving through the air. This airspeed component is added to, or subtracted from, the rotational relative wind depending on whether the blade is advancing or retreating in relation to helicopter movement. Introduction of airspeed relative wind also modifies induced flow. Generally, the downward velocity of induced flow is reduced. The pattern of air circulation through the disk changes when the aircraft has horizontal motion. As the helicopter gains airspeed, the addition of forward velocity results in decreased induced flow velocity. This change results in an improved efficiency (additional lift) being produced from a given blade pitch setting.

Induced Flow (Downwash)

At flat pitch, air leaves the trailing edge of the rotor blade in the same direction it moved across the leading edge; no lift or induced flow is being produced. As blade pitch angle is increased, the rotor disk induces a downward flow of air through the rotor blades creating a downward component of air that is added to the rotational relative wind. Because the blades are moving horizontally, some of the air is displaced downward. The blades travel along the same path and pass a given point in rapid succession. Rotor blade action changes the still air to a column of descending air. Therefore, each blade has a decreased AOA due to the downwash. This downward flow of air is called induced flow (downwash). It is most pronounced at a hover under no-wind conditions. *[Figure 2-21]*

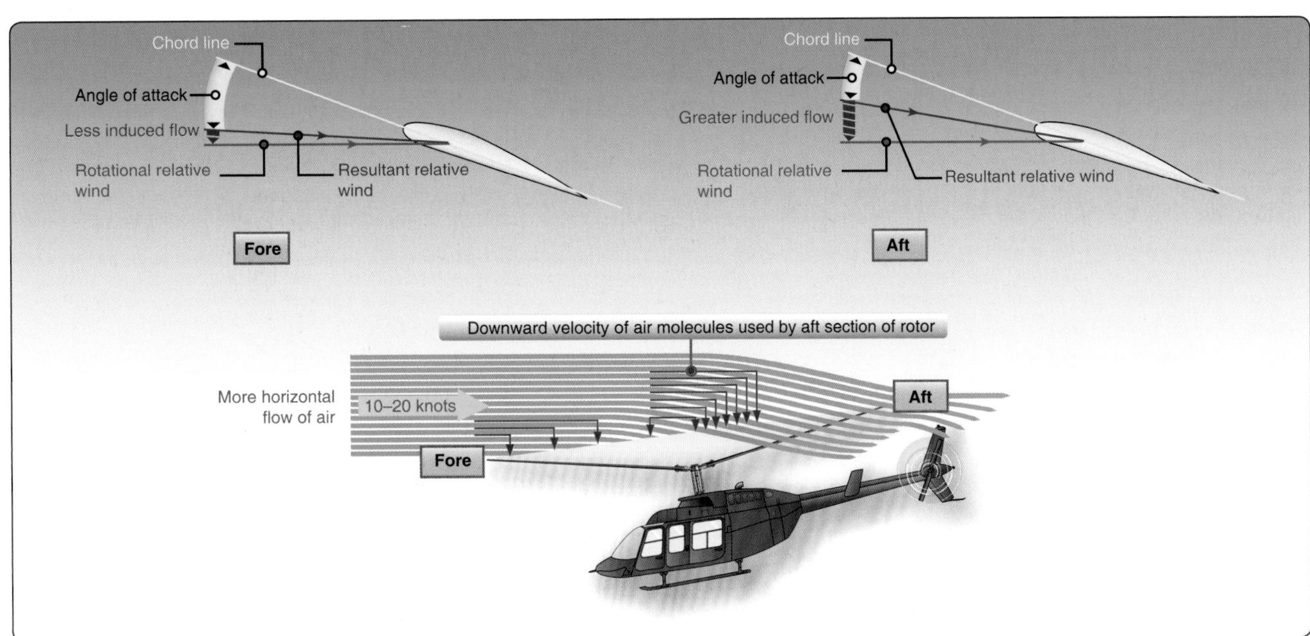

Figure 2-21. *A helicopter in forward flight, or hovering with a headwind or crosswind, has more molecules of air entering the aft portion of the rotor disk. Therefore, at the rear of the rotor disk, the angle of attack is less and the induced flow is greater.*

In Ground Effect (IGE)

Ground effect is the increased efficiency of the rotor disk caused by interference of the airflow when near the ground. The air pressure or density is increased, which acts to decrease the downward velocity of air. Ground effect permits relative wind to be more horizontal, lift vector to be more vertical, and induced drag to be reduced. These conditions allow the rotor disk to be more efficient. Maximum ground effect is achieved when hovering over smooth hard surfaces. When hovering over surfaces as tall grass, trees, bushes, rough terrain, and water, maximum ground effect is reduced. Rotor efficiency is increased by ground effect to a height of about one rotor diameter (measured from the ground to the rotor disk) for most helicopters. Since the induced flow velocities are decreased, the AOA is increased, which requires a reduced blade pitch angle and a reduction in induced drag. This reduces the power required to hover IGE. *[Figure 2-22]*

Out of Ground Effect (OGE)

The benefit of placing the helicopter near the ground is lost above IGE altitude. Above this altitude, the power required to hover remains nearly constant, given similar conditions (such as wind). Induced flow velocity is increased, resulting in a decrease in AOA and a decrease in lift. Under the correct circumstances, this downward flow can become so localized that the helicopter and locally disturbed air will sink at alarming rates. This effect is called vortex ring state (formerly referenced as settling-with-power) and is discussed at length in Chapter 11, Helicopter Emergencies and Hazards. A higher blade pitch angle is required to maintain the same AOA as in IGE hover. The increased pitch angle also creates more drag. This increased pitch angle and drag requires more power to hover OGE than IGE. *[Figure 2-23]*

Rotor Blade Angles

There are two angles that enable a rotor disk to produce the lift required for a helicopter to fly: angle of incidence and angle of attack.

Angle of Incidence

Angle of incidence is the angle between the chord line of a main or tail rotor blade and its rotor disk. It is a mechanical angle rather than an aerodynamic angle and is sometimes referred to as blade pitch angle. *[Figure 2-24]* In the absence of induced flow, AOA and angle of incidence are the same. Whenever induced flow, up flow (inflow), or airspeed modifies the relative wind, the AOA is different from the angle of incidence. Collective input and cyclic feathering (see page 2-12) change the angle of incidence. A change in the angle of incidence changes the AOA, which changes the coefficient of lift, thereby changing the lift produced by the airfoil.

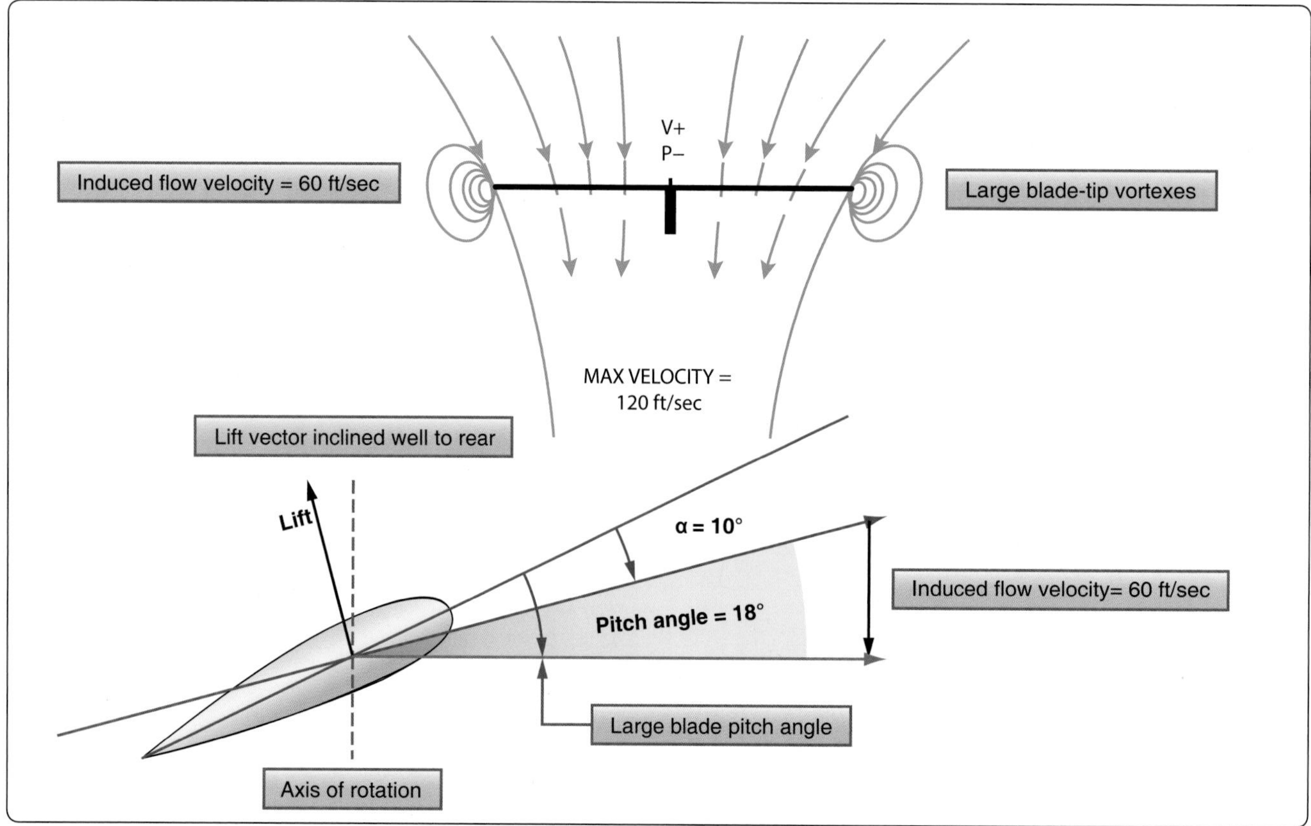

Figure 2-22. *In ground effect (IGE).*

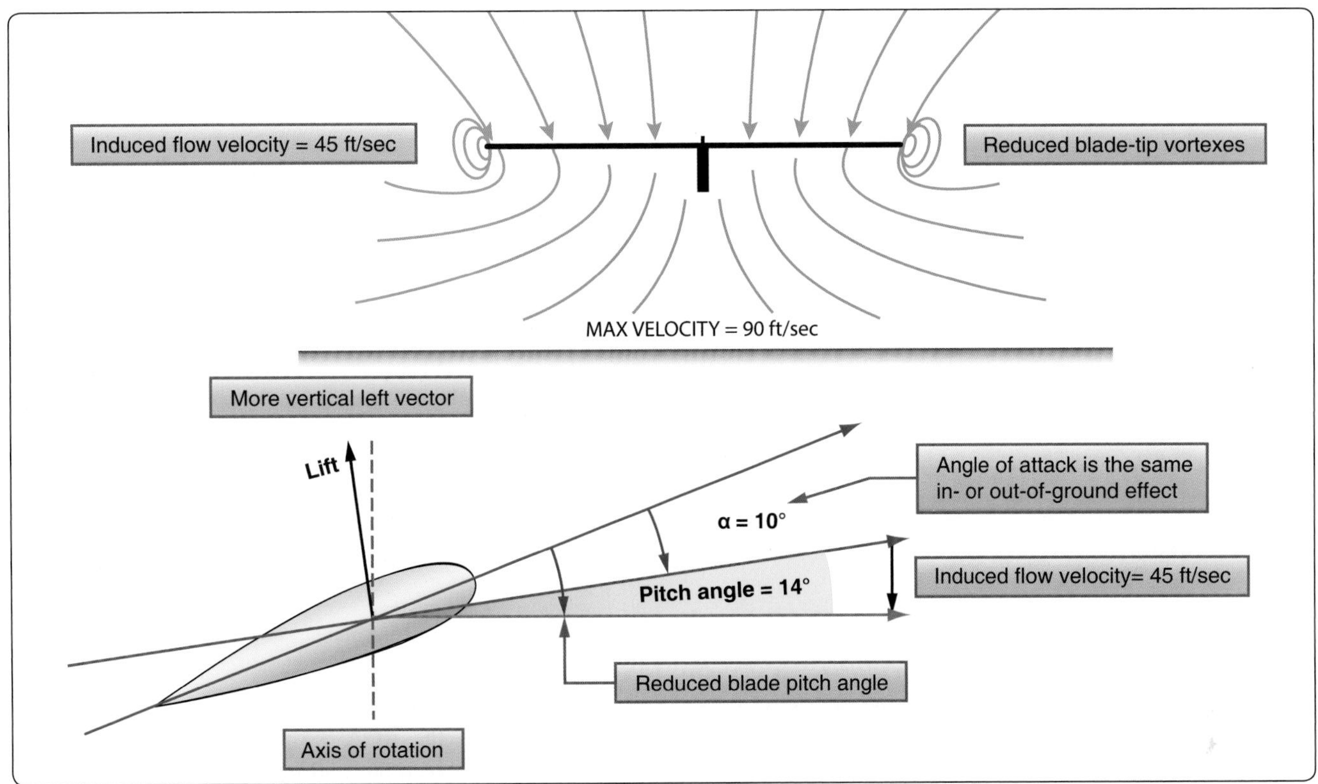

Figure 2-23. *Out of ground effect (OGE).*

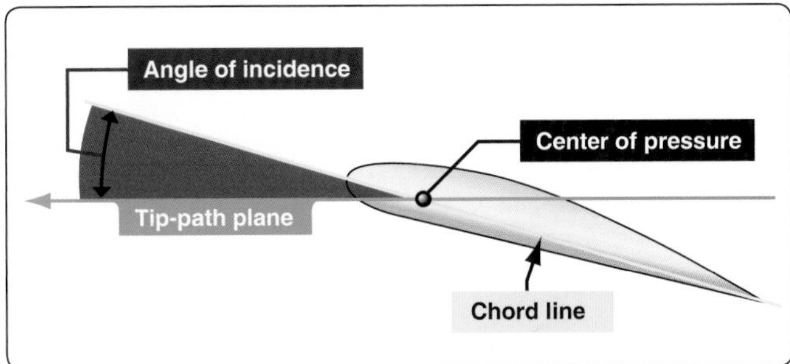

Figure 2-24. *Angle of incidence.*

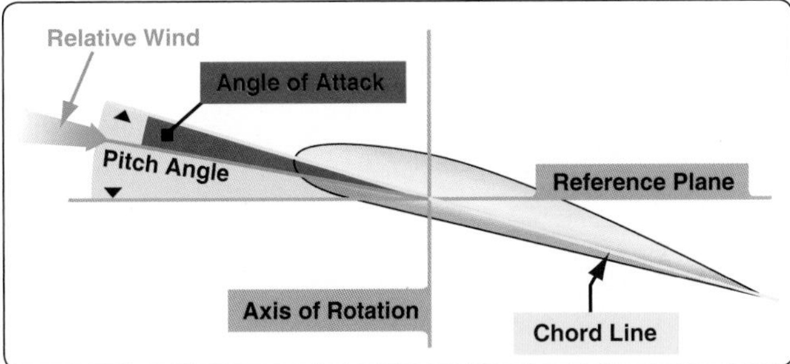

Figure 2-25. *The AOA is the angle between the airfoil chord line and resultant relative wind.*

Angle of Attack

AOA is the angle between the airfoil chord line and resultant relative wind. *[Figure 2-25]* It is an aerodynamic angle and not easy to measure. It can change with no change in the blade pitch angle (angle of incidence, discussed earlier).

When the AOA is increased, air flowing over the airfoil is diverted over a greater distance, resulting in an increase of air velocity and more lift. As the AOA is increased further, it becomes more difficult for air to flow smoothly across the top of the airfoil. At this point, the airflow begins to separate from the airfoil and enters a burbling or turbulent pattern. The turbulence results in a large increase in drag and loss of lift in the area where it is taking place. Increasing the AOA increases lift until the critical angle of attack is reached. Any increase in the AOA beyond this point produces a stall and a rapid decrease in lift (refer to the Low Rotor RPM and Rotor Stall section of Chapter 11, Helicopter Emergencies and Hazards).

Several factors may change the rotor blade AOA. The pilot has little direct control over AOA except indirectly through the flight control input. Collective and cyclic feathering help to make these changes. Feathering is the rotation of the blade about its longitudinal axis by collective/cyclic inputs causing changes in blade pitch angle. Collective feathering changes angle of incidence equally and in the same direction on all rotor blades simultaneously. This action changes AOA, which changes coefficient of lift (CL), and affects overall lift of the rotor disk.

Cyclic feathering changes the blade's AOA differentially around the rotor disk and creates a differential lift. Aviators use cyclic feathering to control attitude of the rotor disk. It is the means to control rearward tilt of the rotor (blowback) caused by flapping action and (along with blade flapping) counteract dissymmetry of lift (discussed in chapter 3). Cyclic feathering causes attitude of the rotor disk to change but does not change the amount of net lift the rotor disk is producing.

Most of the changes in AOA come from change in airspeed and rate of climb or descent; others such as flapping occur automatically due to the rotor system design. Flapping is the up and down movement of rotor blades about a hinge on a fully articulated rotor system. A semi-rigid system does not have a hinge but flap as a unit. A rigid rotor system has no vertical or horizontal hinges, so the blades cannot flap or drag, but they can flex. By flexing, the blades themselves compensate for the forces which previously required rugged hinges. It occurs in response to changes in lift due to changing velocity or cyclic feathering. No flapping occurs when the tip-path plane is perpendicular to the mast. The flapping action alone, or along with cyclic feathering, controls dissymmetry of lift. Flapping is the primary means of compensating for dissymmetry of lift.

Pilots adjust AOA through normal control manipulation of the pitch angle of the blades. If the pitch angle is increased, the AOA increases; if the pitch angle is reduced, the AOA is reduced.

Powered Flight

In powered flight (hovering, vertical, forward, sideward, or rearward), the total lift and thrust forces of a rotor are perpendicular to the rotor disk.

Hovering Flight

Hovering is the most challenging part of flying a helicopter. This is because a helicopter generates its own gusty air while in a hover, which acts against the fuselage and flight control surfaces. The end result is constant control inputs and corrections by the pilot to keep the helicopter where it is required to be. Despite the complexity of the task, the control inputs in a hover are simple. The cyclic is used to eliminate drift in the horizontal plane, controlling forward, backward, right and left movement or travel. The throttle, if not governor controlled, is used to control revolutions per minute (rpm). The collective is used to maintain altitude. The pedals are used to control nose direction or heading. It is the interaction of these controls that makes hovering difficult, since an adjustment in any one control requires an adjustment of the other two, creating a cycle of constant correction. During hovering flight, a helicopter maintains a constant position over a selected point, usually a few feet above the ground. The ability of the helicopter to hover comes from the both the lift component, which is the force developed by the main rotor(s) to overcome gravity and aircraft weight, and the thrust component, which acts horizontally to accelerate or decelerate the helicopter in the desired direction. Pilots direct the thrust of the rotor disk by using the cyclic to rotate the rotor disk plane relative to the horizon. They do this in order to induce travel or compensate for the wind and hold a position. At a hover in a no-wind condition, all opposing forces (lift, thrust, drag, and weight) are in balance; they are equal and opposite. Therefore, lift and weight are equal, resulting in the helicopter remaining at a stationary hover. *[Figure 2-26]*

While hovering, the amount of main rotor thrust can be adjusted to maintain the desired hovering height. This is done by changing the angle of incidence (by moving the collective) of the rotor blades, and hence their AOA. Changing the AOA changes the drag on the rotor blades, and the power delivered by the engine must change as well to keep the rotor speed constant.

The weight that must be supported is the total weight of the helicopter and its occupants. If the amount of lift is greater than the actual weight, the helicopter accelerates upwards until the lift force equals the weight of the helicopter; if lift is less than weight, the helicopter accelerates downward.

The drag of a hovering helicopter is mainly induced drag incurred while the blades are producing lift. There is, however, some profile drag on the blades as they rotate through the air and a small amount of parasite drag from the non-lift-producing surfaces of the helicopter, such as the rotor hub, cowlings, and

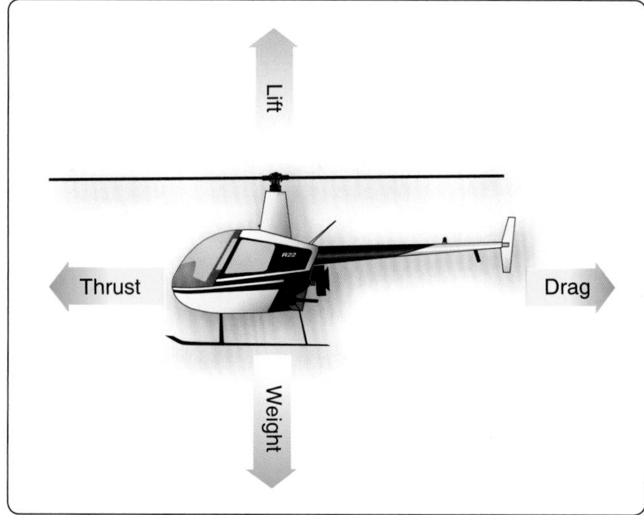

Figure 2-26. *To maintain a hover at a constant altitude, the lift must equal the weight of the helicopter. Thrust must equal any wind and tail rotor thrust to maintain position. The power must be sufficient to turn the rotors and overcome the various drags and frictions involved.*

landing gear. Throughout the rest of this discussion, the term "drag" includes induced, profile and parasite drag.

An important consequence of producing thrust is torque. As discussed earlier, Newton's Third Law states: for every action there is an equal and opposite reaction. Therefore, as the engine turns the main rotor disk in a counterclockwise direction, the helicopter fuselage wants to turn clockwise. The amount of torque is directly related to the amount of engine power being used to turn the main rotor disk. As power changes, torque changes.

To counteract this torque-induced turning tendency, an antitorque rotor or tail rotor is incorporated into most helicopter designs. A pilot can vary the amount of thrust produced by the tail rotor in relation to the amount of torque produced by the engine. As the engine supplies more power to the main rotor, the tail rotor must produce more thrust to overcome the increased torque effect. This control change is accomplished through the use of antitorque pedals (See page 3-4).

Translating Tendency (Drift)

During hovering flight, a single main rotor helicopter tends to move in the direction of tail rotor thrust. This lateral (or sideward) movement is called translating tendency. *[Figure 2-27]*

To counteract this tendency, one or more of the following features may be used. All examples are for a counterclockwise rotating main rotor disk.

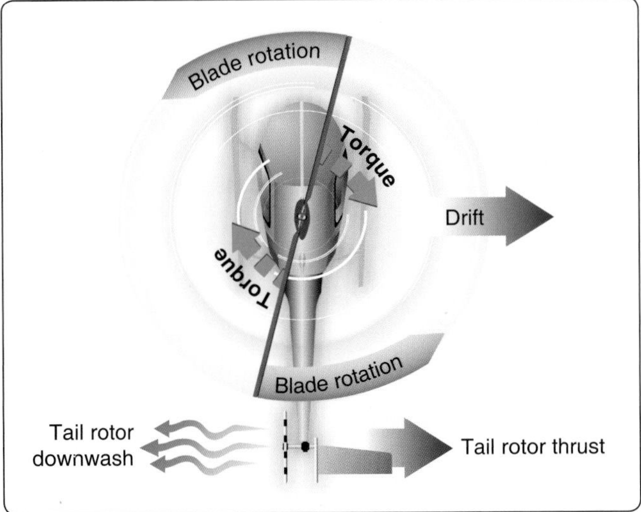

Figure 2-27. *A tail rotor is designed to produce thrust in a direction opposite torque. The thrust produced by the tail rotor is sufficient to move the helicopter laterally.*

- The main transmission is mounted at a slight angle to the left (when viewed from behind) so that the rotor mast has a built-in tilt to oppose the tail rotor thrust.
- Flight controls can be rigged so that the rotor disk is tilted to the left slightly when the cyclic is centered. Whichever method is used, the tip-path plane is tilted slightly to the left in the hover.
- The transmission is mounted so the rotor shaft is vertical with respect to the fuselage, the helicopter "hangs" left skid low in the hover. (The opposite is true for rotor disks turning clockwise when viewed from above.)
- The helicopter fuselage will also be tilted when the tail rotor is below the main rotor disk and supplying antitorque thrust. The fuselage tilt is caused by the imperfect balance of the tail rotor thrust against the main rotor torque in the same plane. The helicopter tilts due to two separate forces, the main rotor disk tilt to neutralize the translating tendency and the lower tail rotor thrust below the plane of the torque action.
- In forward flight, the tail rotor continues to push to the right, and the helicopter makes a small angle with the wind when the rotors are level and the slip ball is in the middle (See page 12-2). This is called inherent sideslip. For some larger helicopters, the vertical fin or stabilizer is often designed with the tail rotor mounted on them to correct this side slip and to eliminate some of the tilting at a hover. (By mounting the tail rotor on top of the vertical fin or pylon, the antitorque is more in line with or closer to the horizontal plane of torque, resulting in less airframe (or body) lean from the tail rotor.) Also, having the tail rotor higher off the ground reduces the risk of objects coming in contact with the blades, but at the cost of increased weight and complexity.

Pendular Action

Since the fuselage of the helicopter, with a single main rotor, is suspended from a single point and has considerable mass, it is free to oscillate either longitudinally or laterally in the same way as a pendulum. This pendular action can be exaggerated by overcontrolling; therefore, control movements should be smooth and not exaggerated. *[Figure 2-28]*

The horizontal stabilizer helps to level the helicopter in forward flight. However, in rearward flight, the horizontal stabilizer can press the tail downward, resulting in a tail strike if the helicopter is moved rearward into the wind. Normally, with the helicopter mostly into the wind, the horizontal stabilizer experiences less headwind component as the helicopter begins rearward travel (downwind). When

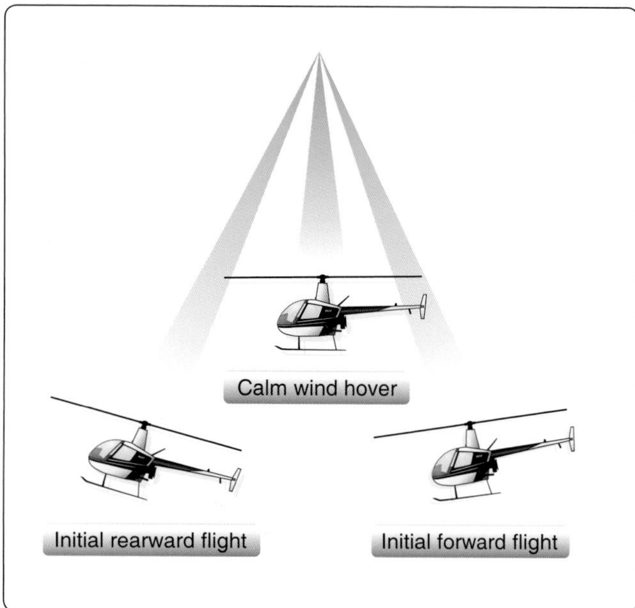

Figure 2-28. *Because the helicopter's body has mass and is suspended from a single point (the rotor mast head), it tends to act much like a pendulum.*

rearward flight groundspeed equals the windspeed, then the helicopter is merely hovering in a no-wind condition. However, rearward hovering into the wind requires considerable care and caution to prevent tail strikes.

It is important to note that there is a difference in the amount of pendular action between a semirigid system and a fully articulated system. Because of the hard connection (offset) of the latter, the centrifugal force pulling out on the blades is transferred to the fuselage, and the fuselage tends to follow the rotor attitude. The semirigid system is a true pendulum, with thrust required to create a moment around the fuselage CG to allow for control of the fuselage. This comes into play later when mast bumping is discussed.

Coning

In order for a helicopter to generate lift, the rotor blades must be turning. Rotor disk rotation drives the blades into the air, creating a relative wind component without having to move the airframe through the air as with an airplane or glider. Depending on the motion of the blades and helicopter airframe, many factors cause the relative wind direction to vary. The rotation of the rotor disk creates centrifugal force (inertia), which tends to pull the blades straight outward from the main rotor hub: the faster the rotation, the greater the centrifugal force, the slower the rotation, the smaller the centrifugal force. This force gives the rotor blades their rigidity and, in turn, the strength to support the weight of the helicopter. The maximum centrifugal force generated is determined by the maximum operating rotor revolutions per minute (rpm).

As lift on the blades is increased (in a takeoff, for example), two major forces are acting at the same time—centrifugal force acting outward, and lift acting upward. The result of these two forces is that the blades assume a conical path instead of remaining in the plane perpendicular to the mast. This can be seen in any helicopter when it takes off; the rotor disk changes from flat to a slight cone shape. *[Figure 2-29]*

If the rotor rpm is allowed to go too low (below the minimum power-on rotor rpm, for example), the centrifugal force becomes smaller and the coning angle becomes much larger. In other words, should the rpm decrease too much, at some point the rotor blades fold up with no chance of recovery.

Coriolis Effect (Law of Conservation of Angular Momentum)

The Coriolis Effect is also referred to as the law of conservation of angular momentum. It states that the value of angular momentum of a rotating body does not change unless an external force is applied. In other words, a rotating body continues to rotate with the same rotational velocity until some external force is applied to change the speed of rotation. Angular momentum is the moment of inertia (mass times distance from the center of rotation squared) multiplied by the speed of rotation.

Changes in angular velocity, known as angular acceleration and deceleration, take place as the mass of a rotating body is moved closer to or farther away from the axis of rotation. The speed of the rotating mass varies proportionately with the square of the radius.

An excellent example of this principle in action is a figure skater performing a spin on ice skates. The skater begins rotation on one foot, with the other leg and both arms extended. The rotation of the skater's body is relatively slow. When a skater draws both arms and one leg inward, the moment of inertia (mass times radius squared) becomes

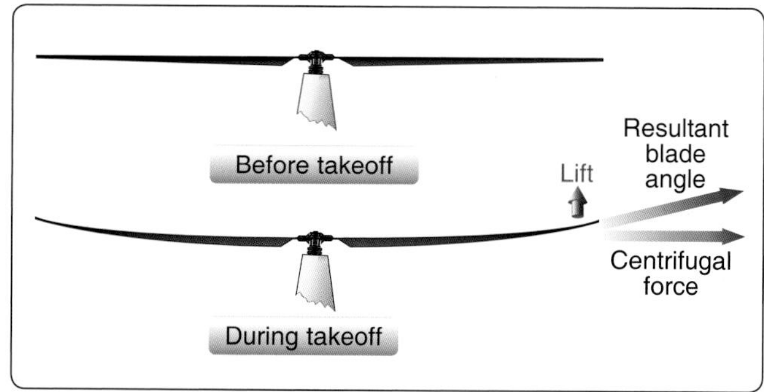

Figure 2-29. *During takeoff, the combination of centrifugal force and lift cause the rotor disk to cone upward.*

much smaller and the body is rotating almost faster than the eye can follow. Because the angular momentum must, by law of nature, remain the same (no external force applied), the angular velocity must increase.

The rotor blade rotating about the rotor hub possesses angular momentum. As the rotor begins to cone due to G-loading maneuvers, the diameter of the rotor disk shrinks. Due to conservation of angular momentum, the blades increase speed even though the blade tips have a shorter distance to travel due to reduced disk diameter. The action results in an increase in rotor rpm which causes a slight increase in lift. Most pilots arrest this increase of rpm with an increase in collective pitch. This increase in blade rpm lift is somewhat negated by the slightly smaller disk area as the blades cone upward.

Gyroscopic Precession

The spinning main rotor of a helicopter acts like a gyroscope. As such, it has the properties of gyroscopic action, one of which is precession. Gyroscopic precession is the resultant action or deflection of a spinning object when a force is applied to this object. This action occurs approximately 90° in the direction of rotation from the point where the force is applied (or 90° later in the rotation cycle). *[Figure 2-30]*

Examine a two-bladed rotor disk to see how gyroscopic precession affects the movement of the tip-path plane. Moving the cyclic pitch control increases the angle of incidence of one rotor blade with the result of a greater lifting force being applied at that point in the plane of rotation. This same control movement simultaneously decreases the angle of incidence of the other blade the same amount, thus decreasing the lifting force applied at that point in the plane of rotation. The blade with the increased angle of incidence tends to flap up; the blade with the decreased angle of incidence tends to flap down. Because the rotor disk acts like a gyro, the blades reach maximum deflection at a point approximately 90° later in the plane of rotation. *Figure 2-31* illustrates the result of a forward cyclic input. The retreating blade angle of incidence is increased, and the advancing blade angle of incidence is decreased resulting in a tipping forward of the tip-path plane, since maximum deflection takes place 90° later when the blades are at the rear and front, respectively. In a rotor disk using three or more blades, the movement of the cyclic pitch control changes the angle of incidence of each blade an appropriate amount so that the end result is the same.

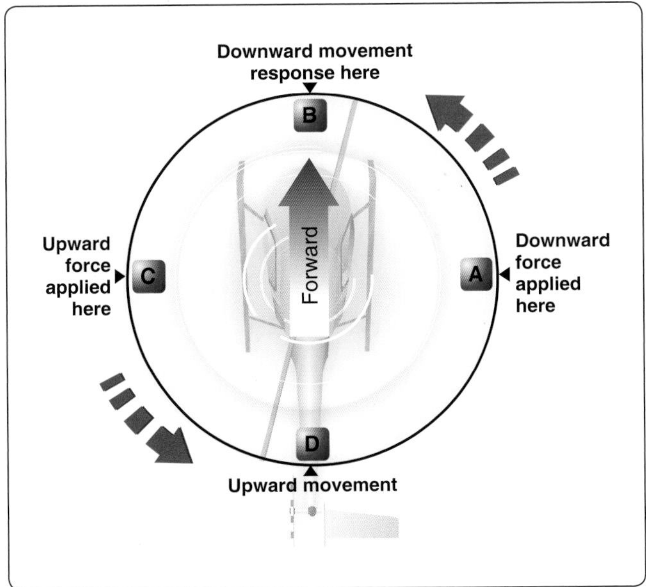

Figure 2-30. *Gyroscopic precession.*

Vertical Flight

Hovering is actually an element of vertical flight. Increasing the angle of incidence of the rotor blades (pitch) while keeping their rotation speed constant generates additional lift and the helicopter ascends. Decreasing the pitch causes the helicopter to descend. In a no-wind condition in which lift and thrust are less than weight and drag, the helicopter descends vertically. If lift and thrust are greater than weight and drag, the helicopter ascends vertically. *[Figure 2-32]*

Forward Flight

In steady forward flight, with no change in airspeed or vertical speed, the four forces of lift, thrust, drag, and weight must be in balance. Once the tip-path plane is tilted forward, the total lift-thrust force is also tilted forward. This resultant lift-thrust force can be resolved into two components—lift acting vertically upward and thrust acting horizontally in the direction of flight. In addition to lift and thrust, there is weight (the downward acting force) and drag (the force opposing the motion of an airfoil through the air). *[Figure 2-33]*

In straight-and-level, unaccelerated forward flight (straight-and-level flight is flight with a constant heading and at a constant altitude), lift equals weight and thrust equals drag. If lift exceeds weight, the helicopter accelerates vertically until the forces are in balance; if thrust is less than drag, the helicopter slows down until the forces are in balance. As a helicopter initiates a move forward, it begins to lose altitude because lift is lost as thrust is diverted forward. However, as the helicopter begins to accelerate from a hover, the rotor disk becomes more efficient due to translational lift (see translational lift on page 2-19). The result is excess power over that which is required to hover. Continued acceleration causes an even larger increase in airflow through the rotor disk (up to a maximum determined by drag and the engine's limit of power), and more efficient flight. In order to maintain unaccelerated flight, the pilot must understand that with

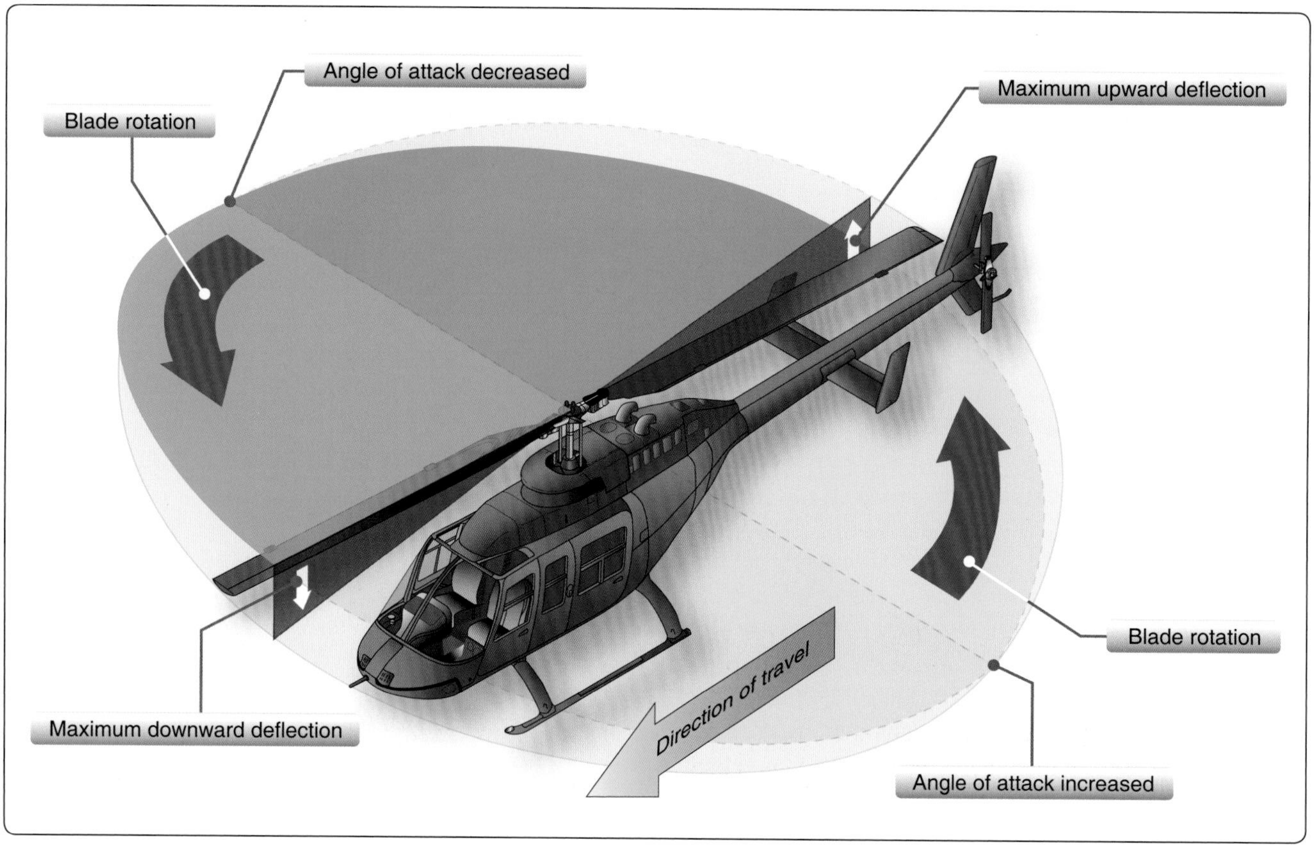

Figure 2-31. *As each blade passes the 90° position on the left in a counterclockwise main rotor blade rotation, the maximum increase in angle of incidence occurs. As each blade passes the 90° position to the right, the maximum decrease in angle of incidence occurs. Maximum deflection takes place 90° later—maximum upward deflection at the rear and maximum downward deflection at the front—and the tip-path plane tips forward.*

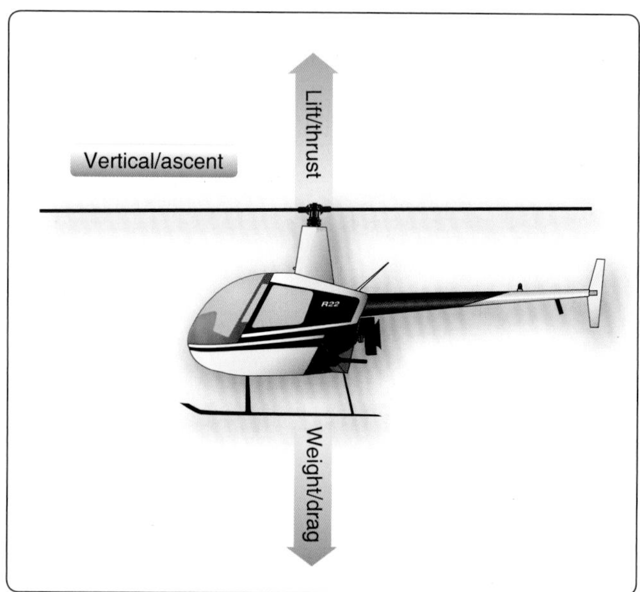

Figure 2-32. *Balanced forces: hovering in a no-wind condition.*

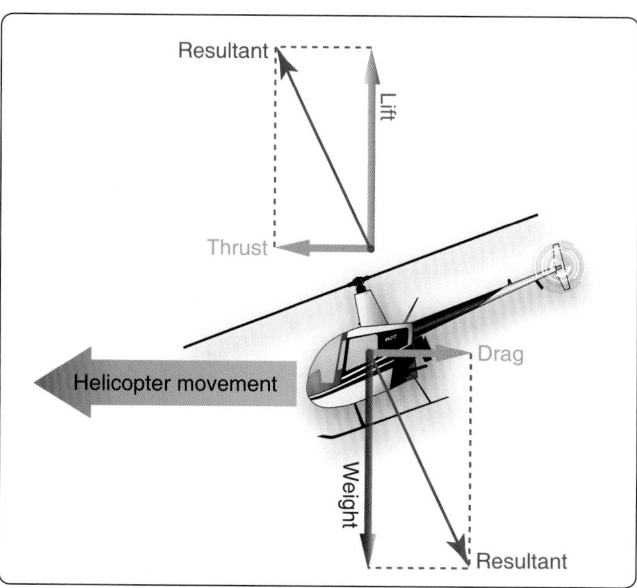

Figure 2-33. *To transition to forward flight, more lift and thrust must be generated to overcome the forces of weight and drag.*

any changes in power or in cyclic movement, the helicopter begins either to climb or to descend. Once straight-and-level flight is obtained, the pilot should make note of the power (torque setting) required and not make major adjustments to the flight controls. *[Figure 2-34]*

Airflow in Forward Flight

Airflow across the rotor disk in forward flight varies from airflow at a hover. In forward flight, air flows opposite the aircraft's flightpath. The velocity of this air flow equals the helicopter's forward speed. Because the rotor blades turn in a circular pattern, the velocity of airflow across a blade depends on the position of the blade in the plane of rotation at a given instant, its rotational velocity, and airspeed of the helicopter. Therefore, the airflow meeting each blade varies continuously as the blade rotates. The highest velocity of airflow occurs over the right side (3 o'clock position) of the helicopter (advancing blade in a rotor disk that turns counterclockwise) and decreases to rotational velocity over the nose. It continues to decrease until the lowest velocity of airflow occurs over the left side (9 o'clock position) of the helicopter (retreating blade). As the blade continues to rotate, velocity of the airflow then increases to rotational velocity over the tail. It continues to increase until the blade is back at the 3 o'clock position.

The advancing blade in *Figure 2-35,* position A, moves in the same direction as the helicopter. The velocity of the air meeting this blade equals rotational velocity of the blade plus wind velocity resulting from forward airspeed. The retreating blade (position C) moves in a flow of air moving in the opposite direction of the helicopter. The velocity of airflow meeting this blade equals rotational velocity of the blade minus wind velocity resulting from forward airspeed. The blades (positions B and D) over the nose and tail move essentially at right angles to the airflow created by forward airspeed; the velocity of airflow meeting these blades equals the rotational velocity. This results in a change to velocity of airflow all across the rotor disk and a change to the lift pattern of the rotor disk.

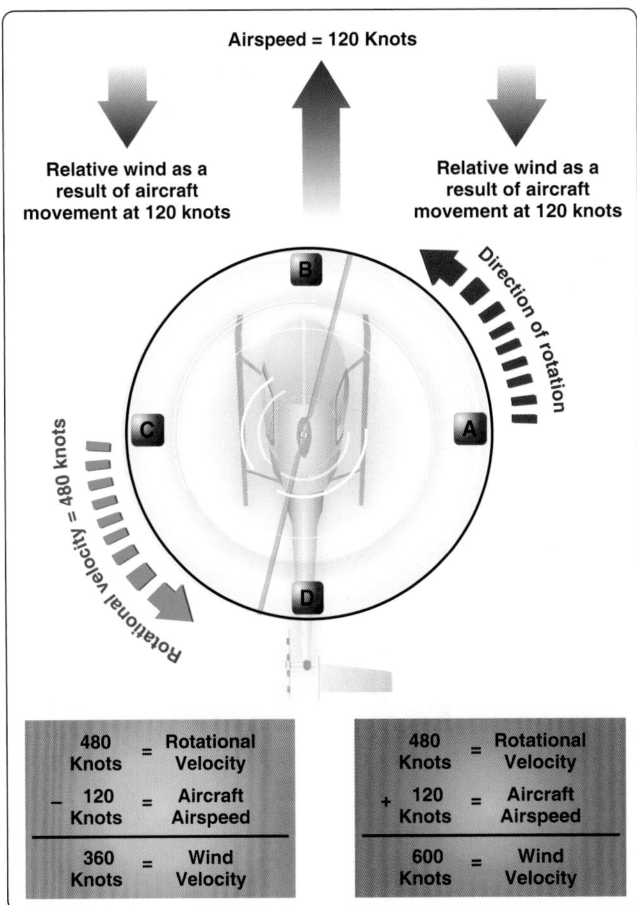

Figure 2-35. *Airflow in forward flight.*

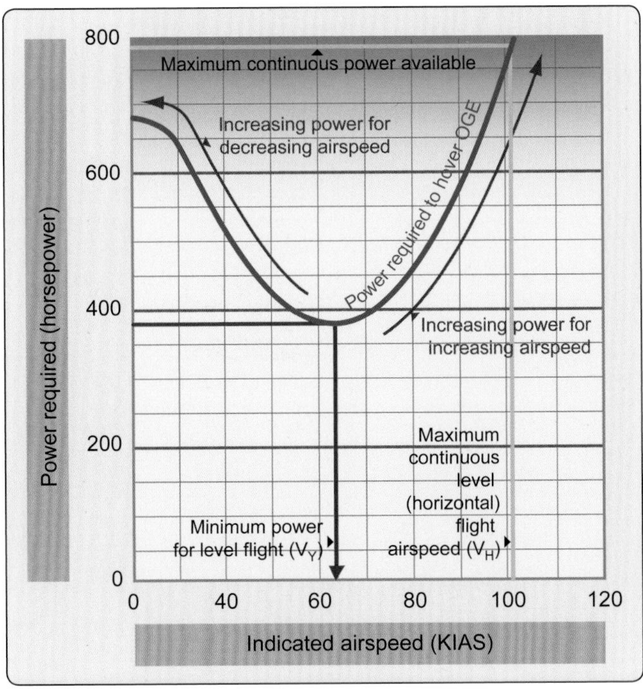

Figure 2-34. *Power versus airspeed chart.*

Advancing Blade

As the relative wind speed of the advancing blade increases, the blade gains lift and begins to flap up. It reaches its maximum upflap velocity at the 3 o'clock position, where the wind velocity is the greatest. This upflap creates a downward flow of air and has the same effect as increasing the induced flow velocity by imposing a downward vertical velocity vector to the relative wind which decreases the AOA.

Retreating Blade

As relative wind speed of the retreating blade decreases, the blade loses lift and begins to flap down. It reaches its maximum downflap velocity at the 9 o'clock position, where wind velocity is the least. This downflap creates an upward flow of air and has the same effect as decreasing the induced flow velocity by imposing an upward velocity vertical vector to the relative wind which increases the AOA.

Dissymmetry of Lift

Dissymmetry of lift is the differential (unequal) lift between advancing and retreating halves of the rotor disk caused by the different wind flow velocity across each half. This difference in lift would cause the helicopter to be uncontrollable in any situation other than hovering in a calm wind. There must be a means of compensating, correcting, or eliminating this unequal lift to attain symmetry of lift.

When the helicopter moves through the air, the relative airflow through the main rotor disk is different on the advancing side from the retreating side. The relative wind encountered by the advancing blade is increased by the forward speed of the helicopter, while the relative wind speed acting on the retreating blade is reduced by the helicopter's forward airspeed. Therefore, as a result of the relative wind speed, the advancing blade side of the rotor disk can produce more lift than the retreating blade side. *[Figure 2-36]*

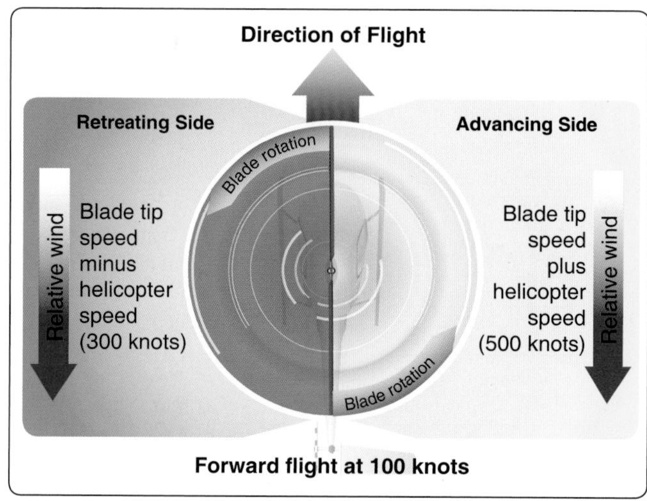

Figure 2-36. *The blade tip speed of this helicopter is approximately 400 knots. If the helicopter is moving forward at 100 knots, the relative windspeed on the advancing side is 500 knots. On the retreating side, it is only 300 knots. This difference in speed causes a dissymmetry of lift.*

If this condition were allowed to exist, a helicopter with a counterclockwise main rotor blade rotation would roll to the left because of the difference in lift. In reality, the main rotor blades flap and feather automatically to equalize lift across the rotor disk. Articulated rotor disks, usually with three or more blades, incorporate a horizontal hinge (flapping hinge) to allow the individual rotor blades to move, or flap up and down as they rotate. A semi-rigid rotor disk (two blades) utilizes a teetering hinge, which allows the blades to flap as a unit. When one blade flaps up, the other blade flaps down.

As shown in *Figure 2-37,* as the rotor blade reaches the advancing side of the rotor disk (A), it reaches its maximum up flap velocity. When the blade flaps upward, the angle between the chord line and the resultant relative wind decreases. This decreases the AOA, which reduces the amount of lift produced by the blade. At position (C), the rotor blade is now at its maximum down flapping velocity. Due to down flapping, the angle between the chord line and the resultant relative wind increases. This increases the AOA and thus the amount of lift produced by the blade.

The combination of blade flapping and slow relative wind acting on the retreating blade normally limits the maximum forward speed of a helicopter. At a high forward speed, the retreating blade stalls because of a high AOA and slow relative wind speed. This situation is called retreating blade stall and is evidenced by a nose pitch up, vibration, and a rolling tendency—usually to the left in helicopters with counterclockwise blade rotation.

Pilots can avoid retreating blade stall by not exceeding the never-exceed speed. This speed is designated V_{NE} and is indicated on a placard and marked on the airspeed indicator by a red line.

Blade flapping compensates for dissymmetry of lift in the following way. At a hover, equal lift is produced around the rotor disk with equal pitch (AOI) on all the blades and at all points in the rotor disk (disregarding compensation for translating tendency). The rotor disk is parallel to the horizon. To develop a thrust force, the rotor disk must be tilted in the desired direction of movement. Cyclic feathering changes the angle of incidence differentially around the rotor disk. For a counterclockwise rotation, forward cyclic movement decreases the angle of incidence on the right of the rotor disk and increases it on the left.

When transitioning to forward flight either from a hover or taking off from the ground, pilots must be aware that as the helicopter speed increases, translational lift becomes more effective and causes the nose to rise or pitch up (sometimes referred to as blowback). This tendency is caused by the combined effects of dissymmetry of lift and transverse flow. Pilots must correct for this tendency by maintaining a constant rotor disk attitude that will move the helicopter through the speed range in which blowback occurs. If the nose is permitted to pitch up while passing through this speed range, the aircraft may also tend to roll to the right. To correct for this tendency, the pilot must continuously move the cyclic forward as velocity of the helicopter increases until the takeoff is complete, and the helicopter has transitioned into forward flight.

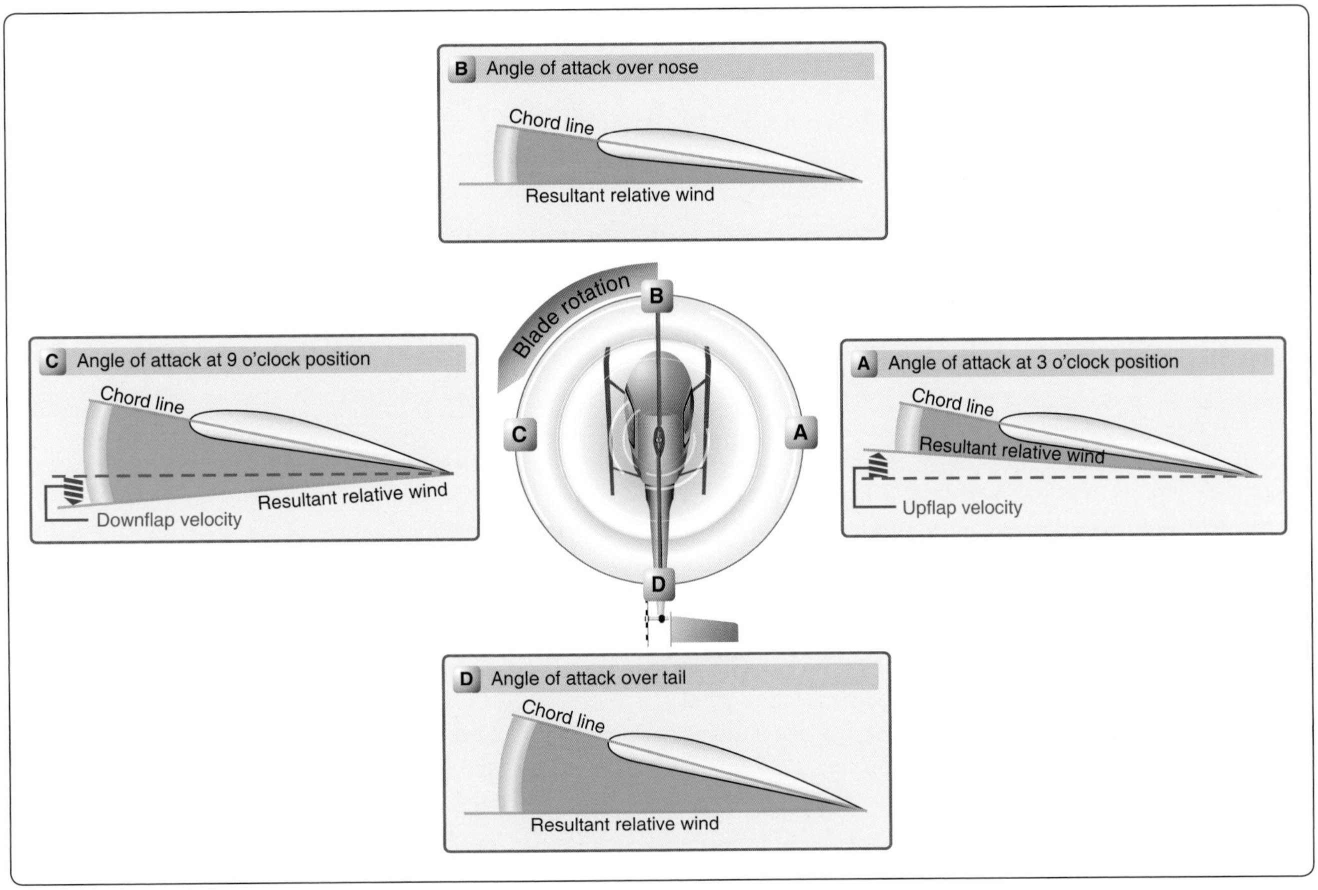

Figure 2-37. *The combined upward flapping (reduced lift) of the advancing blade and downward flapping (increased lift) of the retreating blade equalizes lift across the main rotor disk, counteracting dissymmetry of lift.*

Figure 2-38 illustrates the tilting forward of the rotor disk, which is the result of a change in pitch angle with forward cyclic. At a hover, the cyclic is centered and the pitch angle on the advancing and retreating blades is the same. At low forward speeds, moving the cyclic forward reduces pitch angle on the advancing blade and increases pitch angle on the retreating blade. This causes a slight rotor disk tilt. At higher forward speeds, the pilot must continue to move the cyclic forward. This further reduces pitch angle on the advancing blade and further increases pitch angle on the retreating blade. As a result, there is even more tilt to the rotor disk than at lower speeds.

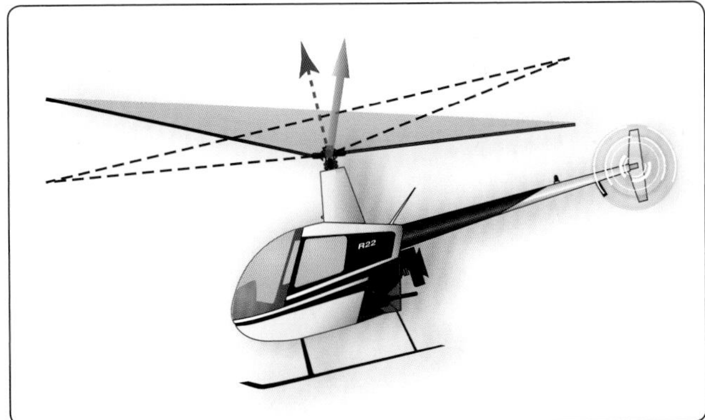

Figure 2-38. *To compensate for blowback, you must move the cyclic forward.*

A horizontal lift component (thrust) generates higher helicopter airspeed. The higher airspeed induces blade flapping to maintain symmetry of lift. The combination of flapping and cyclic feathering maintains symmetry of lift and desired attitude on the rotor disk and helicopter.

Translational Lift

Improved rotor efficiency resulting from directional flight is called translational lift. The efficiency of the hovering rotor disk is greatly improved with each knot of incoming wind gained by horizontal movement of the aircraft or surface wind. As the incoming wind produced by aircraft movement or surface wind enters the rotor disk, turbulence and vortices are left behind and the flow of air becomes more horizontal. In addition, the tail rotor becomes more aerodynamically efficient during the transition from hover to forward flight. *Figures 2-39* and *2-40* show the different airflow patterns at different speeds and how airflow affects the efficiency of the tail rotor.

Effective Translational Lift (ETL)

While transitioning to forward flight at about 16 to 24 knots, the helicopter goes through effective translational lift (ETL). As mentioned earlier in the discussion on translational lift, the rotor blades become more efficient as forward airspeed increases. Between 16 and 24 knots, the rotor disk completely outruns the recirculation of old vortices and begins to work in relatively undisturbed air. The flow of air through the rotor disk is more horizontal, which reduces induced flow and drag with a corresponding increase in angle of attach and lift. The additional lift available at this speed is referred to as the ETL, which makes the rotor disk operate more efficiently. This increased efficiency continues with increased airspeed until the best climb airspeed is reached, and total drag is at its lowest point.

As speed increases, translational lift becomes more effective, nose rises or pitches up, and aircraft rolls to the right. The combined effects of dissymmetry of lift, gyroscopic precession, and transverse flow effect cause this tendency. It is important to understand these effects and anticipate correcting for them. Once the helicopter is transitioning through ETL, the pilot needs to apply forward and left lateral cyclic input to maintain a constant rotor-disk attitude. *[Figure 2-41]*

Translational Thrust

Translational thrust occurs when the tail rotor becomes more aerodynamically efficient during the transition from hover to forward flight. As the tail rotor works in progressively less turbulent air, this improved efficiency produces more antitorque thrust, causing the nose of the aircraft to yaw left

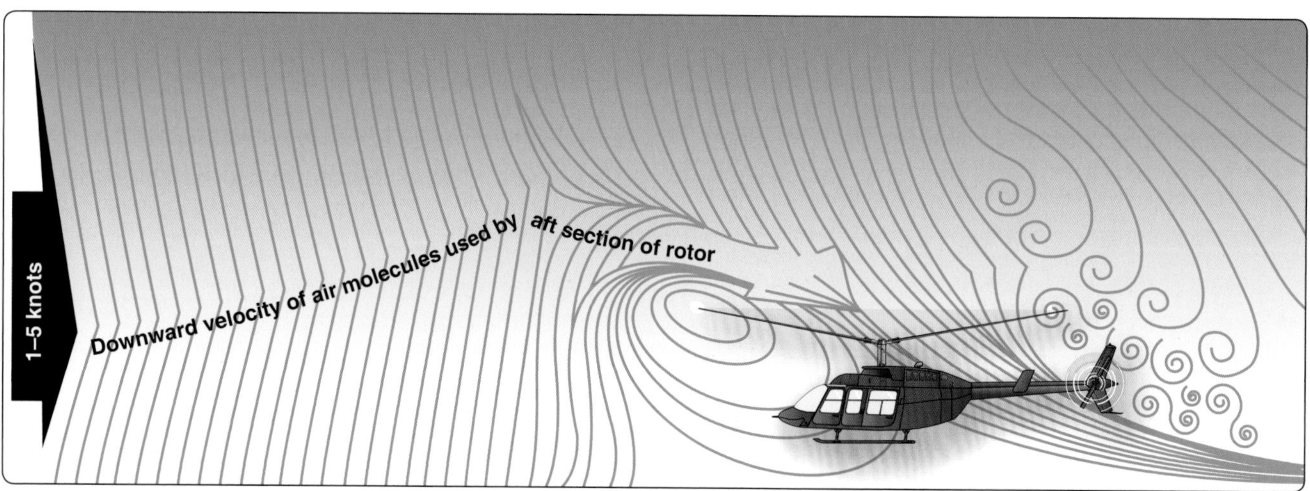

Figure 2-39. *The airflow pattern for 1–5 knots of forward airspeed. Note how the downwind vortex is beginning to dissipate and induced flow down through the rear of the rotor disk is more horizontal.*

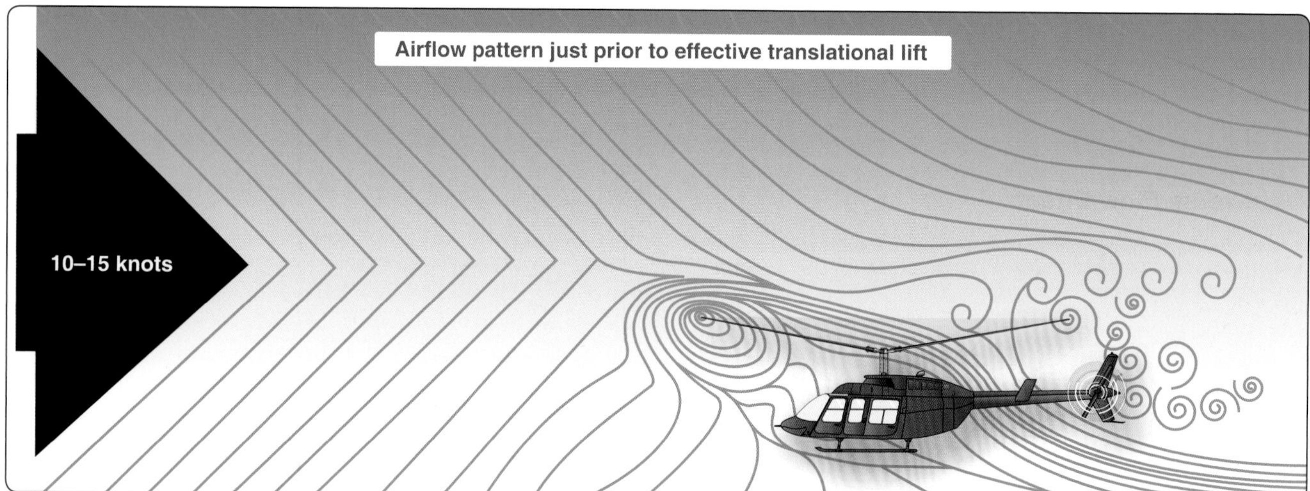

Figure 2-40. *An airflow pattern at a speed of 10–15 knots. At this increased airspeed, the airflow continues to become more horizontal. The leading edge of the downwash pattern is being overrun and is well back under the nose of the helicopter.*

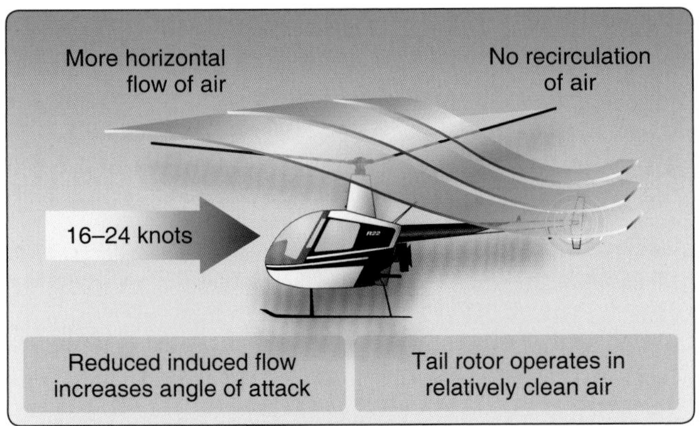

Figure 2-41. *Effective translational lift is easily recognized in actual flight by a transient induced aerodynamic vibration and increased performance of the helicopter.*

(with a main rotor turning counterclockwise) and forces the pilot to apply right pedal (decreasing the AOA in the tail rotor blades) in response. In addition, during this period, the airflow affects the horizontal components of the stabilizer found on most helicopters which tends to bring the nose of the helicopter to a more level attitude.

Induced Flow

As the rotor blades rotate, they generate what is called rotational relative wind. This airflow is characterized as flowing parallel and opposite the rotor's plane of rotation and striking perpendicular to the rotor blade's leading edge. This rotational relative wind is used to generate lift. As rotor blades produce lift, air is accelerated over the foil and projected downward. Anytime a helicopter is producing lift, it moves large masses of air vertically and down through the rotor disk. This downwash or induced flow can significantly change the efficiency of the rotor disk. Rotational relative wind combines with induced flow to form the resultant relative wind. As induced flow increases, resultant relative wind becomes less horizontal. Since AOA is determined by measuring the difference between the chord line and the resultant relative wind, as the resultant relative wind becomes less horizontal, AOA decreases. *[See Figure 2-21]*

Transverse Flow Effect

As the helicopter accelerates in forward flight, induced flow drops to near zero at the forward disk area and increases at the aft disk area. These differences in lift between the fore and aft portions of the rotor disk are called transverse flow effect. *[Figure 2-41]* This increases the AOA at the front disk area causing the rotor blade to flap up and reduces AOA at the aft disk area causing the rotor blade to flap down. Because the rotor acts like a gyro, maximum displacement occurs 90° in the direction of rotation. The result is a tendency for the helicopter to roll slightly to the right as it accelerates through approximately 20 knots or if the headwind is approximately 20 knots.

Transverse flow effect is recognized by increased vibrations of the helicopter at airspeeds around 12 to 15 knots and can be produced by forward flight or from the wind while in a hover. This vibration happens at an airspeed just below ETL on takeoff and after passing through ETL during landing. The vibration happens close to the same airspeed as ETL because that is when the greatest lift differential exists between the front and rear portions of the rotor system. As such, some pilots confuse the vibration felt by transverse flow effect with passing through ETL. To counteract transverse flow effect, a cyclic input to the left may be needed.

Sideward Flight

In sideward flight, the tip-path plane is tilted in the direction that flight is desired. This tilts the total lift-thrust vector sideward. In this case, the vertical or lift component is still straight up and weight straight down, but the horizontal or thrust component now acts sideward with drag acting to the opposite side. *[Figure 2-42]*

Sideward flight can be a very unstable condition due to the parasitic drag of the fuselage combined with the lack of horizontal stabilizer for that direction of flight. Increased altitudes help with control and the pilot must always scan in the direction of flight. Movement of the cyclic in the intended direction of flight causes the helicopter to move, controls the rate of speed, and ground track, but the collective and pedals are key to successful sideward flight. Just as in forward flight, the collective keeps the helicopter from contacting the ground and the pedals help maintain the correct heading; even in sideward flight, the tail of the helicopter should remain behind you. Inputs to the cyclic should be smooth and controlled, and the pilot should always be aware of the tip-path plane in relation to the ground. *[Figure 2-43]*

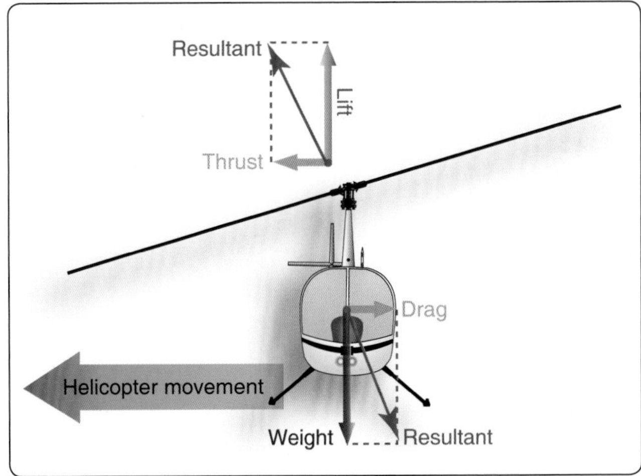

Figure 2-42. *Forces acting on the helicopter during sideward flight.*

Contacting the ground with the skids during sideward flight will most likely result in a dynamic rollover event before the pilot has a chance to react. Extreme caution should be used when maneuvering the helicopter sideways to avoid such hazards from happening. Refer to Chapter 11, Helicopter Hazards and Emergencies.

Rearward Flight

For rearward flight, the tip-path plane is tilted rearward, which, in turn, tilts the lift-thrust vector rearward. Drag now acts forward with the lift component straight up and weight straight down. *[Figure 2-44]*

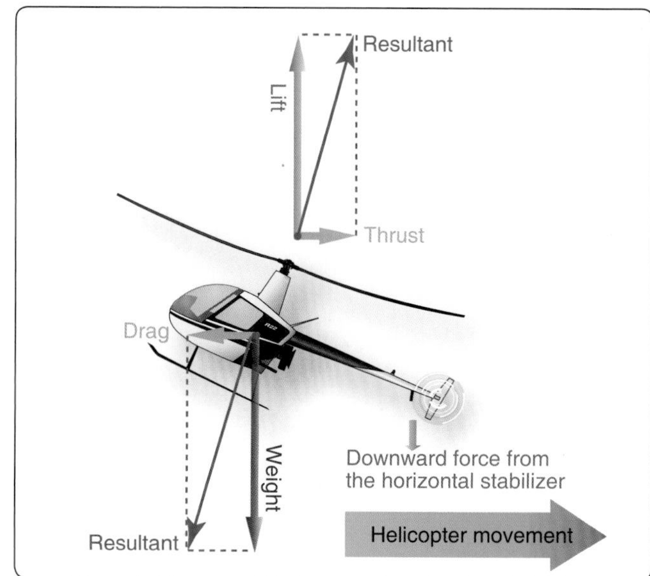

Figure 2-44. *Forces acting on the helicopter during rearward flight.*

Pilots must be aware of the hazards of rearward flight. Because of the position of the horizontal stabilizer, the tail end of the helicopter tends to pitch downward in rearward flight, causing the probability of hitting the ground to be greater than in forward flight. Another factor to consider in rearward flight is skid design. Most helicopter skids are not turned upward in the back, and any contact with the ground during rearward flight can put the helicopter in an uncontrollable position leading to tail rotor contact with the ground. Pilots must do a thorough scan of the area before attempting to hover rearward, looking for obstacles and terrain changes. Slower airspeeds can help mitigate risk and maintain a higher-than-normal hover altitude.

Turning Flight

In forward flight, the rotor disk is tilted forward, which also tilts the total lift-thrust force of the rotor disk forward. When the helicopter is banked, the rotor disk is tilted sideward resulting in lift being separated into two components. Lift acting upward and opposing weight is called the vertical component of lift. Lift acting horizontally and opposing inertia (centrifugal force) is the horizontal component of lift (centripetal force). *[Figure 2-45]*

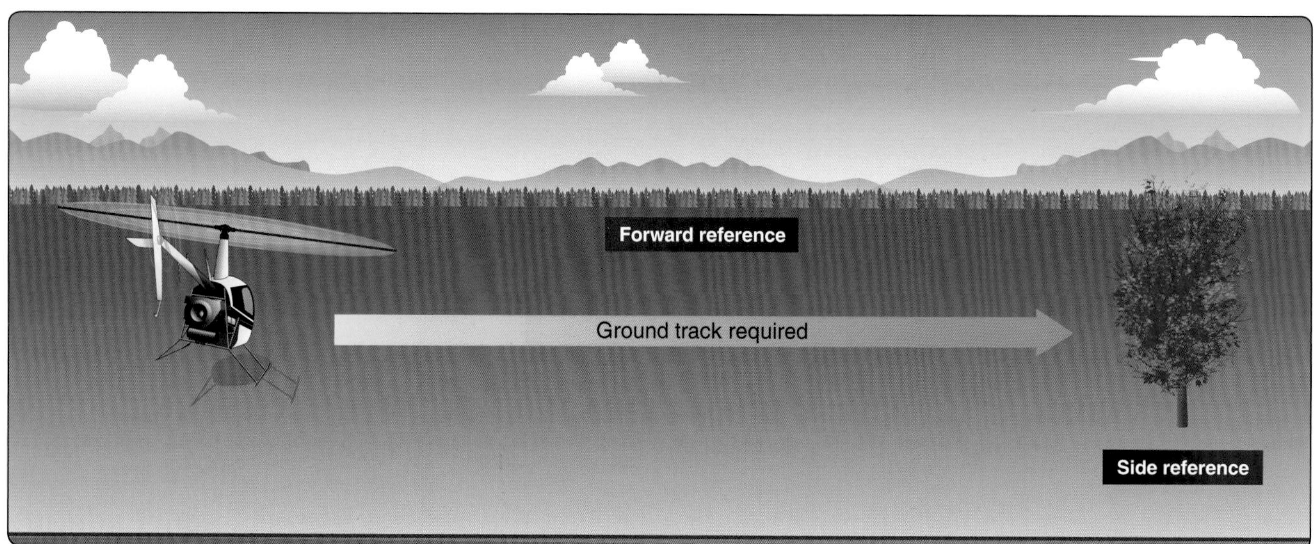

Figure 2-43. *Forces acting on the helicopter during sideward flight.*

As the angle of bank increases, the total lift force is tilted more toward the horizontal, thus causing the rate of turn to increase because more lift is acting horizontally. Since the resultant lifting force acts more horizontally, the effect of lift acting

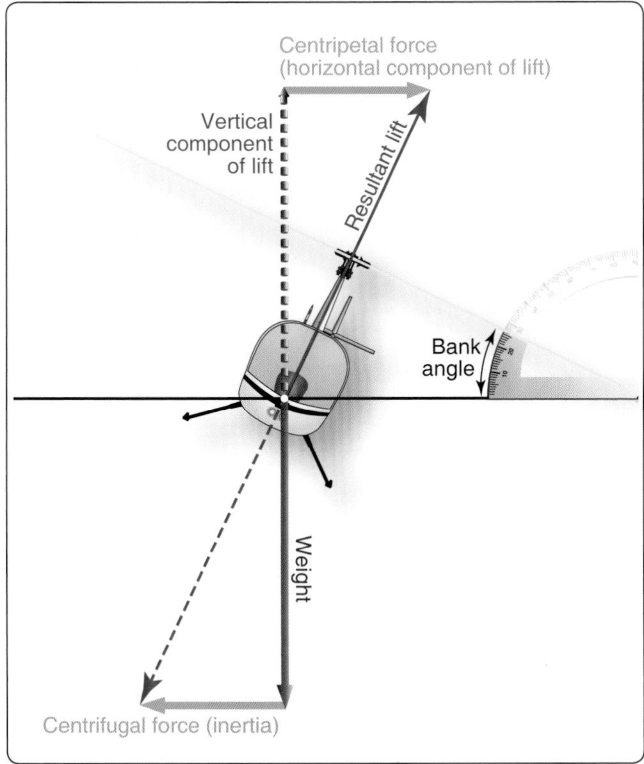

Figure 2-45. *Forces acting on the helicopter during turning flight.*

vertically is decreased. To compensate for this decreased vertical lift, the AOA of the rotor blades must be increased in order to maintain altitude. The steeper the angle of bank is, the greater the AOA of the rotor blades required to maintain altitude. Thus, with an increase in bank and a greater AOA, the resultant lifting force increases, and the rate of turn is higher. Simply put, collective pitch must be increased in order to maintain altitude and airspeed while turning. Collective pitch controls the angle of incidence and along with other factors, determines the overall AOA in the rotor disk.

Autorotation

Autorotation is the state of flight where the main rotor disk of a helicopter is being turned by the action of air moving up through the rotor rather than engine power driving the rotor. In normal, powered flight, air is drawn into the main rotor disk from above and exhausted downward, but during autorotation, air moves up into the rotor disk from below as the helicopter descends. Autorotation is permitted mechanically by a freewheeling unit, which is a special clutch mechanism that allows the main rotor to continue turning even if the engine is not running. If the engine fails, the freewheeling unit automatically disengages the engine from the main rotor allowing the main rotor to rotate freely. It is the means by which a helicopter can be landed safely in the event of an engine failure; consequently, all helicopters must demonstrate this capability in order to be certified. *[Figure 2-46]* If a decision is made to attempt an engine restart in flight (the parameters for this emergency procedure will be different for each helicopter and must be precisely followed) the pilot must reengage the engine starter switch to start the engine. Once the engine is started, the freewheeling unit will reengage the engine with the main rotor.

Vertical Autorotation

Most autorotations are performed with forward speed. For simplicity, the following aerodynamic explanation is based on a vertical autorotative descent (no forward speed) in still air. Under these conditions, the forces that cause the blades to turn are similar for all blades regardless of their position in the plane of rotation. Therefore, dissymmetry of lift resulting from helicopter airspeed is not a factor.

During vertical autorotation, the rotor disk is divided into three regions (as illustrated in *Figure 2-47)*: driven region,

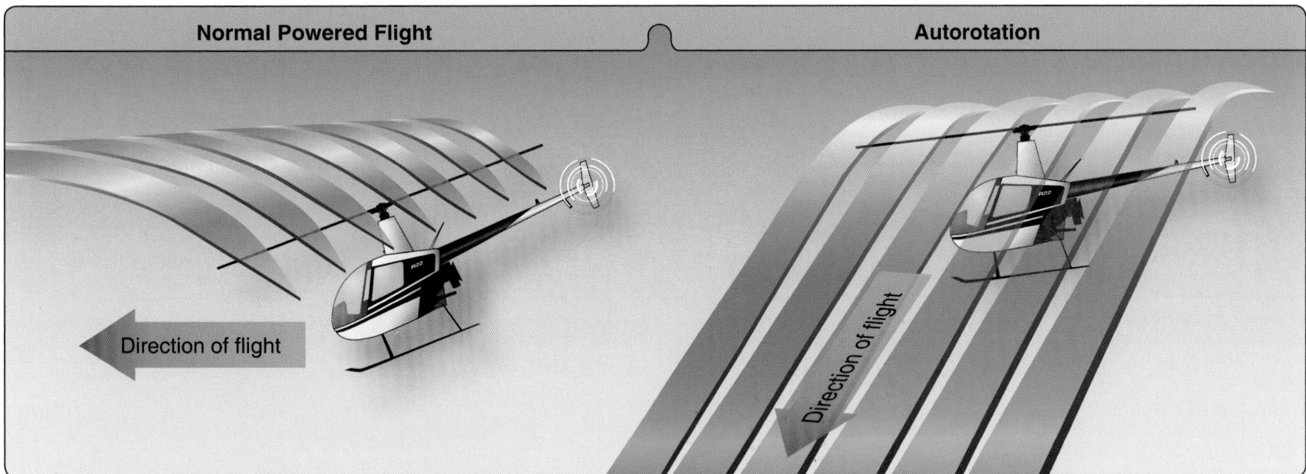

Figure 2-46. *During an autorotation, the upward flow of relative wind permits the main rotor blades to rotate at their normal speed. In effect, the blades are "gliding" in their rotational plane.*

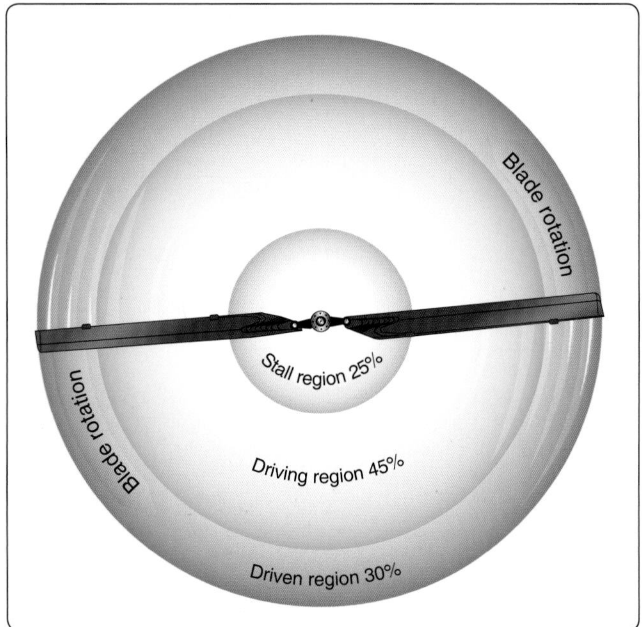

Figure 2-47. *Blade regions during autorotational descent.*

driving region, and stall region. *Figure 2-48* shows three blade sections that illustrate force vectors. Part A is the driven region, B and D are points of equilibrium, part C is the driving region, and part E is the stall region. Force vectors are different in each region because rotational relative wind is slower near the blade root and increases continually toward the blade tip. Also, blade twist gives a more positive AOA in the driving region than in the driven region. The combination of the inflow up through the rotor with rotational relative wind produces different combinations of aerodynamic force at every point along the blade.

The driven region, also called the propeller region, is nearest the blade tips. Normally, it consists of about 30 percent of the radius. In the driven region, part A of *Figure 2-48,* the TAF acts behind the axis of rotation, resulting in an overall drag force. The driven region produces some lift, but that lift is offset by drag. The overall result is a deceleration in the rotation of the blade. The size of this region varies with the blade pitch, rate of descent, and rotor rpm. When changing autorotative rpm blade pitch, or rate of descent, the size of the driven region in relation to the other regions also changes.

There are two points of equilibrium on the blade—one between the driven region and the driving region, and one between the driving region and the stall region. At points of equilibrium, TAF is aligned with the axis of rotation. Lift and drag are produced, but the total effect produces neither acceleration nor deceleration.

The driving region, or autorotative region, normally lies between 25 to 70 percent of the blade radius. Part C of *Figure 2-48* shows the driving region of the blade, which produces the forces needed to turn the blades during autorotation. Total aerodynamic force in the driving region is inclined slightly forward of the axis of rotation, producing a continual acceleration force. This inclination supplies thrust, which tends to accelerate the rotation of the blade. Driving region size varies with blade pitch setting, rate of descent, and rotor rpm.

By controlling the size of this region, a pilot can adjust autorotative rpm. For example, if the collective pitch is raised, the pitch angle increases in all regions. This causes the point of equilibrium to move inboard along the blade's span, thus increasing the size of the driven region. The stall region also becomes larger while the driving region becomes smaller. Reducing the size of the driving region causes the acceleration force of the driving region and rpm to decrease. A constant rotor rpm is achieved by adjusting the collective pitch so blade acceleration forces from the driving region are balanced with the deceleration forces from the driven and stall regions.

The inner 25 percent of the rotor blade is referred to as the stall region and operates above its maximum AOA (stall angle), causing drag, which tends to slow rotation of the blade. Part E of *Figure 2-48* depicts the stall region.

Autorotation (Forward Flight)

Autorotative force in forward flight is produced in exactly the same manner as when the helicopter is descending vertically in still air. However, because forward speed changes the inflow of air up through the rotor disk, all three regions move outboard along the blade span on the retreating side of the disk where AOA is larger. *[Figure 2-49]* With lower AOA on the advancing side blade, more of the blade falls in the driven region. On the retreating side, more of the blade is in the stall region. A small section near the root experiences a reversed flow; therefore, the size of the driven region on the retreating side is reduced.

Prior to landing from an autorotative descent (or autorotation), the pilot must flare the helicopter in order to decelerate. The pilot initiates the flare by applying aft cyclic. As the helicopter flares back, the airflow patterns change around the blades causing the rpm to increase. Pilots must adjust the collective as necessary to keep the rpm within operating limits.

Chapter Summary

This chapter introduced the basics of aerodynamic fundamentals and theory and how they relate to flying a helicopter. This chapter also explained how aerodynamics affect helicopter flight and how important it is for pilots to understand aerodynamic principles and be prepared to react to these effects. For additional information on aerodynamics, refer to the aerodynamics of flight portion of the Pilot's Handbook of Aeronautical Knowledge.

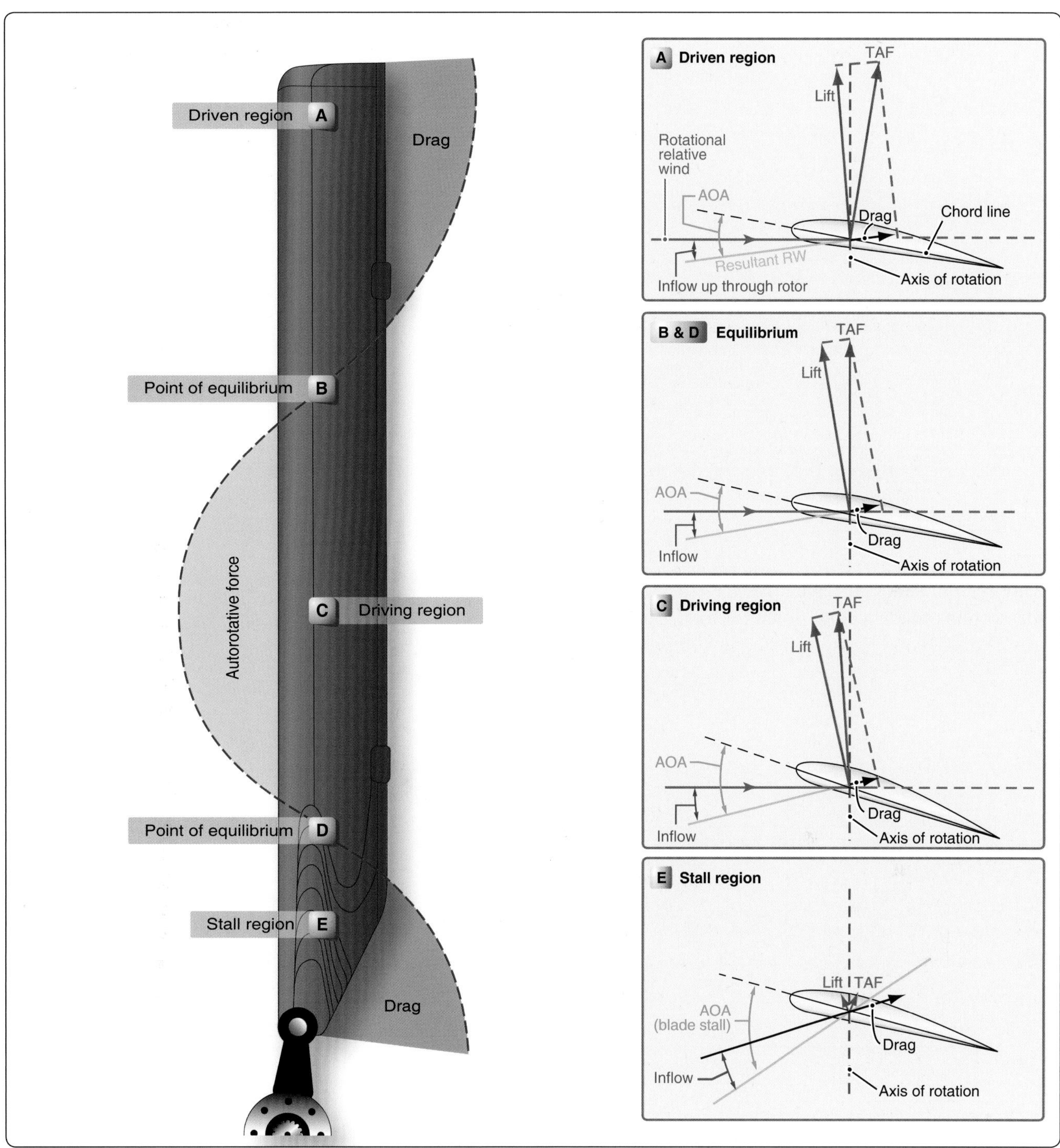

Figure 2-48. *Force vectors in vertical autorotation descent.*

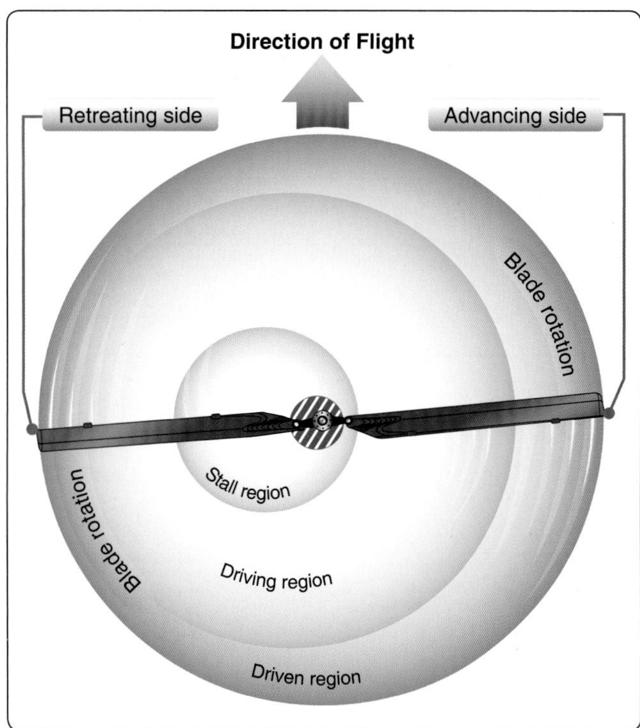

Figure 2-49. *Blade regions in forward autorotation descent.*

Helicopter Flight Controls

Introduction

There are three major controls in a helicopter that the pilot must use during flight. They are the collective pitch control, the cyclic pitch control, and the antitorque pedals or tail rotor control. In addition to these major controls, the pilot must also use the throttle control, which is usually mounted directly to the collective pitch control in order to fly the helicopter.

In this chapter, the control systems described are not limited to the single main rotor type helicopter but are employed in one form or another in most helicopter configurations. All examples in this chapter refer to a counterclockwise main rotor blade rotation as viewed from above. If flying a helicopter with a clockwise rotation, left and right references must be reversed, particularly in the areas of rotor blade pitch change, antitorque pedal movement, and tail rotor thrust.

Collective Pitch Control

The collective pitch control (or simply "collective" or "thrust lever") is located on the left side of the pilot's seat and is operated with the left hand. The collective is used to make changes to the pitch angle of the main rotor blades and does this simultaneously, or collectively, as the name implies. As the collective pitch control is raised, there is a simultaneous and equal increase in pitch angle of all main rotor blades; as it is lowered, there is a simultaneous and equal decrease in pitch angle. This is done through a series of mechanical linkages and the amount of movement in the collective lever determines the amount of blade pitch change. *[Figure 3-1]* An adjustable friction control helps prevent inadvertent collective pitch movement.

Changing the pitch angle on the blades changes the angle of incidence on each blade. With a change in angle of incidence comes a change in drag, which affects the speed or revolutions per minute (rpm) of the main rotor. As the pitch angle increases, angle of incidence increases, drag increases, and rotor rpm decreases. Decreasing pitch angle decreases both angle of incidence and drag, while rotor rpm increases. In order to maintain a constant rotor rpm, which is essential in helicopter operations, a proportionate change in power is required to compensate for the change in drag. This is accomplished with the throttle control or governor, which automatically adjusts engine power.

Throttle Control

The function of the throttle is to regulate engine rpm. If the correlator or governor system does not maintain the desired rpm when the collective is raised or lowered, or if those systems are not installed, the throttle must be moved manually with the twist grip in order to maintain rpm. In most helicopters, rotating the twist-grip throttle away from the pilot (counter-clockwise), increases engine rpm; rotating the twist-grip throttle towards the pilot (clockwise) decreases engine rpm. [Figure 3-2]

Governor/Correlator

A governor is a sensing device that senses rotor and engine rpm and makes the necessary adjustments in order to keep rotor rpm constant. In normal operations, once the rotor rpm is set, the governor keeps the rpm constant, and there is no need to make any throttle adjustments. Governors are common on all turbine helicopters (as it is a function of the fuel control system of the turbine engine) and used on some piston powered helicopters.

A correlator is a mechanical connection between the collective lever and the engine throttle. When the collective lever is raised, power is automatically increased; when lowered, power is decreased. This system maintains rpm close to the desired value, but still requires adjustment of the throttle for fine tuning.

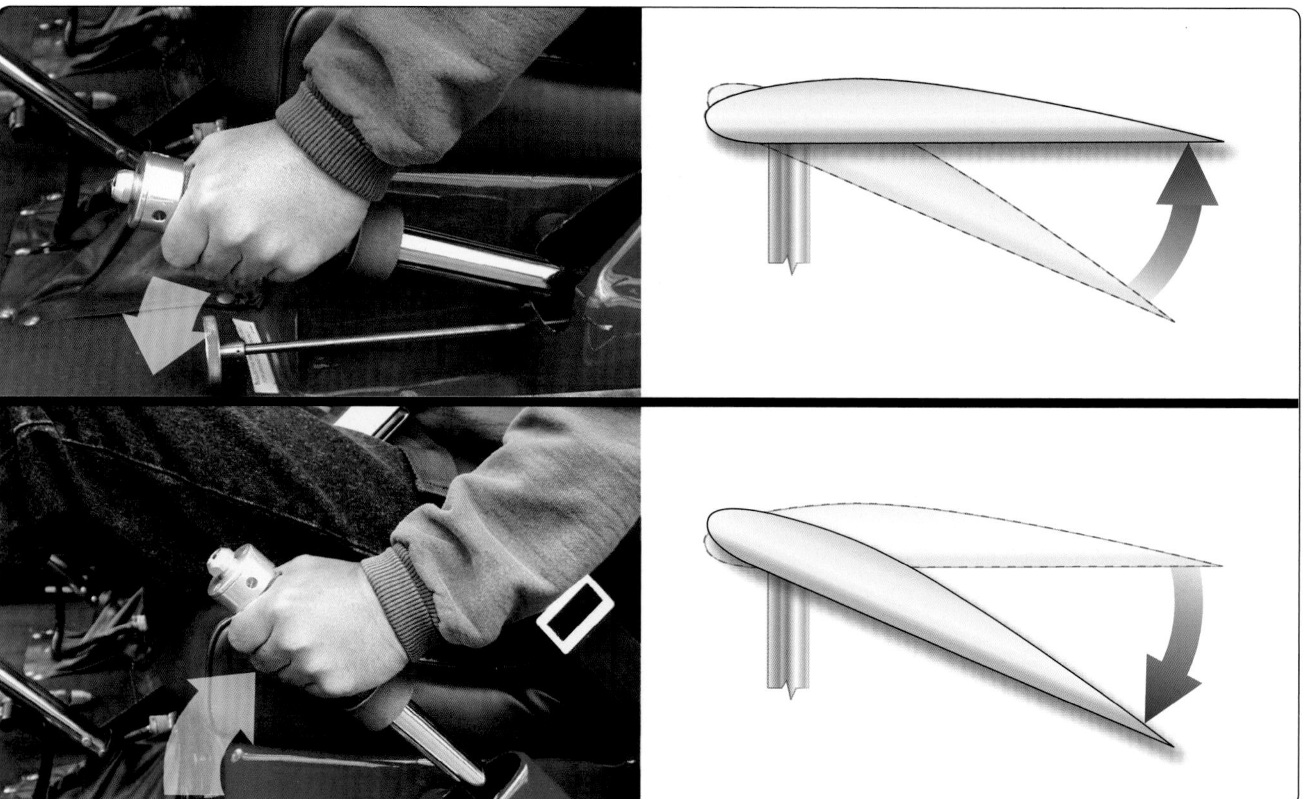

Figure 3-1. *Raising the collective pitch control increases the pitch angle, or angle of incidence, by the same amount on all blades.*

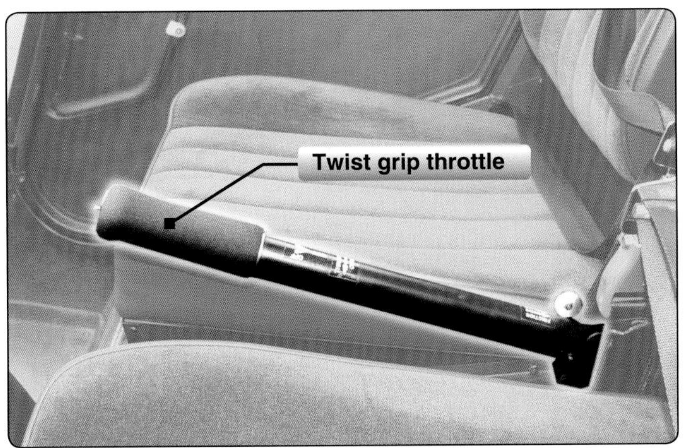

Figure 3-2. *A twist grip throttle is usually mounted on the end of the collective lever. The throttles on some turbine helicopters are mounted on the overhead panel or on the floor in the cockpit.*

Some helicopters do not have correlators or governors and require coordination of all collective and throttle movements. When the collective is raised, the throttle must be increased; when the collective is lowered, the throttle must be decreased. As with any aircraft control, large adjustments of either collective pitch or throttle should be avoided. All corrections should be made through the use of smooth pressure.

In piston helicopters, the collective pitch is the primary control for manifold pressure, and the throttle is the primary control for rpm. However, the collective pitch control also influences rpm, and the throttle also influences manifold pressure; therefore, each is considered to be a secondary control of the other's function. Both the tachometer (rpm indicator) and the manifold pressure gauge must be analyzed to determine which control to use. *Figure 3-3* illustrates this relationship.

if rpm is	and manifold pressure is	Solution
LOW	LOW	Increasing the throttle increases manifold pressure and rpm
LOW	HIGH	Lowering the collective pitch decreases manifold pressure and increases rpm
HIGH	LOW	Raising the collective pitch increases manifold pressure and decreases rpm
HIGH	HIGH	Reducing the throttle decreases manifold pressure and rpm

Figure 3-3. *Relationship between rpm, manifold pressure, collective, and throttle.*

Cyclic Pitch Control

The cyclic pitch control (or simply "cyclic") is usually projected upward from the cockpit floor, between the pilot's legs or between the two pilot seats in some models. *[Figure 3-4]* This primary flight control allows the pilot to fly the helicopter in any direction of travel: forward, rearward, left, and right. As discussed in Chapter 2, Aerodynamics of Flight, the total lift force is always perpendicular to the tip-path plane of the main rotor. The purpose of the cyclic pitch control is to tilt the tip-path plane in the direction of the desired horizontal direction. The cyclic controls the rotor disk tilt versus the horizon, which directs the rotor disk thrust to enable the pilot to control the direction of travel of the helicopter.

The rotor disk tilts in the same direction the cyclic pitch control is moved. If the cyclic is moved forward, the rotor disk tilts forward; if the cyclic is moved aft, the disk tilts aft, and so on. Because the rotor disk acts like a gyro, the mechanical linkages for the cyclic control rods are rigged in such a way that they

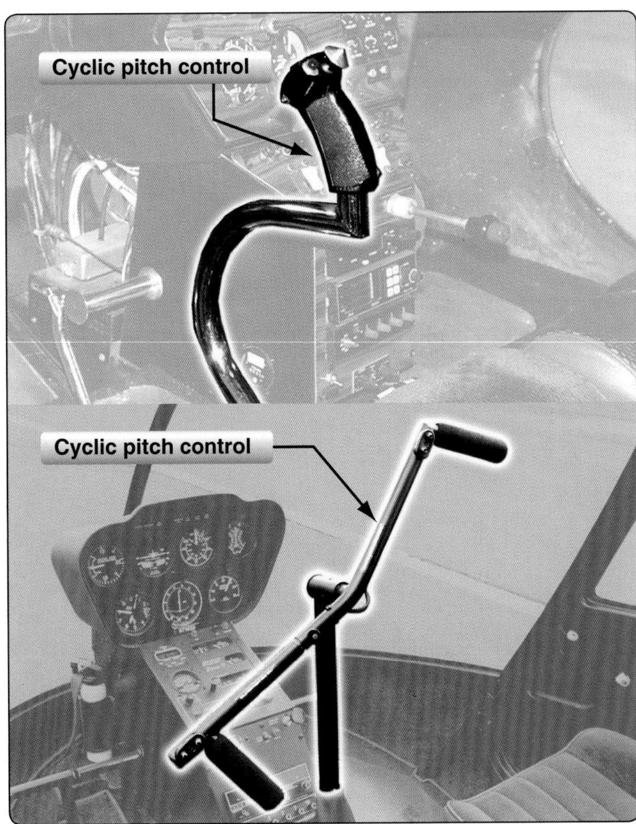

Figure 3-4. *The cyclic pitch control may be mounted vertically between the pilot's knees or on a teetering bar from a single cyclic located in the center of the helicopter. The cyclic can pivot in all directions.*

decrease the pitch angle of the rotor blade approximately 90° before it reaches the direction of cyclic displacement and increase the pitch angle of the rotor blade approximately 90° after it passes the direction of displacement. An increase in pitch angle increases AOA; a decrease in pitch angle decreases AOA. For example, if the cyclic is moved forward, the AOA decreases as the rotor blade passes the right side of the helicopter and increases on the left side. This results in maximum downward deflection of the rotor blade in front of the helicopter and maximum upward deflection behind it, causing the rotor disk to tilt forward.

Antitorque Pedals

The antitorque pedals, located on the cabin floor by the pilot's feet, control the pitch and therefore the thrust of the tail rotor blades or other antitorque system. See Chapter 5, Helicopter Components, Sections, and Systems, for a discussion on these other systems. *[Figure 3-5]* Newton's Third Law was discussed in Chapter 2, General Aerodynamics, stating that for every action there is an equal and opposite reaction. This law applies to the helicopter fuselage and its rotation in the opposite direction of the main rotor blades unless counteracted and controlled. To make flight possible and to compensate for this torque, most helicopter designs incorporate an antitorque rotor or tail rotor. The antitorque pedals allow the pilot to control the pitch angle of the tail rotor blades, which in forward flight puts the helicopter in longitudinal trim and, while at a hover, enables the pilot to turn the helicopter 360°. The antitorque pedals are connected to the pitch change mechanism on the tail rotor gearbox and allow the pitch angle on the tail rotor blades to be increased or decreased.

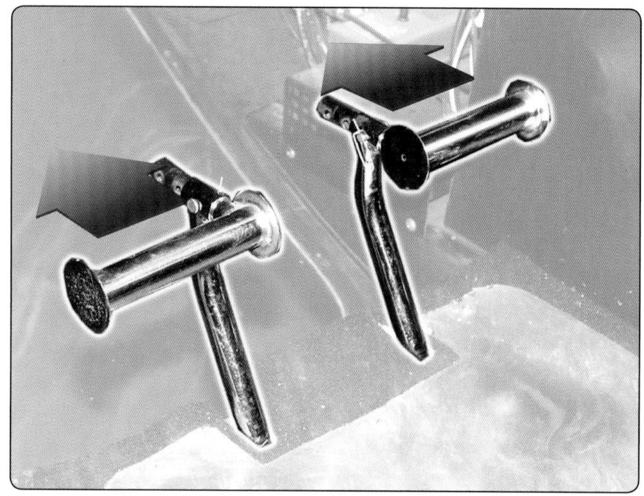

Figure 3-5. *Antitorque pedals compensate for changes in torque and control heading in a hover.*

Heading Control

The tail rotor is used to control the heading of the helicopter while hovering or when making hovering turns, as well as counteracting the torque of the main rotor. Hovering turns are commonly referred to as "pedal turns."

At speeds above translational lift, the pedals are used to compensate for torque to put the helicopter in longitudinal trim, so that coordinated flight can be maintained. The cyclic control is used to change heading by making a turn to the desired direction.

The thrust of the tail rotor depends on the pitch angle of the tail rotor blades. This pitch angle can be positive, negative, or zero. A positive pitch angle tends to move the tail to the right. A negative pitch angle moves the tail to the left, while no thrust is produced with a zero pitch angle. The maximum positive pitch angle of the tail rotor is generally greater than the maximum negative pitch angle available. This is because the primary purpose of the tail rotor is to counteract the torque of the main rotor. The capability for tail rotors to produce thrust to the left (negative pitch angle) is necessary, because during autorotation the drag of the transmission tends to yaw the nose to the left, or in the same direction the main rotor is turning.

From the neutral position, applying right pedal causes the nose of the helicopter to yaw right and the tail to swing to the left. Pressing on the left pedal has the opposite effect: the nose of the helicopter yaws to the left and the tail swings right. *[Figure 3-6]*

With the antitorque pedals in the neutral position, the tail rotor has a medium positive pitch angle. In medium positive pitch, the tail rotor thrust approximately equals the torque of the main rotor during cruise flight, so the helicopter maintains a constant heading in level flight.

A vertical fin or stabilizer is used in many single-rotor helicopters to help aid in heading control. The fin is designed to optimize directional stability in flight with a zero tail rotor thrust setting. The size of the fin is crucial to this design. If the surface is too large, the tail rotor thrust may be blocked. Heading control would be more difficult at slower airspeeds and at a hover and the vertical fin would then weathervane.

Helicopters that are designed with tandem rotors do not have an antitorque rotor. The helicopter is designed with both rotor systems rotating in opposite directions to counteract the torque rather than a tail rotor. Directional antitorque pedals

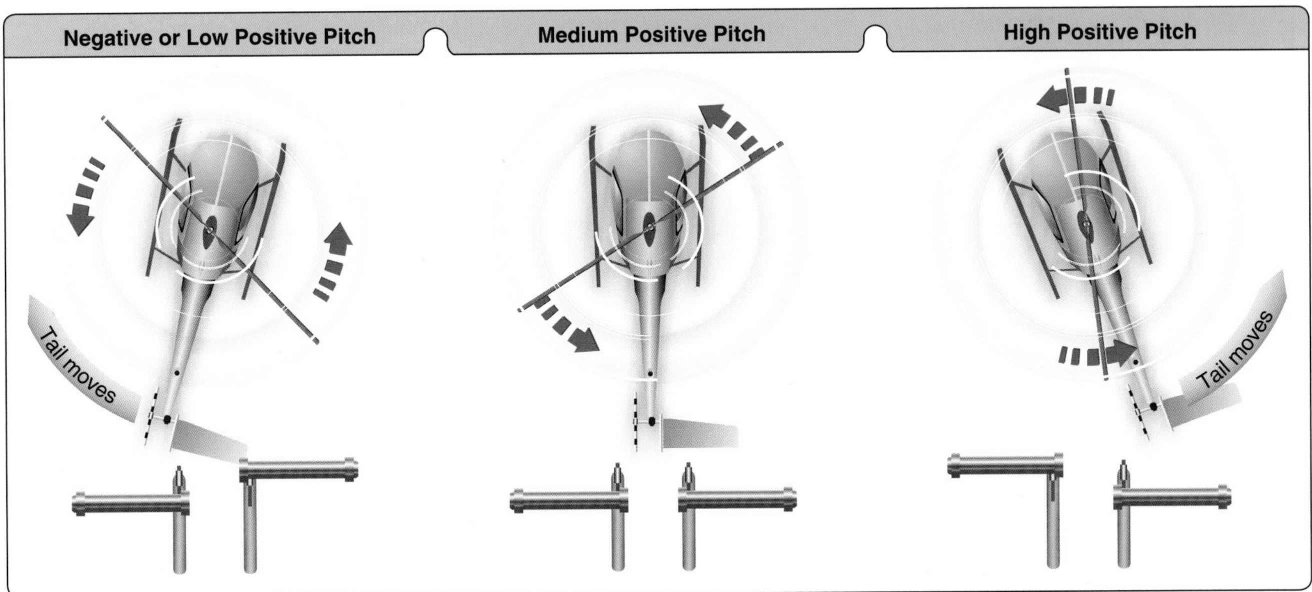

Figure 3-6. *Tail rotor pitch angle and thrust in relation to pedal positions during cruising flight.*

are used for directional control of the aircraft while in flight, as well as while taxiing with the forward gear off the ground.

In intermeshing rotor systems, which are a set of two rotors turning in opposite directions with each rotor mast mounted on the helicopter with a slight angle to the other so that the blades intermesh without colliding, and a coaxial rotor systems, which are a pair of rotors mounted one above the other on the same shaft and turning in opposite directions, the heading pedals control the heading of the helicopter while at a hover by imbalancing torque between the rotors, allowing for the torque to turn the helicopter.

Chapter Summary

This chapter introduced the pilot to the major flight controls and how they work in relation to each other. The chapter also correlates the use of flight controls and aerodynamics and how the two work together to make flight possible.

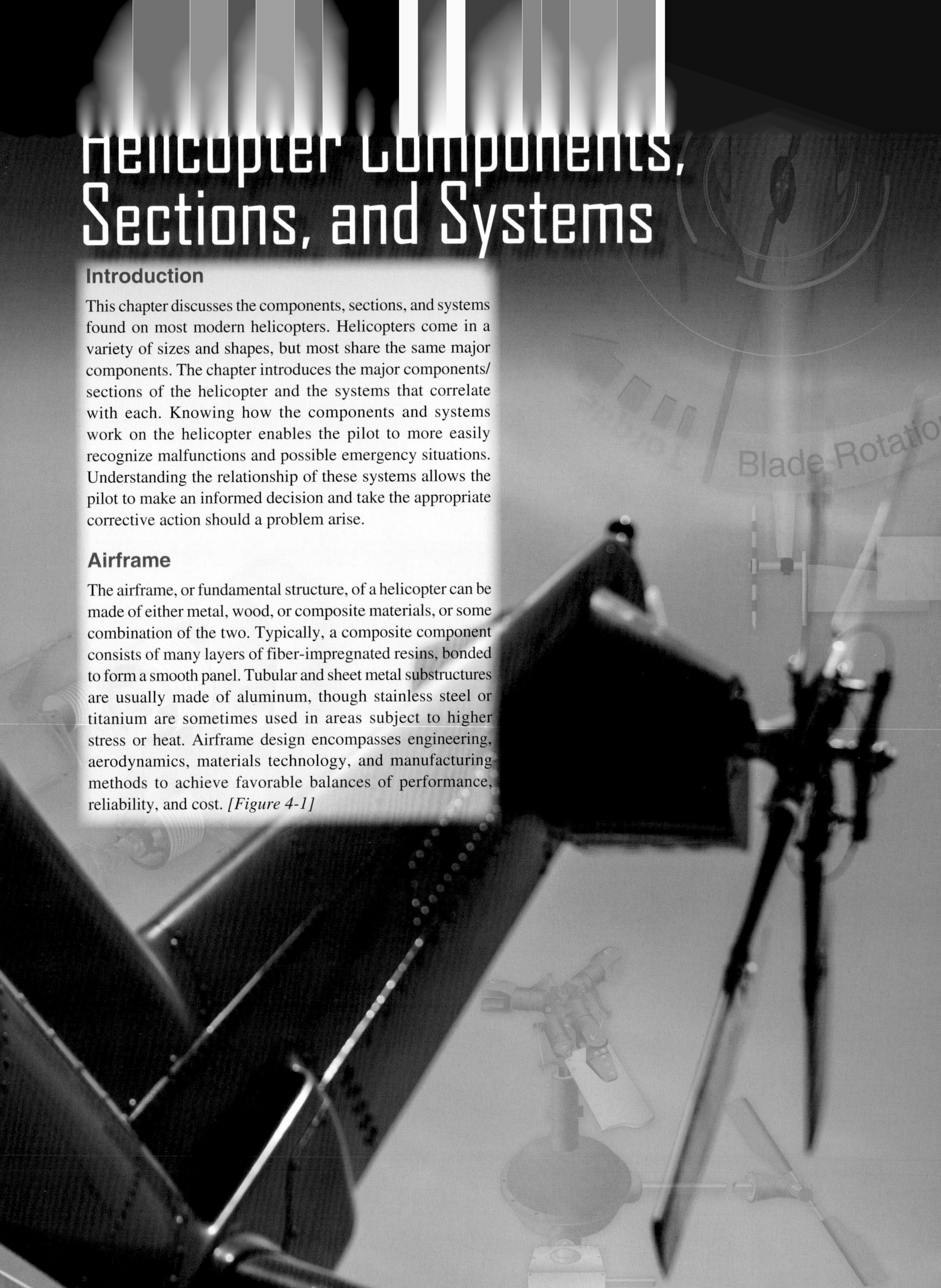

Helicopter Components, Sections, and Systems

Introduction

This chapter discusses the components, sections, and systems found on most modern helicopters. Helicopters come in a variety of sizes and shapes, but most share the same major components. The chapter introduces the major components/sections of the helicopter and the systems that correlate with each. Knowing how the components and systems work on the helicopter enables the pilot to more easily recognize malfunctions and possible emergency situations. Understanding the relationship of these systems allows the pilot to make an informed decision and take the appropriate corrective action should a problem arise.

Airframe

The airframe, or fundamental structure, of a helicopter can be made of either metal, wood, or composite materials, or some combination of the two. Typically, a composite component consists of many layers of fiber-impregnated resins, bonded to form a smooth panel. Tubular and sheet metal substructures are usually made of aluminum, though stainless steel or titanium are sometimes used in areas subject to higher stress or heat. Airframe design encompasses engineering, aerodynamics, materials technology, and manufacturing methods to achieve favorable balances of performance, reliability, and cost. *[Figure 4-1]*

Fuselage

The fuselage, the outer core of the airframe, is an aircraft's main body section that houses the cabin that holds the crew, passengers, and cargo. Helicopter cabins have a variety of seating arrangements. Most have the pilot seated on the right side, although there are some with the pilot seated on the left side or center. The fuselage also houses the engine, the transmission, avionics, flight controls, and the powerplant. *[Figure 4-1]*

Main Rotor System

The rotor system is the rotating part of a helicopter which generates lift. The rotor consists of a mast, hub, and rotor blades. The mast is a hollow cylindrical metal shaft which extends upwards from and is driven and sometimes supported by the transmission. At the top of the mast is the attachment point for the rotor blades called the hub. The rotor blades are then attached to the hub by any number of different methods. Main rotor systems are classified according to how the main rotor blades are attached and move relative to the main rotor hub. There are three basic classifications: semirigid, rigid, or fully articulated. Some modern rotor systems, such as the bearingless rotor system, use an engineered combination of these types.

Semirigid Rotor System

A semirigid rotor system is usually composed of two blades that are rigidly mounted to the main rotor hub. The main rotor hub is free to tilt with respect to the main rotor shaft on what is known as a teetering or flapping hinge. This allows the blades to flap together as a unit. As one blade flaps up, the other flaps down. Since there is no vertical drag hinge, lead/lag forces are absorbed and mitigated by blade bending. The semirigid rotor is also capable of feathering, which means that the pitch angle of the blade changes. This is made possible by the feathering hinge. *[Figure 4-2]*

If the semirigid rotor system is an underslung rotor, the center of gravity (CG) is below where it is attached to the mast. This underslung mounting is designed to align the blade's center of mass with a common flapping hinge so that both blades' centers of mass vary equally in distance from the center of rotation during flapping. The rotational speed of the system tends to change, but this is restrained by the inertia of the engine and flexibility of the drive system. Only a moderate amount of stiffening at the blade root is necessary to handle this restriction. Simply put, underslinging effectively eliminates geometric imbalance. *[Figure 4-3]*

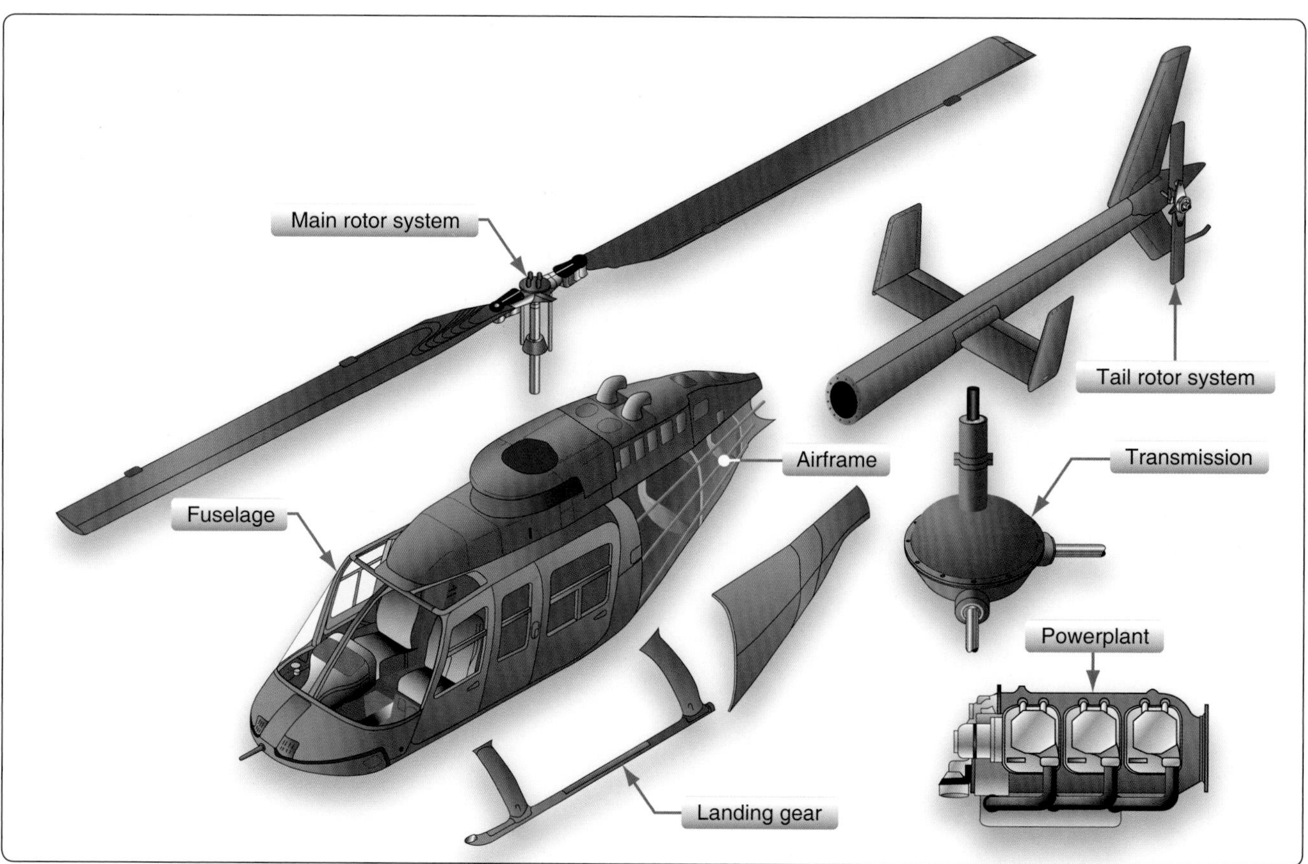

Figure 4-1. *The major components of a helicopter are the airframe, fuselage, landing gear, powerplant, transmission, main rotor system,and tail rotor system.*

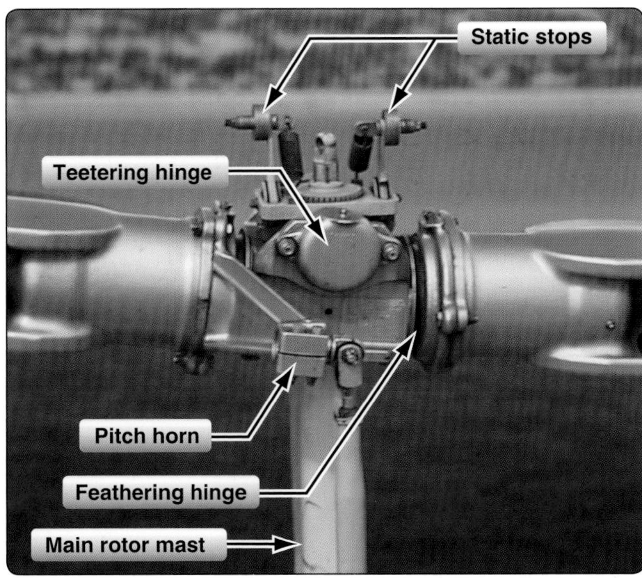

Figure 4-2. *The teetering hinge allows the main rotor hub to tilt, and the feathering hinge enables the pitch angle of the blades to change.*

The underslung rotor system mitigates the lead/lag forces by mounting the blades slightly lower than the usual plane of rotation, so the lead/lag forces are minimized. As the blades cone upward, the center of pressures of the blades are almost in the same plane as the hub. Whatever stresses are remaining bend the blades for compliance.

Helicopters with semirigid rotors are vulnerable to a condition known as mast bumping which can cause the rotor flap stops to shear the mast. The mechanical design of the semirigid rotor system dictates downward flapping of the blades must have some physical limit. Mast bumping is the result of excessive rotor flapping. Each rotor system design has a maximum flapping angle. If flapping exceeds the design value, the static stop will contact the mast. The static stop is a component of the main rotor providing limited movement of strap fittings and a contoured surface between the mast and hub. It is the violent contact between the static stop and the mast during flight that causes mast damage or separation. This contact must be avoided at all costs.

Mast bumping is directly related to how much the blade system flaps. In straight and level flight, blade flapping is minimal, perhaps 2° under usual flight conditions. Flapping angles increase slightly with high forward speeds, at low rotor rpm, at high-density altitudes, at high gross weights, and when encountering turbulence. Maneuvering the aircraft in a sideslip or during low-speed flight at extreme CG positions can induce larger flapping angles.

Rigid Rotor System

The rigid rotor system shown in *Figure 4-4* is mechanically simple, but structurally complex because operating loads must be absorbed in bending rather than through hinges. In this system, the blade roots are rigidly attached to the rotor hub. Rigid rotor systems tend to behave like fully articulated systems through aerodynamics, but lack flapping or lead/lag hinges. Instead, the blades accommodate these motions by bending. They cannot flap or lead/lag, but they can be feathered. As advancements in helicopter aerodynamics

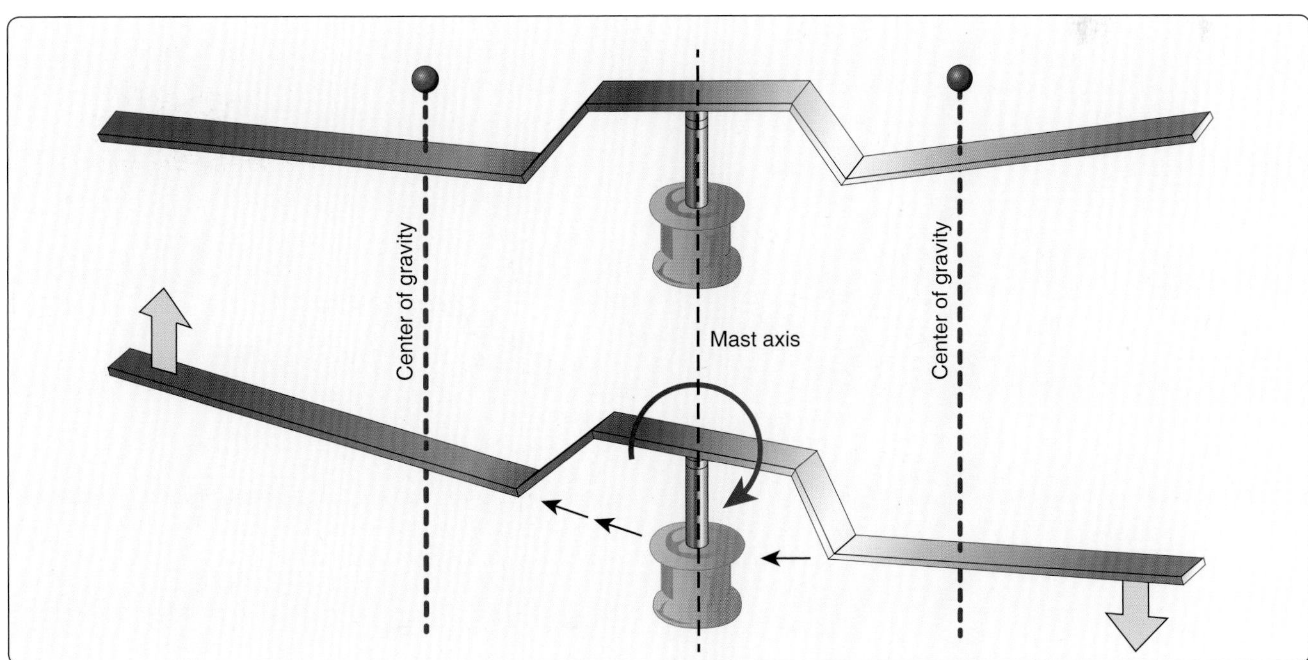

Figure 4-3. *With an underslung rotor, the center of gravity (CG) remains in the same approximate location relative to the mast before and after rotor tilt.*

Figure 4-4. *Four-blade hingeless (rigid) main rotor. Rotor blades are comprised of glass fiber reinforced material. The hub is a single piece of forged rigid titanium.*

and materials continue to improve, rigid rotor systems may become more common because the system is fundamentally easier to design and offers the best properties of both semirigid and fully articulated systems.

The rigid rotor system is very responsive and is usually not susceptible to mast bumping like the semirigid systems because the rotor hubs are mounted solid to the main rotor mast. This allows the rotor and fuselage to move together as one entity and eliminates much of the oscillation usually present in the other rotor systems. Other advantages of the rigid rotor include a reduction in the weight and drag of the rotor hub and a larger flapping arm, which significantly reduces control inputs. Without the complex hinges, the rotor system becomes much more reliable and easier to maintain than the other rotor configurations. A disadvantage of this system is the quality of ride in turbulent or gusty air. Because there are no hinges to help absorb the larger loads, vibrations are felt in the cabin much more than with other rotor head designs.

There are several variations of the basic three rotor head designs. The bearingless rotor system is closely related to the articulated rotor system but has no bearings or hinges. This design relies on the structure of blades and hub to absorb stresses. The main difference between the rigid rotor system and the bearingless system is that the bearingless system has no feathering bearing—the material inside the cuff is twisted by the action of the pitch change arm. Nearly all bearingless rotor hubs are made of fiber-composite materials. The differences in handling between the types of rotor system are summarized in *Figure 4-5.*

Fully Articulated Rotor System

Fully articulated rotor systems allow each blade to lead/lag (move back and forth in plane), flap (move up and down about an inboard mounted hinge) independent of the other blades, and feather (rotate about the pitch axis to change lift). *[Figures 4-6* and *4-7]* Each of these blade motions is related

System Type	Advantages	Disadvantages
Articulated	Good control response	High aerodynamic drag. *More complex, greater cost.*
Semirigid (Teetering, Underslung, or See-Saw)	Simple, easy to hangar due to two blades	Reaction to control input not as quick as articulated head. Vibration can be higher than *multi-bladed articulated* systems.
Rigid	Simple design, crisp response	Higher vibration than articulated rotor.

Figure 4-5. *Differences in handling between the types of rotor systems.*

Figure 4-6. *Lead/lag hinge allows the rotor blade to move back and forth in plane.*

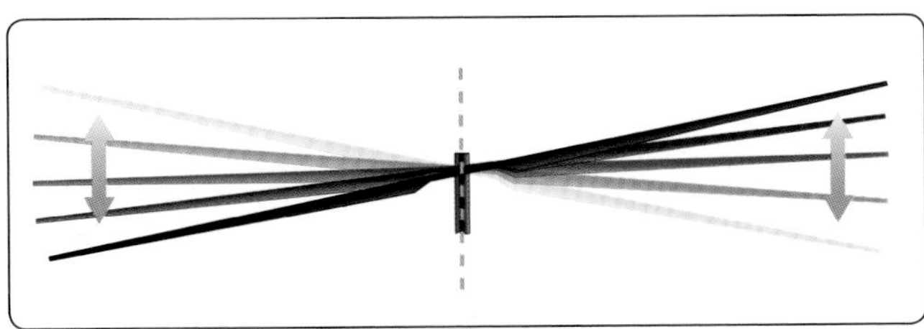

Figure 4-7. *Fully articulated flapping hub.*

to the others. Fully articulated rotor systems are found on helicopters with more than two main rotor blades.

As the rotor spins, each blade responds to inputs from the control system to enable aircraft control. The center of lift on the whole rotor system moves in response to these inputs to effect pitch, roll, and upward motion. The magnitude of this lift force is based on the collective input, which changes pitch on all blades in the same direction at the same time. The location of this lift force is based on the pitch and roll inputs from the pilot. Therefore, the feathering angle of each blade (proportional to its own lifting force) changes as it rotates with the rotor, hence the name "cyclic control."

As the lift on a given blade increases, it tends to flap upwards. The flapping hinge for the blade permits this motion and is balanced by the centrifugal force of the weight of the blade, which tries to keep it in the horizontal plane. *[Figure 4-8]*

Either way, some motion must be accommodated. The centrifugal force is nominally constant; however, the flapping force is affected by the severity of the maneuver (rate of climb, forward speed, aircraft gross weight). As the blade flaps, its CG changes. This changes the local moment of inertia of the blade with respect to the rotor system and it speeds up or slows down with respect to the rest of the blades and the whole rotor system. This is accommodated by the lead/lag or drag hinge, shown in *Figure 4-9,* and is easier to visualize with the classical 'ice skater doing a spin' image. As the skater moves her arms in, she spins faster because her inertia changes but her total energy remains constant (neglect friction for purposes of this explanation). Conversely, as her arms extend, her spin slows. This is also known as the conservation of angular momentum. An in-plane damper typically moderates lead/lag motion.

Following a single blade through a single rotation beginning at some neutral position, as load increases from increased feathering, it flaps up and leads forward. As it continues

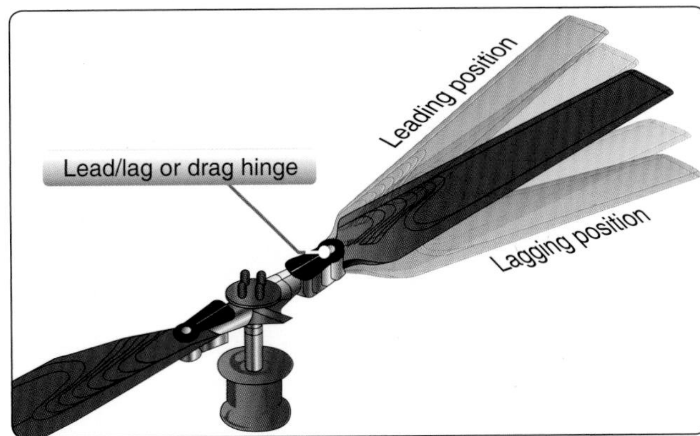

Figure 4-8. *Fully articulated rotor blade with flapping hinge.*

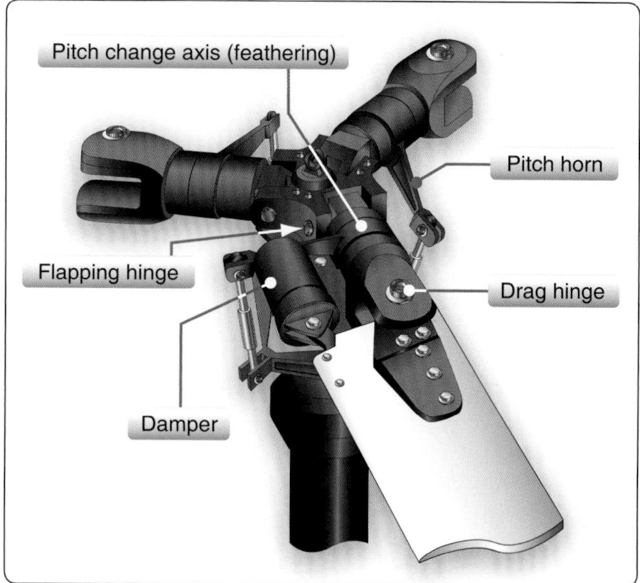

Figure 4-9. *Drag hinge.*

around, it flaps down and lags backward. At the lowest point of load, it is at its lowest flap angle and also at its most 'rearward' lag position. Because the rotor is a large, rotating mass, it behaves somewhat like a gyroscope. The effect of this is that a control input is usually realized on the attached body at a position 90° prior to the control input displacement in the axis of rotation. This is accounted for by the designers through placement of the control input to the rotor system so that a forward input of the cyclic control stick results in a nominally forward motion of the aircraft. The effect is made transparent to the pilot.

Older hinge designs relied on conventional metal bearings. By basic geometry, this precludes a coincident flapping and lead/lag hinge and is cause for recurring maintenance. Newer rotor systems use elastomeric bearings, arrangements of rubber and steel that can permit motion in two axes. Besides solving some of the above-mentioned kinematic issues, these bearings are usually in compression, can be readily inspected, and eliminate the maintenance associated with metallic bearings.

Elastomeric bearings are naturally fail-safe, and their wear is gradual and visible. The metal-to-metal contact of older bearings and the need for lubrication is eliminated in this design.

Tandem Rotor

Tandem rotor (sometimes referred to as dual rotor) helicopters have two large horizontal rotor assemblies; a twin rotor system, instead of one main assembly, and a smaller tail rotor. *[Figure 4-10]* Single rotor helicopters need an anti-torque system to neutralize the twisting momentum produced by the single large rotor. Tandem rotor helicopters, however, use counter-rotating rotors, with each canceling

Figure 4-10. *Tandem rotor heads.*

out the other's torque. Counter-rotating rotor blades will not collide with and destroy each other if they flex into the other rotor's pathway. This configuration also has the advantage of being able to hold more weight with shorter blades, since there are two sets. Also, all of the power from the engines can be used for lift, whereas a single rotor helicopter uses power to counter the torque.

Coaxial Rotors

A coaxial rotor system is a pair of rotors mounted on the same shaft but turning in opposite directions. This design eliminates the need for a tail rotor or other antitorque mechanisms, and since the blades turn in opposite directions, the effects of dissymmetry of lift are avoided. The main disadvantage of coaxial rotors is the increased mechanical complexity of the rotor system. Numerous Russian helicopters, such as the Kaman Ka-31 and Ka-50, along with the Sikorsky experimental X2 use a coaxial rotor design.

Intermeshing Rotors

An intermeshing rotor system is a set of two rotors turning in the opposite directions with each rotor mast mounted on the helicopter with a slight angle, so the blades intermesh without colliding. This design also eliminates the need for an antitorque system, which provides more engine power for lift. However, neither rotor lifts directly vertical which reduces each rotor's efficiency. The Kaman HH-43, which was used by the USAF in a firefighting role and the Kaman K-MAX are examples of an intermeshing rotor systems.

Swash Plate Assembly

The purpose of the swash plate is to convert stationary control inputs from the pilot into rotating inputs which can be connected to the rotor blades or control surfaces. It consists of two main parts: stationary swash plate and rotating swash plate. *[Figure 4-11]*

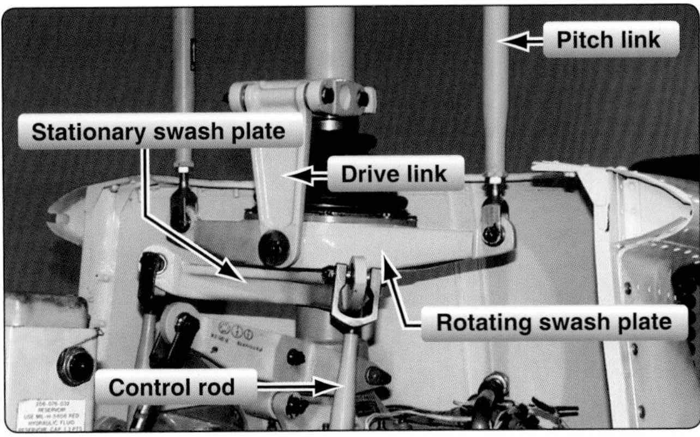

Figure 4-11. *Stationary and rotating swash plate.*

The stationary swash plate is mounted around the main rotor mast and connected to the cyclic and collective controls by a series of pushrods. It is restrained from rotating by an anti-drive link but can tilt in all directions and move vertically. The rotating swash plate is mounted to the stationary swash plate by means of a uniball sleeve. It is connected to the mast by drive links and must rotate in constant relationship with the main rotor mast. Both swash plates tilt and slide up and down as one unit. The rotating swash plate is connected to the pitch horns by the pitch links.

Freewheeling Unit

Since lift in a helicopter is provided by rotating airfoils, these airfoils must be free to rotate if the engine fails. The freewheeling unit automatically disengages the engine from the main rotor when engine revolutions per minute (rpm) is less than main rotor rpm. *[Figure 4-12]* This allows the main rotor and tail rotor to continue turning at normal in-flight speeds. The most common freewheeling unit assembly consists of a one-way sprag clutch located between the engine and main rotor transmission. This is usually in the upper pulley in a piston helicopter or mounted on the accessory gearbox in a turbine helicopter. When the engine is driving the rotor, inclined surfaces in the sprag clutch force rollers against an outer drum. This prevents the engine from exceeding transmission rpm. If the engine fails, the rollers move inward, allowing the outer drum to exceed the speed of the inner portion. The transmission can then exceed the speed of the engine. In this condition, engine speed is less than that of the drive system, and the helicopter is in an autorotative state.

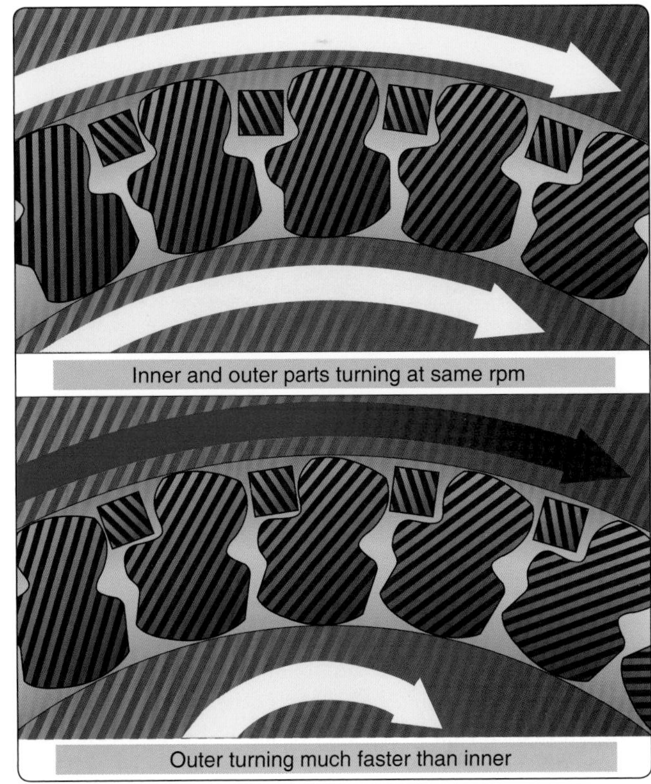

Figure 4-12. *Freewheeling unit in normal drive position and freewheeling position. Note that in the top example, the engine output shaft (inner part) drives the rotor shaft (outer part) at the same speed (normal flight). In the bottom example, the rotor shaft (outer part) breaks free under autorotation, as it turns faster than the driver shaft (inner part).*

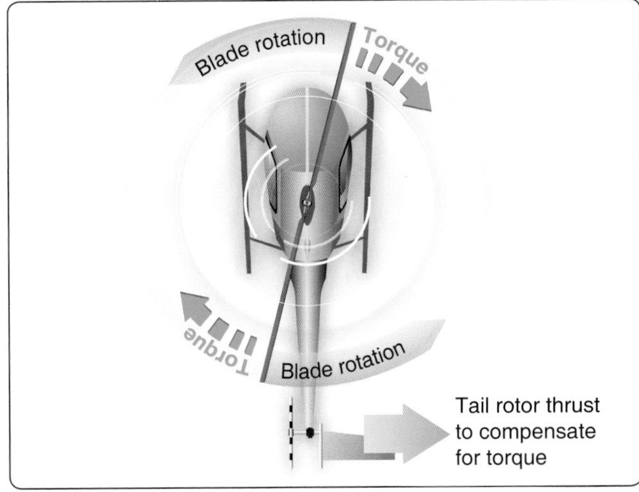

Figure 4-13. *Antitorque rotor produces thrust to oppose torque.*

Antitorque System

Helicopters with a single, main rotor system require a separate antitorque system. This is most often accomplished through a variable pitch, antitorque rotor or tail rotor. *[Figure 4-13]* Pilots vary the thrust of the antitorque system to maintain directional control whenever the main rotor torque changes, or to make heading changes while hovering. Most helicopters drive the tail rotor shaft from the transmission to ensure tail rotor rotation (and hence control) in the event that the engine quits. Usually, negative antitorque thrust is needed in autorotations to overcome transmission friction.

Fenestron

Another form of antitorque system is the Fenestron or "fan-in-tail" design. This system uses a series of rotating blades shrouded within a vertical tail. Because the blades are located within a circular duct, they are less likely to come into contact with people or objects. *[Figure 4-14]*

NOTAR®

Using the natural characteristics of helicopter aerodynamics, the NOTAR® antitorque system provides safe, quiet, responsive, foreign object damage (FOD) resistant directional control. The enclosed variable-pitch composite blade fan produces a low pressure, high volume of ambient air to pressurize the composite tailboom. The air is expelled through two slots which run the length of the tailboom on the right side, causing a boundary-layer control called the Coanda effect. The result is that the tailboom becomes a "wing," flying in the downwash of the rotor system, producing up to 60 percent of the antitorque required in a hover. The balance of the directional control is accomplished by a rotating direct jet thruster. In forward flight, the vertical stabilizers provide the majority of the antitorque; however, directional control remains a function of the direct jet thruster. The NOTAR® antitorque system eliminates some of the mechanical disadvantages of a tail rotor, including long drive shafts, hanger bearings, intermediate gearboxes and 90° gearboxes. *[Figure 4-15]*

Antitorque Drive Systems

The antitorque drive system consists of an antitorque drive shaft and a antitorque gearbox mounted at the end of the tail boom. The drive shaft may consist of one long shaft or a series of shorter shafts connected at both ends with flexible couplings. This allows the drive shaft to flex with the tail boom. The tail rotor gearbox provides a right-angle drive for the tail rotor and may also include gearing to adjust the output to optimum tail rotor rpm. *[Figure 4-16]* Tail rotors may also have an intermediate gearbox to turn the power up a pylon or vertical fin.

Figure 4-14. *Fenestron or "fan-in-tail" antitorque system. This design provides an improved margin of safety during ground operations.*

Engines

Reciprocating Engines

Reciprocating engines, also called piston engines, are generally used in smaller helicopters. Most training helicopters use reciprocating engines because they are relatively simple and inexpensive to operate. Refer to the Pilot's Handbook of Aeronautical Knowledge for a detailed explanation and illustrations of the piston engine.

Turbine Engines

Turbine engines are more powerful and are used in a wide variety of helicopters. They produce a tremendous amount of power for their size but are generally more expensive to operate. The turbine engine used in helicopters operates differently from those used in airplane applications. In most applications, the exhaust outlets simply release expended gases and do not contribute to the forward motion of the helicopter. Approximately 75 percent of the incoming airflow is used to cool the engine.

The gas turbine engine mounted on most helicopters is made up of a compressor, combustion chamber, turbine, and accessory gearbox assembly. The compressor draws filtered air into the plenum chamber and compresses it. Common type filters are centrifugal swirl tubes where debris is ejected outward and blown overboard prior to entering the compressor, or engine barrier filters (EBF), similar to the K&N filter element used in automotive applications.

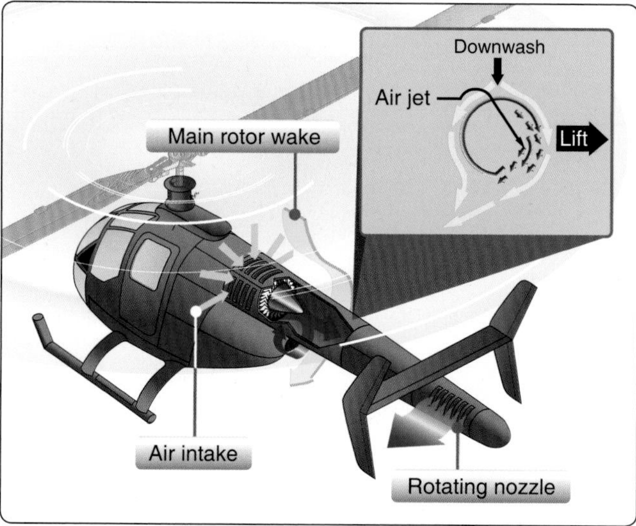

Figure 4-15. *While in a hover, Coanda effect supplies approximately two-thirds of the lift necessary to maintain directional control. The rest is created by directing the thrust from the controllable rotating nozzle.*

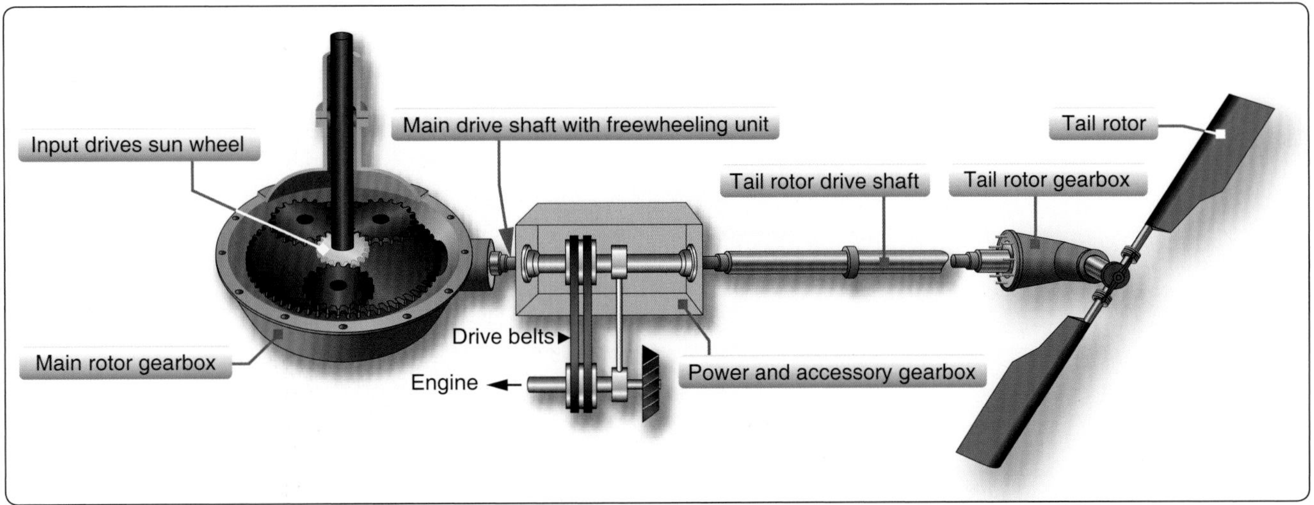

Figure 4-16. *The tail rotor driveshaft is connected to both the main transmission and the tail rotor transmission.*

Although this design significantly reduces the ingestion of foreign objects into the engine, it is important for pilots to be aware of how much debris is actually being filtered. Operating in the sand, dust, or even in grassy type materials can choke an engine in just minutes. The compressed air is directed to the combustion section through discharge tubes where atomized fuel is injected into it. The fuel/air mixture is ignited and allowed to expand. This combustion gas is then forced through a series of turbine wheels causing them to turn. These turbine wheels provide power to both the engine compressor and the accessory gearbox. Depending on model and manufacturer, the rpm can vary from 20,000 to 51,600.

Power is provided to the main rotor and tail rotor systems through the freewheeling unit which is attached to the accessory gearbox power output gear shaft. The combustion gas is finally expelled through an exhaust outlet. The temperature of gas is measured at different locations and is referenced differently by each manufacturer. Some common terms are inter-turbine temperature (ITT), exhaust gas temperature (EGT), measured gas temperature (MGT), or turbine outlet temperature (TOT). TOT is used throughout this discussion for simplicity. *[Figure 4-17]*

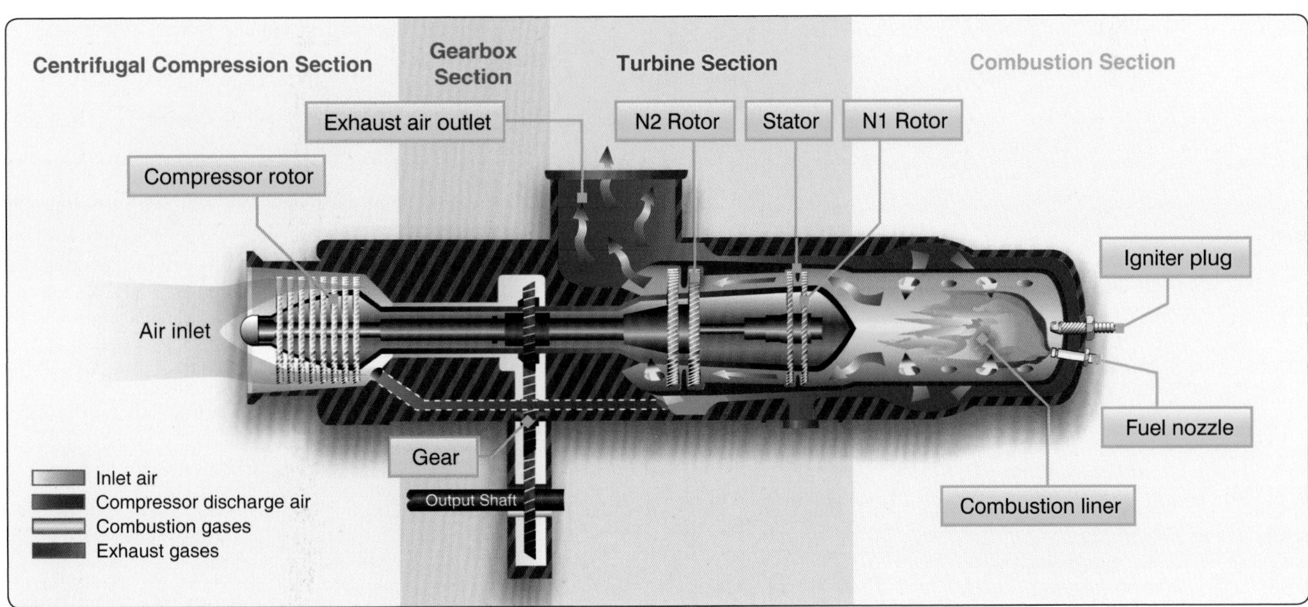

Figure 4-17. *Many helicopters use a turboshaft engine as shown above to drive the main transmission and rotor systems. The main difference between a turboshaft and a turbojet engine is that most of the energy produced by the expanding gases is used to drive a turbine rather than producing thrust through the expulsion of exhaust gases.*

Compressor

The compressor may consist of an axial compressor, a centrifugal compressor, or combination of the two.

An axial compressor consists of two main elements: the rotor and the stator. The rotor consists of a number of blades fixed on a rotating spindle and resembles a fan. As the rotor turns, air is drawn inward. Stator vanes are arranged in fixed rows between the rotor blades and act as a diffuser at each stage to decrease air velocity and increase air pressure. There may be a number of rows of rotor blades and stator vanes. Each row constitutes a pressure stage, and the number of stages depends on the amount of air and pressure rise required for the particular engine.

A centrifugal compressor consists of an impeller, diffuser, and a manifold. The impeller, which is a forged disc with integral blades, rotates at a high speed to draw air in and expel it at an accelerated rate. The air then passes through the diffuser, which slows the air down. When the velocity of the air is slowed, static pressure increases, resulting in compressed, high pressure air. The high-pressure air then passes through the compressor manifold where it is distributed to the combustion chamber via discharge tubes.

If the airflow through the compressor is disturbed, a condition called surge, or compressor stall, may take effect. This phenomenon is a periodic stalling of the compressor blades. When this occurs, the pressure at the compressor is reduced and the combustion pressure may cause reverse flow into the compressor output. As the airflow through the compressor is reduced, the air pressure then increases temporarily correcting the condition until it occurs again. This is felt throughout the airframe as vibrations and is accompanied by power loss and an increase in TOT as the fuel control adds fuel in an attempt to maintain power. This condition may be corrected by activating the bleed air system which vents excess pressure to the atmosphere and allows a larger volume of air to enter the compressor to unstall the compressor blades.

Combustion Chamber

Unlike a piston engine, the combustion in a turbine engine is continuous. An igniter plug serves only to ignite the fuel/air mixture when starting the engine. Once the fuel/air mixture is ignited, it continues to burn as long as the fuel/air mixture continues to be present. If there is an interruption of fuel, air, or both, combustion ceases. This is known as a "flameout," and the engine must be restarted or re-lit. Some helicopters are equipped with auto-relight, which automatically activates the igniters to start combustion if the engine flames out.

Turbine

The two-stage turbine section consists of a series of turbine wheels that are used to drive the compressor section and other components attached to the accessory gearbox. Both stages may consist of one or more turbine wheels. The first stage is usually referred to as the gas producer (N1 or NG) while the second stage is commonly called the power turbine (N2 or NP). (The letter N is used to denote rotational speed.)

If the first and second stage turbines are mechanically coupled to each other, the system is said to be a fixed turbine (turboshaft). These engines share a common shaft, which means the first and second stage turbines, and thus the compressor and output shaft, are connected.

On most turbine assemblies used in helicopters, the first stage and second stage turbines are not mechanically connected to each other. Rather, they are mounted on independent shafts, one inside the other, and can turn freely with respect to each other. This is referred to as a "free turbine." When a free turbine engine is running, the combustion gases pass through the first stage turbine (N1) to drive the compressor and other components, and then past the independent second stage turbine (N2), which turns the power and accessory gearbox to drive the output shaft, as well as other miscellaneous components.

Accessory Gearbox

The accessory gearbox of the engine houses all of the necessary gears to drive the numerous components of the helicopter. Power is provided to the accessory gearbox through the independent shafts connected to the N1 and N2 turbine wheels. The N1 stage drives the components necessary to complete the turbine cycle, making the engine self-sustaining. Common components driven by the N1 stage are the compressor, oil pump, fuel pump, and starter/generator. The N2 stage is dedicated to driving the main rotor and tail rotor drive systems and other accessories such as generators, alternators, and air conditioning.

Transmission System

The transmission system transfers power from the engine to the main rotor, tail rotor, and other accessories during normal flight conditions. The main components of the transmission system are the main rotor transmission, tail rotor drive system, clutch, and freewheeling unit. The freewheeling unit or autorotative clutch allows the main rotor transmission to drive the tail rotor drive shaft during autorotation. In some helicopter designs, such as the Bell BH-206, the freewheeling unit is located in the accessory gearbox. Because it is part of the transmission system, the transmission lubricates it to ensure free rotation. Helicopter transmissions are normally lubricated and cooled with their own oil supply. A sight gauge is provided to check the oil level. Some transmissions have chip detectors located in the sump, to detect loose pieces of metal. These detectors are wired to warning lights located on the pilot's instrument panel that illuminate in the event of an internal problem. Some chip detectors on modern helicopters have a "burn off" capability and attempt to correct the situation without pilot action. If the problem cannot be corrected on its own, the pilot must refer to the emergency procedures for that particular helicopter.

Main Rotor Transmission

The primary purpose of the main rotor transmission is to reduce engine output rpm to optimum rotor rpm. This reduction is different for the various helicopters. As an example, suppose the engine rpm of a specific helicopter is 2,700. A rotor speed of 450 rpm would require a 6:1 reduction. A 9:1 reduction would mean the rotor would turn at 300 rpm.

Dual Tachometers

Most helicopters use a dual-needle tachometer or a vertical scale instrument to show both engine and rotor rpm or a percentage of engine and rotor rpm. The rotor rpm indicator is used during clutch engagement to monitor rotor acceleration, and in autorotation to maintain rpm within prescribed limits. It is vital to understand that rotor rpm is paramount, and that engine rpm is secondary. If the rotor tachometer fails, rotor rpm can still be determined indirectly by the engine rpm during powered flight, because the engine drives the rotor at a fixed, one-to-one ratio (by virtue of the sprag clutch). There have been many accidents where the pilot responded to the rotor rpm tachometer failure and entered into autorotation while the engine was still operating.

Look closer at the markings on the gauges in *Figure 4-18.* All gauges shown are dual tachometer gauges. The two on the left have two needles each, one marked with the letter 'T' (turbine) the other marked with the letter 'R' (rotor). The lower left gauge shows two arced areas within the same needle location. In this case, both needles should be nearly together or superimposed during normal operation. Note the top left gauge shows two numerical arcs. The outer arc, with larger numbers, applies one set of values to engine rpm. The inner arc, or smaller numbers, represents a separate set of values for rotor rpm. Normal operating limits are shown when the needles are married or appear superimposed. The top right gauge shows independent needles, focused toward the middle of the gauge, with colored limitation areas respective to the needle head. The left side represents engine operational parameters; the right, rotor operational parameters.

In normal conditions when the rotor is coupled to the engine, both needles move together in the same direction. However, with a sudden loss in engine power the needles "split" showing that the engine and rotor are no longer coupled as the clutch has disconnected. *[Figure 4-19]*

Many newer aircraft have what is referred to as a glass cockpit, meaning the instrumentation is digital and displayed

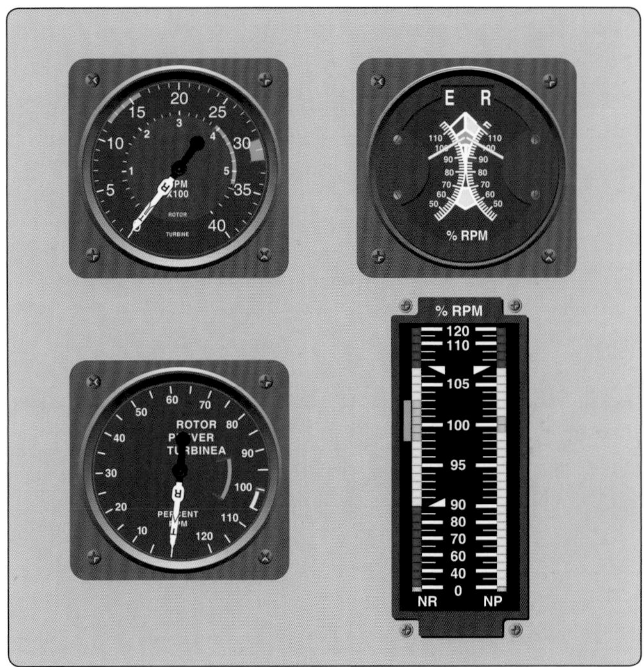

Figure 4-18. *Various types of dual-needle tachometers.*

Figure 4-19. *A "split" or divided needle condition is a result of a sudden loss of engine power.*

to the pilot on digital screens and vertical scale instruments. The bottom right gauge in *Figure 4-18* replicates a vertical scale instrument. The dual tachometer shown displays rotor rpm (NR) on the left and engine rpm (NP) on the right side of the vertical scale. Corresponding color limits are present for each component parameter.

Structural Design

In helicopters with horizontally mounted engines, another purpose of the main rotor transmission is to change the axis of rotation from the horizontal axis of the engine to the vertical axis of the rotor shaft. *[Figure 4-20]* This differs from airplanes, which have their propellers mounted directly to the crankshaft or to a shaft that is geared to the crankshaft.

Maintaining main rotor rpm is essential for adequate lift. RPM within normal limits produces adequate lift for normal maneuvering. Therefore, it is imperative not only to know the location of the tachometers, but also to understand the information they provide. If rotor rpm is allowed to go below normal limits, the outcome could be catastrophic.

Clutch

In a conventional airplane, the engine and propeller are permanently connected. However, in a helicopter they are not. Because of the greater weight of a rotor in relation to the power of the engine, as compared to the weight of a propeller and the power in an airplane, the rotor must be disconnected from the engine when the starter is engaged. A clutch allows the engine to be started and then gradually pick up the load of the rotor.

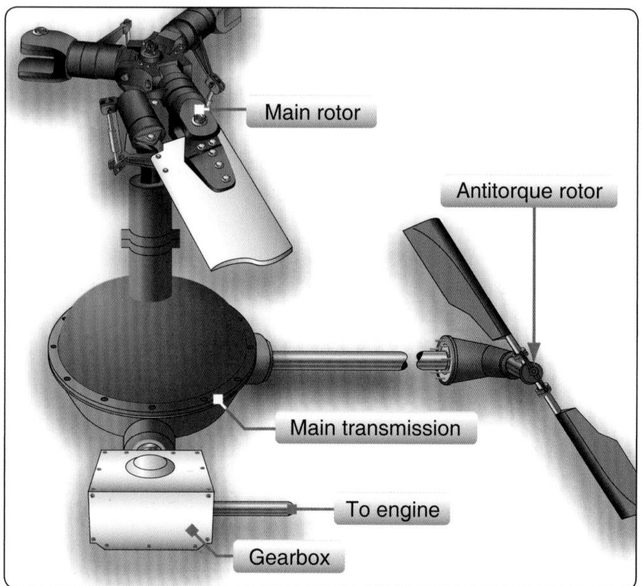

Figure 4-20. *The main rotor transmission reduces engine output rpm to optimum rotor rpm.*

Freewheeling turbine engines do not require a separate clutch since the air coupling between the gas producer turbine and the power (takeoff) turbine functions as an air clutch for starting purposes. When the engine is started, there is little resistance from the power turbine. This enables the gas-producer turbine to accelerate to normal idle speed without the load of the transmission and rotor system dragging it down. As the gas pressure increases through the power turbine, the rotor blades begin to turn, slowly at first and then gradually accelerate to normal operating rpm.

On reciprocating and fixed turbine engines, a clutch is required to enable engine start. Air, or windmilling starts, are not possible. The two main types of clutches are the centrifugal clutch and the idler or manual clutch.

How the clutch engages the main rotor system during engine start differs between helicopter design. Piston-powered helicopters have a means of engaging the clutch manually just as a manual clutch in an automobile. This may be by means of an electric motor that positions a pulley when the engine is at the proper operating condition (oil temperature and pressure in the appropriate range), but which is controlled by a cockpit mounted switch.

Belt Drive Clutch

Some helicopters utilize a belt drive to transmit power from the engine to the transmission. A belt drive consists of a lower pulley attached to the engine, an upper pulley attached to the transmission input shaft, a belt or a set of V-belts, and some means of applying tension to the belts. The belts fit loosely over the upper and lower pulley when there is no tension on the belts. *[Figure 4-21]*

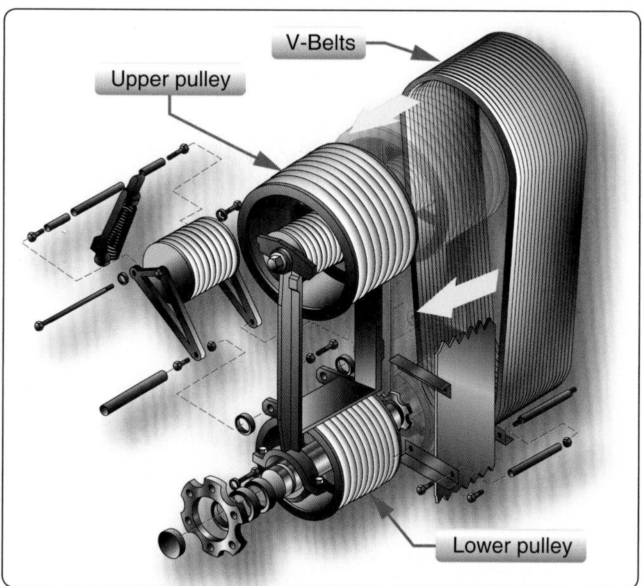

Figure 4-21. *Idler or manual clutch.*

Some aircraft utilize a clutch for starting. This allows the engine to be started without requiring power to turn the transmission. One advantage this concept has is that without a load on the engine starting may be accomplished with minimal throttle application. However, caution should also be used during starting, since rapid or large throttle inputs may cause overspeeds.

Once the engine is running, tension on the belts is gradually increased. When the rotor and engine tachometer needles are superimposed, the rotor and the engine are synchronized, and the clutch is then fully engaged. Advantages of this system include vibration isolation, simple maintenance. When the clutch is not engaged, engines are very easy to overspeed, resulting in costly inspections and maintenance. Power, or throttle control, is very important in this phase of engine operation.

Centrifugal Clutch

A centrifugal clutch is made up of an inner assembly and an outer drum. The inner assembly, which is connected to the engine driveshaft, consists of shoes lined with material similar to automotive brake linings. At low engine speeds, springs hold the shoes in, so there is no contact with the outer drum, which is attached to the transmission input shaft. As engine speed increases, centrifugal force causes the clutch shoes to move outward and begin sliding against the outer drum. The transmission input shaft begins to rotate, causing the rotor to turn slowly at first, but increasing as the friction increases between the clutch shoes and transmission drum.

As rotor speed increases, the rotor tachometer needle shows an increase by moving toward the engine tachometer needle. When the two needles are superimposed (in the case of a coaxial-type gage), the engine and the rotor are synchronized, indicating the clutch is fully engaged and there is no further slippage of the clutch shoes.

The turbine engine engages the clutch through centrifugal force, as stated above. Unless a rotor brake is used to separate the automatic engagement of the main driveshaft and subsequently the main rotor, the drive shaft turns at the same time as the engine and the inner drum of the freewheeling unit engages gradually to turn the main rotor system.

Fuel Systems

The fuel system in a helicopter is made up of two components: supply and control.

Fuel Supply System

The supply system consists of a fuel tank or tanks, fuel quantity gauges, a shut-off valve, fuel filter, a fuel line to the engine, and possibly a primer and fuel pumps. *[Figure 4-22]* The fuel tanks are usually mounted to the airframe as close as possible to the CG. This way, as fuel is burned off, there is a negligible effect on the CG. A drain valve located on the bottom of the fuel tank allows the pilot to drain water and sediment that may have collected in the tank. A fuel vent prevents the formation of a vacuum in the tank, and an overflow drain allows fuel to expand without rupturing the tank.

The fuel travels from the fuel tank through a shut-off valve, which provides a means to completely stop fuel flow to the engine in the event of an emergency or fire. The shut-off valve remains in the open position for all normal operations.

Most non-gravity feed fuel systems contain both an electric pump and a mechanical engine-driven pump. The electrical pump is used to maintain positive fuel pressure to the engine pump and may also serve as a backup in the event of mechanical pump failure. The electrical pump is controlled by a switch in the cockpit. The engine driven pump is the primary pump that supplies fuel to the engine and operates any time the engine is running. A fuel filter removes moisture and other sediment from the fuel before it reaches the engine. These contaminants are usually heavier than fuel and settle to the bottom of the fuel filter sump where they can be drained out by the pilot.

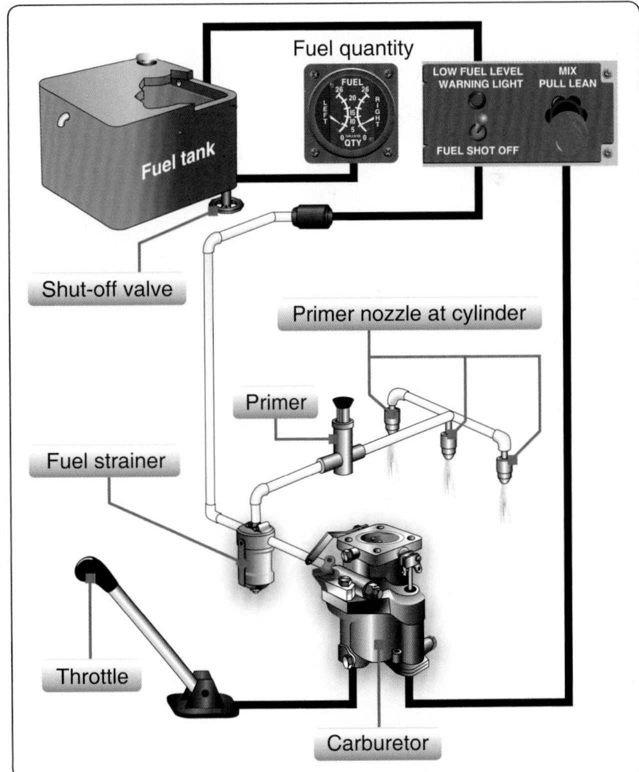

Figure 4-22. *A typical gravity feed fuel system, in a helicopter with a reciprocating engine, contains the components shown here.*

Some fuel systems contain a small hand-operated pump called a primer. A primer allows fuel to be pumped directly into the intake port of the cylinders prior to engine start. The primer is useful in cold weather when fuel in the carburetor is difficult to vaporize.

A fuel quantity gauge located on the pilot's instrument panel shows the amount of fuel measured by a sensing unit inside the tank. Most fuel gauges will indicate in gallons or pounds and must be accurate only when empty.

It is worth noting that in accordance with Title 14 of the Code of Federal Regulations (14 CFR) section 27.1337(b)(1), fuel quantity indicators "must be calibrated to read 'zero' during level flight when the quantity of fuel remaining in the tank is equal to the unusable fuel supply." Therefore, it is of the utmost importance that the pilot or operator determine an accurate means of verifying partial or full fuel loads. It is always a good habit, if possible, to visually verify the fuel on board prior to flight and determine if adequate fuel is present for the duration of the flight.

Additionally, 14 CFR section 27.1305(l)(1) requires newer helicopters to have warning systems "provide a warning to the flight crew when approximately 10 minutes of usable fuel remains in the tank." Caution should be used to eliminate unnecessary or erratic maneuvering that could cause interruption of fuel flow to the engine. Although these systems must be calibrated, never assume the entire amount is available. Many pilots have not reached their destinations due to poor fuel planning or faulty fuel indications.

Engine Fuel Control System

Regardless of the device, the reciprocating engine and the turbine engine both use the ignition and combustion of the fuel/air mix to provide the source of their power. Engine fuel control systems utilize several components to meter the proper amount of fuel necessary to produce the required amount of power. The fuel control system, in concert with the air induction components, combines the proper amount of fuel and air to be ignited in the combustion chamber. Refer to the Pilot's Handbook of Aeronautical Knowledge for a detailed explanation and illustration.

Carburetor Ice

The effect of fuel vaporization and/or a decrease of air pressure in the venturi causes a rapid decrease in air temperature in the carburetor. If the air is moist, the water vapor in the air may condense causing ice to form in the carburetor. If ice is allowed to form inside the carburetor, engine failure is a very real possibility and the ability to restart the engine is greatly reduced. Carburetor icing can occur during any phase of flight but is particularly dangerous when you are using reduced power, such as during a descent. You may not notice it during the descent until you try to add power. Indications of carburetor icing are a decrease in engine rpm or manifold pressure, the carburetor air temperature gauge indicating a temperature outside the safe operating range, and engine roughness. A reciprocating engine with a governor may mask the formation of carburetor ice since it will maintain a constant manifold pressure and rpm.

Since changes in rpm or manifold pressure can occur for a number of reasons, closely check the carburetor air temperature gauge when in possible carburetor icing conditions. Carburetor air temperature gauges are marked with a yellow caution arc or green operating arcs. In most cases, it is best to keep the needle out of the yellow arc or in the green arc. This is accomplished by using a carburetor heat system, which eliminates the ice by routing air across

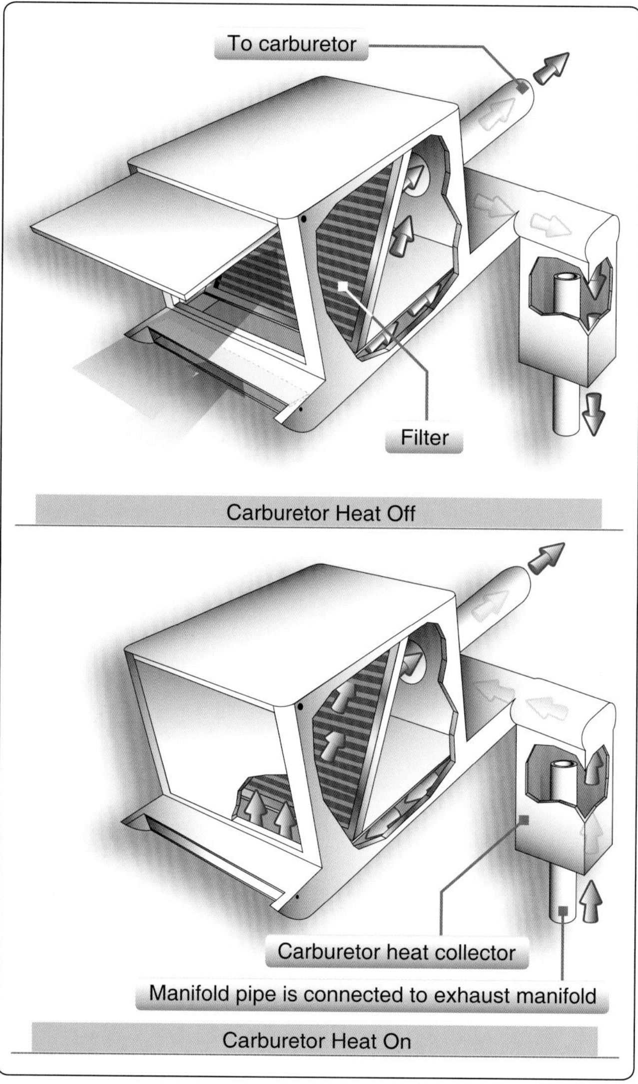

Figure 4-23. *When the carburetor heat is turned ON, normal air flow is blocked, and heated air from an alternate source flows through the filter to the carburetor.*

a heat source, such as an exhaust manifold, before it enters the carburetor. [*Figure 4-23*] Refer to the RFM (see Chapter 5, Rotorcraft Flight Manual) for the specific procedure as to when and how to apply carburetor heat.

Fuel Injection

In a fuel injection system, fuel and air are metered at the fuel control unit but are not mixed. The fuel is injected directly into the intake port of the cylinder where it is mixed with

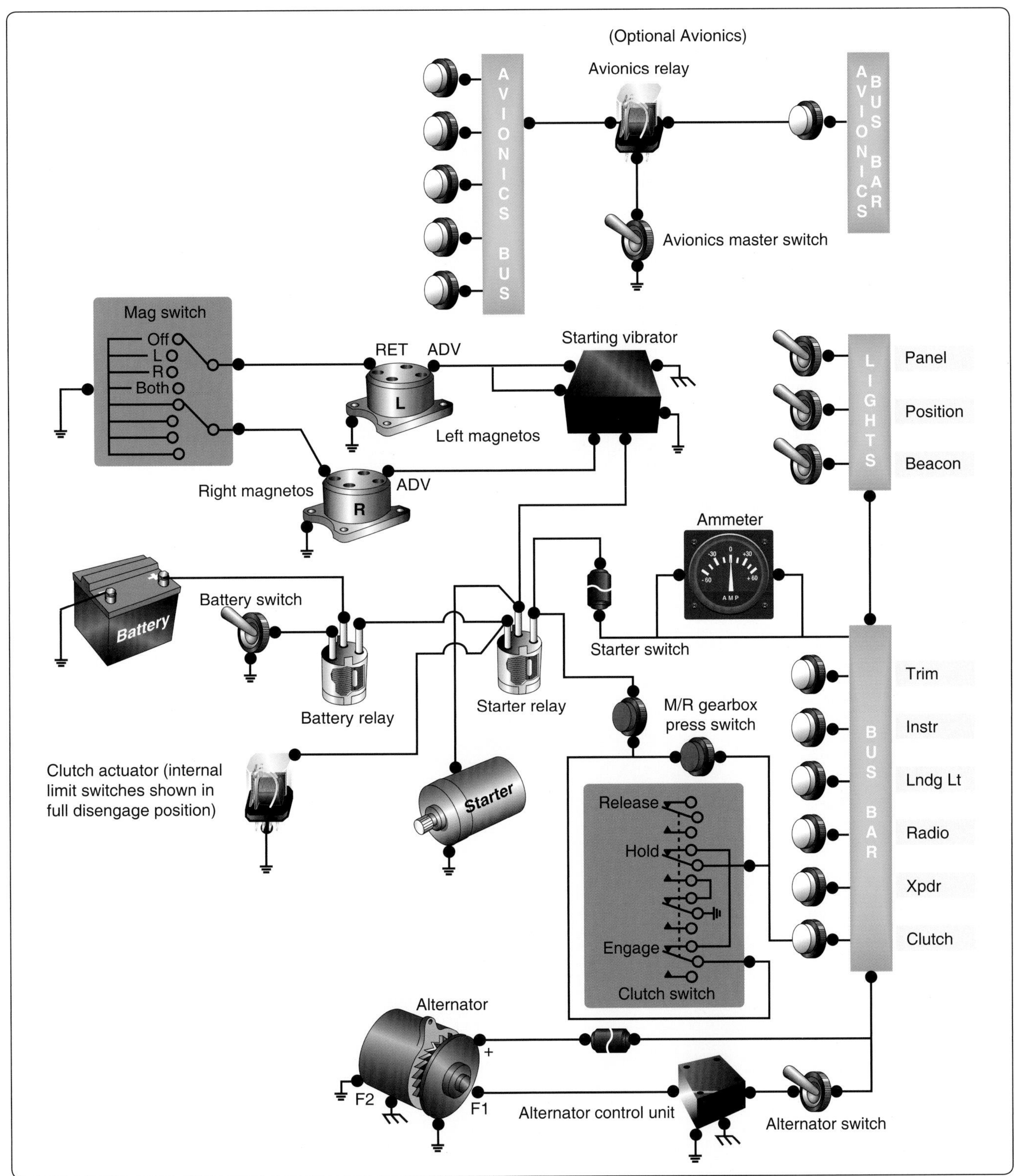

Figure 4-24. *An electrical system schematic like this sample is included in most POHs. Notice that the various bus bar accessories are protected by circuit breakers. However, ensure that all electrical equipment is turned off before starting the engine. This protects sensitive components, particularly the radios, from damage that may be caused by random voltages generated during the starting process.*

the air just before entering the cylinder. This system ensures a more even fuel distribution between cylinders and better vaporization, which in turn promotes more efficient use of fuel. Also, the fuel injection system eliminates the problem of carburetor icing and the need for a carburetor heat system.

Electrical Systems

The electrical systems, in most helicopters, reflect the increased use of sophisticated avionics and other electrical accessories. *[Figure 4-24]* More and more operations in today's flight environment are dependent on the aircraft's electrical system; however, all helicopters can be safely flown without any electrical power in the event of an electrical malfunction or emergency.

Helicopters have either a 14- or 28-volt, direct-current electrical system. On small, piston powered helicopters, electrical energy is supplied by an engine-driven alternator by means of a belt and pulley system similar to that of an automobile. These alternators have advantages over older-style generators as they are lighter in weight, require lower maintenance, and maintain a uniform electrical output even at low engine rpm. (As a reminder, think of volts or voltage as the measure of electrical pressure in the system, analogous to pounds per square inch in water systems. Amperes is the measure of electrical quantity in the system or available. For example, a 100-amp alternator would be analogous to a 100 gallon per hour water pump.)

Turbine-powered helicopters use a starter/generator system. The starter/generator is permanently coupled to the accessory gearbox. When starting the engine, electrical power from the battery is supplied to the starter/generator, which turns the engine over. Once the engine is running, the starter/generator is driven by the engine and then functions as a generator.

Current from the alternator or generator is delivered through a voltage regulator to a bus bar. The voltage regulator maintains the constant voltage required by the electrical system, by regulating the output of the alternator or generator. An over-voltage control may be incorporated to prevent excessive voltage, which may damage the electrical components. The bus bar serves to distribute the current to the various electrical components of the helicopter.

A battery is used mainly for starting the engine. In addition, it permits limited operation of electrical components, such as radios and lights, without the engine running. The battery is also a valuable source of standby or emergency electrical power in the event of alternator or generator failure.

An ammeter (or load meter) is used to monitor the electrical current within the system. The ammeter reflects current flowing to and from the battery. A charging ammeter indicates that the battery is being charged. This is normal after an engine start since the battery power used in starting is being replaced. After the battery is charged, the ammeter should stabilize near zero since the alternator or generator is supplying the electrical needs of the system.

An ammeter showing a discharge means the electrical load is exceeding the output of the alternator or generator, and the battery is helping to supply electrical power. This may mean the alternator or generator is malfunctioning, or the electrical load is excessive. An ammeter displays the load placed on the alternator or generator by the electrical equipment. The RFM (see page 5-1) for a particular helicopter shows the normal load to expect. Loss of the alternator or generator causes the load meter to indicate zero.

Electrical switches are used to select electrical components. Power may be supplied directly to the component or to a relay, which in turn provides power to the component. Relays are used when high current and/or heavy electrical cables are required for a particular component, which may exceed the capacity of the switch.

Circuit breakers or fuses are used to protect various electrical components from overload. A circuit breaker pops out when its respective component is overloaded. The circuit breaker may be reset by pushing it back in, unless a short or the overload still exists. In this case, the circuit breaker continues to pop, indicating an electrical malfunction. A fuse simply burns out when it is overloaded and needs to be replaced. Manufacturers usually provide a holder for spare fuses in the event one has to be replaced in flight. Caution lights on the instrument panel may be installed to show the malfunction of an electrical component.

Hydraulics

Most helicopters, other than smaller piston-powered helicopters, incorporate the use of hydraulic actuators to overcome high control forces. *[Figure 4-25]* A typical hydraulic system consists of actuators, also called servos, on each flight control, a pump which is usually driven by the main rotor transmission and a reservoir to store the hydraulic fluid. Some helicopters have accumulators located on the pressure side of the hydraulic system. This allows for a continuous fluid pressure into the system. A switch in the cockpit can turn the system off, although it is left on under normal conditions. When the pilot places the hydraulic switch/circuit breaker into the on position, the electrical power is being removed from the solenoid valve allowing

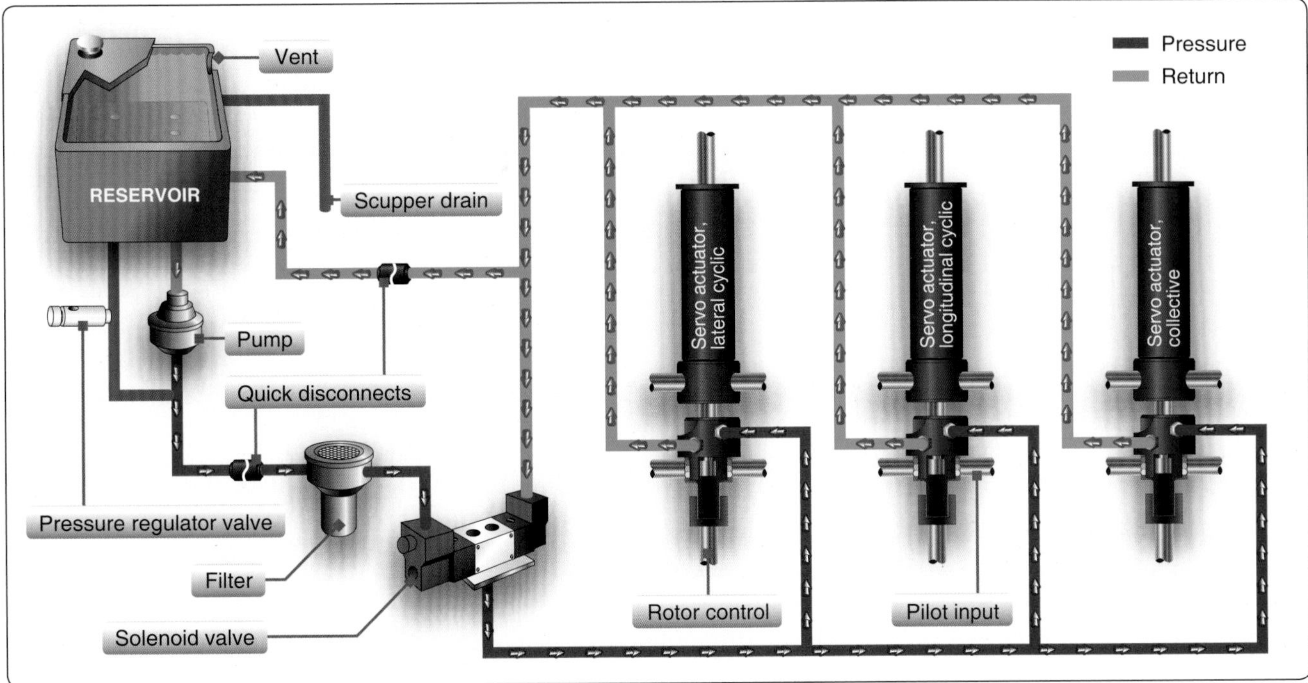

Figure 4-25. *A typical hydraulic system for helicopters in the light to medium range.*

hydraulic fluid to enter the system. When the switch/circuit breaker is put in the off position, the solenoid valve is now de-energized and closes, which then allows the pilot to maintain control of the helicopter with the hydraulic fluid in the actuators. This is known as a failsafe system. If helicopter electrical power is lost in flight, the pilot is still able to maintain control of the hydraulic system. A pressure indicator in the cockpit may also be installed to monitor the system.

When making a control input, the servo is activated and provides an assisting force to move the respective flight control, thus reducing the force the pilot must provide. These boosted flight controls ease pilot workload and fatigue. In the event of hydraulic system failure, a pilot is still able to control the helicopter, but the control forces are very heavy.

In those helicopters in which the control forces are so high that they cannot be moved without hydraulic assistance, two or more independent hydraulic systems may be installed. Some helicopters are designed to use their hydraulic accumulators to store hydraulic pressure for an emergency, allowing for uninterrupted use of the controls for a short period of time following a hydraulic pump failure. This gives you enough time to land the helicopter with normal control.

Stability Augmentations Systems

Some helicopters incorporate a stability augmentation system (SAS) to help stabilize the helicopter in flight and in a hover. The original purpose and design allowed decreased pilot workload and lessened fatigue. It allowed pilots to place an aircraft at a set attitude to accomplish other tasks or simply stabilize the aircraft for long cross-country flights.

Force Trim

Force trim was a passive system that simply held the cyclic in a position that gave a control force to transitioning airplane pilots who had become accustomed to such control forces. The system uses a magnetic clutch and springs to hold the cyclic control in the position where it was released. The system does not use sensor-based data to make corrections, but rather is used by the pilot to "hold" the cyclic in a desired position. The most basic versions only apply to the cyclic requiring the pilot to continue power and tail rotor inputs. With the force trim on or in use, the pilot can override the system by disengaging the system through the use of a force trim release button or, with greater resistance, can physically manipulate the controls. Some recent basic systems are referred to as attitude retention systems.

Active Augmentation Systems

So-called actual augmentation systems use electric actuators that provide input to the hydraulic servos. These servos receive control commands from a computer that senses external environmental inputs, such as wind and turbulence. SAS complexity varies by manufacturer but can be as sophisticated as providing three-axis stability. That is, computer-based inputs adjust attitude, power and aircraft trim for a more stabilized flight.

Once engaged by the pilot, these actual systems use a multitude of sensors, from stabilized gyros to electro-mechanical actuators, which provide instantaneous inputs to all flight controls without pilot assistance. As with all SASs, they may be overridden or disconnected by the pilot at any time. Helicopters with complex Automatic Flight Control Systems (AFCS) and autopilots normally have a trim switch referred to as "beeper trim." This switch is used when minor changes to the trim setting are desired.

Stability augmentation systems reduce pilot workload by improving basic aircraft control harmony and decreasing disturbances. These systems are very useful when the pilot is required to perform other duties, such as sling loading and search-and-rescue operations. Other inputs such as heading, speed, altitude, and navigation information may be supplied to the computer to form a complete autopilot system.

Autopilot

Helicopter autopilot systems are similar to stability augmentation systems, but they have additional features. An autopilot can actually fly the helicopter and perform certain functions selected by the pilot. These functions depend on the type of autopilot and systems installed in the helicopter.

The most common functions are altitude and heading hold. Some more advanced systems include a vertical speed or indicated airspeed (IAS) hold mode, where a constant rate of climb/descent or IAS is maintained by the autopilot. Some autopilots have navigation capabilities, such as very high frequency (VHF) OmniRange Navigation System (VOR), Instrument Landing System (ILS), and global positioning system (GPS) intercept and tracking, which is especially useful in instrument flight rules (IFR) conditions. This is referred to as a coupled system. An additional component, called a flight director (FD), may also be installed. The FD provides visual guidance cues to the pilot to fly selected lateral and vertical modes of operation. The most advanced autopilots can fly an instrument approach to a hover without any additional pilot input once the initial functions have been selected.

The autopilot system consists of electric actuators or servos connected to the flight controls. The number and location of these servos depends on the type of system installed. A two-axis autopilot controls the helicopter in pitch and roll; one servo controls fore and aft cyclic, and another controls left and right cyclic. A three-axis autopilot has an additional servo connected to the antitorque pedals and controls the helicopter in yaw. A four-axis system uses a fourth servo which controls the collective. These servos move the respective flight controls when they receive control commands from a central computer. This computer receives data input from the flight instruments for attitude reference and from the navigation equipment for navigation and tracking reference. An autopilot has a control panel in the cockpit that allows the pilot to select the desired functions, as well as engage the autopilot.

For safety purposes, an automatic disengagement feature is usually included which automatically disconnects the autopilot in heavy turbulence or when extreme flight attitudes are reached. Even though all autopilots can be overridden by the pilot, there is also an autopilot disengagement button located on the cyclic or collective which allows pilots to completely disengage the autopilot without removing their hands from the controls. Because autopilot systems and installations differ from one helicopter to another, it is very important to refer to the autopilot operating procedures located in the RFM.

Environmental Systems

Heating and cooling the helicopter cabin can be accomplished in different ways. The simplest form of cooling is by ram air. Air ducts in the front or sides of the helicopter are opened or closed by the pilot to let ram air into the cabin. This system is limited as it requires forward airspeed to provide airflow and also depends on the temperature of the outside air. Air conditioning provides better cooling, but it is more complex and weighs more than a ram air system.

One of the simplest methods of cooling a helicopter is to remove the doors allowing air to flow through the cockpit and engine compartments. Care must be taken to store the doors properly, whether in a designed door-holding rack in a hangar, or if it is necessary to carry them on the flight, in the helicopter. When storing the doors, care must be taken to not scratch the windows. Special attention should be paid to ensuring that all seat belt cushions and any other loose items are stored away to prevent ingestion into the main or tail rotor. When reattaching the doors, proper care must be taken to ensure that they are fully secured and closed.

Air conditioners or heat exchanges can be fitted to the helicopter as well. They operate by drawing bleed air from the compressor, passing it through the heart exchanger and then releasing it into the cabin. As the compressed air is released, the expansion absorbs heat and cools the cabin. The disadvantage of this type of system is that power is required to compress the air or gas for the cooling function, thus robbing the engine of some of its capability. Some systems are restricted from use during takeoff and landings.

Piston-powered helicopters use a heat exchanger shroud around the exhaust manifold to provide cabin heat. Outside air is piped to the shroud and the hot exhaust manifold heats the air, which is then blown into the cockpit. This warm air is heated by the exhaust manifold but is not exhaust gas. Turbine helicopters use a bleed air system for heat. Bleed air is hot, compressed, discharge air from the engine compressor. Hot air is ducted from the compressor to the bleed air heater assembly where it is combined with ambient air through and induction port mounted to the fuselage. The amount of heat delivered to the helicopter cabin is regulated by a pilot-controlled bleed air mixing valve.

Anti-Icing Systems

Anti-icing is the process of protecting against the formation of frozen contaminant, snow, ice, or slush on a surface.

Engine Anti-Ice

The anti-icing system found on most turbine-powered helicopters uses engine bleed air. Bleed air in turbine engines is compressed air taken from within the engine, after the compressor stage(s) and before the fuel is injected in the burners. The bleed air flows through the inlet guide vanes and to the inlet itself to prevent ice formation on the hollow vanes. A pilot-controlled, electrically operated valve on the compressor controls the air flow. Engine anti-ice systems should be on prior to entry into icing conditions and remain on until exiting those conditions. Use of the engine anti-ice system should always be in accordance with the proper RFM.

Airframe Anti-Ice

Airframe and rotor anti-icing may be found on some larger helicopters, but it is not common due to the complexity, expense, and weight of such systems. The leading edges of rotors may be heated with bleed air or electrical elements to prevent ice formation. Balance and control problems might arise if ice is allowed to form unevenly on the blades. Research is being done on lightweight ice-phobic (anti-icing) materials or coatings. These materials placed in strategic areas could significantly reduce ice formation and improve performance.

The pitot tube on a helicopter is very susceptible to ice and moisture buildup as well. To prevent this, they are usually equipped with a heating system that uses an electrical element to heat the tube.

Deicing

Deicing is the process of removing frozen contaminant, snow, ice, and/or slush from a surface. Deicing of the helicopter fuselage and rotor blades is critical prior to starting. Helicopters that are unsheltered by hangars are subject to frost, snow, freezing drizzle, and freezing rain that can cause icing of rotor blades and fuselages, rendering them unflyable until cleaned. Asymmetrical shedding of ice from the blades can lead to component failure, and shedding ice can be dangerous as it may hit any structures or people that are around the helicopter. The tail rotor is very vulnerable to shedding ice damage. Thorough preflight checks should be made before starting the rotor blades. If any ice was removed prior to starting, ensure that the flight controls move freely. While in flight, for those helicopters that have them, deicing systems should be activated immediately after entry into an icing condition.

Chapter Summary

This chapter discussed all of the common components, sections, and systems of the helicopter. The chapter also explained how each of them work with one another to make flight possible.

Chapter 5

Rotorcraft Flight Manual

Introduction

Title 14 of the Code of Federal Regulations (14 CFR) part 91 requires pilot compliance with the operating limitations specified in approved rotorcraft flight manuals, markings, and placards. Originally, flight manuals were often characterized by a lack of essential information and followed whatever format and content the manufacturer deemed appropriate. This changed with the acceptance of the General Aviation Manufacturers Association (GAMA) specification for a Pilot's Operating Handbook, which established a standardized format for all general aviation airplane and rotorcraft flight manuals. The term "Pilot's Operating Handbook (POH)" is often used in place of "Rotorcraft Flight Manual (RFM)."

However, if "Pilot's Operating Handbook" is used as the main title instead of "Rotorcraft Flight Manual," a statement must be included on the title page indicating that the document is the Federal Aviation Administration (FAA) approved Rotorcraft Flight Manual (RFM). *[Figure 5-1]*

Not including the preliminary pages, an FAA-approved RFM may contain as many as ten sections. These sections are: General Information; Operating Limitations; Emergency Procedures; Normal Procedures; Performance; Weight and Balance; Aircraft and Systems Description; Handling, Servicing, and Maintenance Supplements; and Safety and Operational Tips. Manufacturers have the option of including a tenth section on safety and operational tips and an alphabetical index at the end of the handbook.

Preliminary Pages

While RFMs may appear similar for the same make and model of aircraft, each flight manual is unique since it contains specific information about a particular aircraft, such as the equipment installed, and weight and balance information. Therefore, manufacturers are required to include the serial number and registration on the title page to identify the aircraft to which the flight manual belongs. If a flight manual does not indicate a specific aircraft registration and serial number, it is limited to general study purposes only.

Most manufacturers include a table of contents, which identifies the order of the entire manual by section number and title. In addition, some helicopters may include a log of changes or a revision page to track changes to the manual. Usually, each section also contains its own table of contents. Page numbers reflect the section being read, 1-1, 2-1, 3-1, and so on. If the flight manual is published in looseleaf form, each section is usually marked with a divider tab indicating the section number or title, or both. The emergency procedures section may have a red tab for quick identification and reference.

Figure 5-1. *The RFM is a regulatory document in terms of the maneuvers, procedures, and operating limitations described therein.*

General Information (Section 1)

The general information section provides the basic descriptive information on the rotorcraft and the powerplant. In some manuals there is a three-view drawing of the rotorcraft that provides the dimensions of various components, including the overall length and width, and the diameter of the rotor systems. This is a good place for pilots to quickly familiarize themselves with the aircraft. Pilots need to be aware of the dimensions of the helicopter since they often must decide the suitability of an operations area for themselves, as well as hanger space, landing pad, and ground handling needs.

Pilots can find definitions, abbreviations, explanations of symbology, and some of the terminology used in the manual at the end of this section. At the option of the manufacturer, metric and other conversion tables may also be included.

Operating Limitations (Section 2)

The operating limitations section contains only those limitations required by regulation or that are necessary for the safe operation of the rotorcraft, powerplant, systems, and equipment. It includes operating limitations, instrument markings, color coding, and basic placards. Some of the areas included are: airspeed, altitude, rotor, and powerplant limitations, including fuel and oil requirements; weight and loading distribution; and flight limitations.

Instrument Markings

Instrument markings may include, but are not limited to, green, red, and yellow ranges for the safe operation of the aircraft. The green marking indicates a range of continuous operation. The red range indicates the maximum or minimum operation allowed while the yellow range indicates a caution or transition area.

Airspeed Limitations

Airspeed limitations are shown on the airspeed indicator by color coding and on placards or graphs in the aircraft. A red line on the airspeed indicator shows the airspeed limit beyond which structural damage could occur. This is called the never exceed speed, or V_{NE}. The normal operating speed range is depicted by a green arc. A blue or a red cross-hatched line is sometimes added to show the maximum autorotation speed. *[Figure 5-2]*

Other airspeed limitations may be included in this section of the RFM. Examples include reduced V_{NE} when doors are removed, maximum airspeed for level flight with maximum continuous power (V_H), or restrictions when carrying an external load. Pilots need to understand and adhere to all airspeed limitations appropriate to the make, model, and configuration of the helicopter being flown.

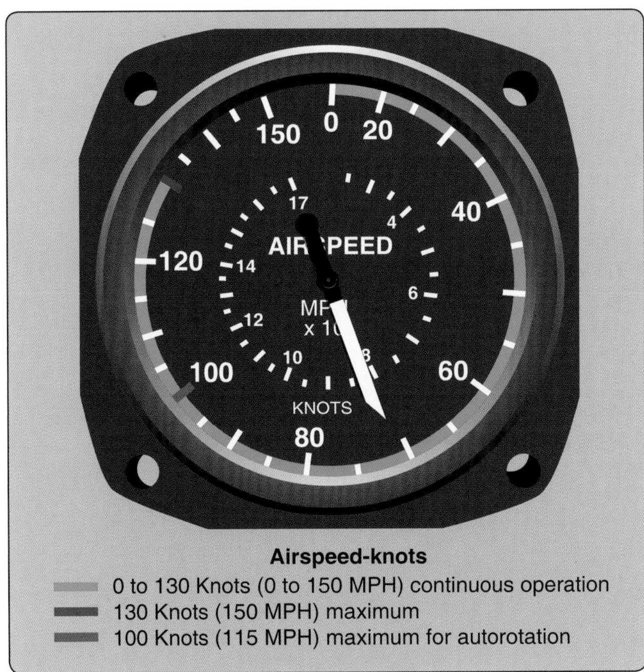

Figure 5-2. *Typical airspeed indicator limitations and markings.*

Figure 5-3. *Markings on a typical dual-needle tachometer in a reciprocating-engine helicopter. The outer band shows the limits of the superimposed needles when the engine is turning the rotor. The inner band indicates the power-off limits.*

Altitude Limitations

If the rotorcraft has a maximum operating density altitude (see page 7-2), it is indicated in this section of the flight manual. Sometimes the maximum altitude varies based on different gross weights.

Rotor Limitations

Low rpm does not produce sufficient lift, and high rpm may cause structural damage, therefore rotor rpm limitations have minimum and maximum values. A green arc depicts the normal operating range with red lines showing the minimum and maximum limits. *[Figure 5-3]*

There are two different rotor rpm limitations: power-on and power-off. Power-on limitations apply anytime the engine is turning the rotor and is depicted by a fairly narrow green band. A yellow arc may be included to show a transition range, which means that operation within this range is limited due to the possibility of increased vibrations or harmonics. This range may be associated with tailboom dynamic modes. Power-off limitations apply anytime the engine is not turning the rotor, such as when in an autorotation. In this case, the green arc is wider than the power-on arc, indicating a larger operating range.

Powerplant Limitations

The powerplant limitations area describes operating limitations on the helicopter's engine including such items as rpm range, power limitations, operating temperatures, and fuel and oil requirements. Most turbine engines and some reciprocating engines have a maximum power and a maximum continuous power rating. The "maximum power" rating is the maximum power the engine can generate and is usually limited by time. The maximum power range is depicted by a yellow arc on the engine power instruments, with a red line indicating the maximum power that must not be exceeded. "Maximum continuous power" is the maximum power the engine can generate continually and is depicted by a green arc. *[Figure 5-4]*

Manifold pressure is a measure of vacuum at the intake manifold. It is the difference between the air pressure (or vacuum) inside the intake manifold and the relative atmospheric pressure of the air around the engine. The red line on a manifold pressure gauge indicates the maximum amount of power. A yellow arc on the gauge warns of pressures approaching the limit of rated power.

Figure 5-4. *Torque and turbine outlet temperature (TOT) gauges are commonly used with turbine-powered aircraft.*

[Figure 5-5] A placard near the gauge lists the maximum readings for specific conditions.

Weight and Loading Distribution

The weight and loading distribution section of the manufacturer's RFM contains the maximum certificated weights, as well as the center of gravity (CG) range. The location of the reference datum used in balance computations should also be included in this section. Weight and balance computations are not provided here, but rather in the weight and balance section of the RFM.

Flight Limitations

This area lists any maneuvers which are prohibited, such as acrobatic flight or flight into known icing conditions. If the rotorcraft can only be flown in visual flight rules (VFR) conditions, it is noted in this area. Also included are the minimum crew requirements, and the pilot seat location, if applicable, from which solo flights must be conducted.

Placards

All rotorcraft generally have one or more placards displayed that have a direct and important bearing on the safe operation of the rotorcraft. These placards are located in a conspicuous place within the cabin and normally appear in the limitations section. Since V_{NE} varies with altitude, this placard can be found in all helicopters. *[Figure 5-6]*

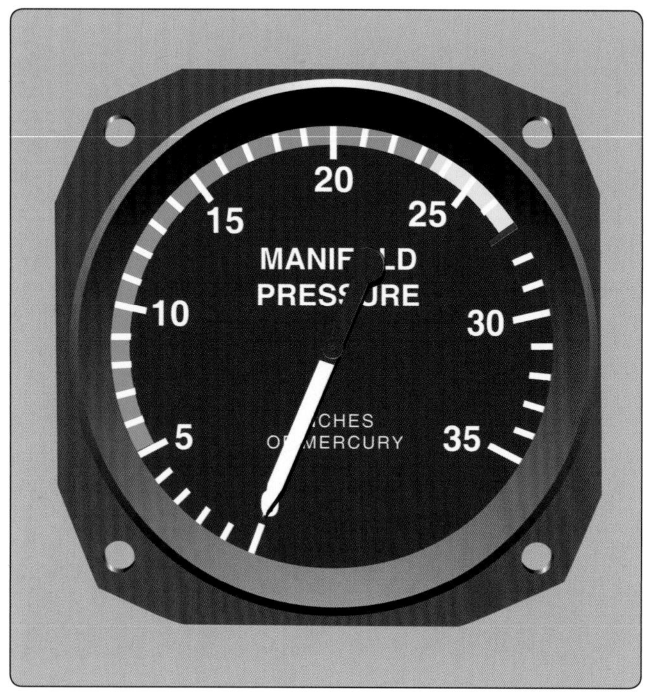

Figure 5-5. *The manifold pressure gauge is an engine instrument typically used in piston aircraft engines to measure the pressure inside the induction system of an engine. Manifold pressure is a measurement of vacuum and the measurement is taken at the intake manifold.*

Press Alt. (1,000 ft.)	V_{NE}—MPH IAS							Gross Weight
F OAT	2	4	6	8	10	12	14	
0	109	109	105	84	61	--	--	More than 1,700 lb
20	109	109	94	72	49	--	--	
40	109	103	81	59	--	--	--	
60	109	91	70	48	--	--	--	
80	109	80	59	--	--	--	--	
100	109	70	48	--	--	--	--	
0	**109**	**109**	**109**	**109**	**98**	**77**	**58**	1,700 lb or less
20	**109**	**109**	**109**	**109**	**85**	**67**	**48**	
40	**109**	**109**	**109**	**96**	**75**	**57**	**--**	
60	**109**	**109**	**108**	**84**	**66**	**48**	**--**	
80	**109**	**109**	**95**	**74**	**57**	**--**	**--**	
100	**109**	**108**	**84**	**66**	**48**	**--**	**--**	
Maximum V_{NE} Doors off—102 MPH IAS								

NEVER EXCEED SPEED

V_{NE} KIAS: 110, 100, 90, 80, 70, 60, 50

-20°C, 0°C, +20°C, +40°C, MAX ALT.

Pressure Altitude (1,000 feet): 0, 2, 4, 6, 8, 10, 12, 14

Figure 5-6. *Various VNE placards.*

Emergency Procedures (Section 3)

Concise checklists describing the recommended procedures and airspeeds for coping with various types of emergencies or critical situations can be found in this section. Some of the emergencies covered include: engine failure in a hover and at altitude, tail rotor failures, fires, and systems failures. The procedures for restarting an engine and for ditching in the water might also be included.

Manufacturers may first show the emergencies checklists in an abbreviated form with the order of items reflecting the sequence of action. This is followed by amplified checklists providing additional information to clarify the procedure. To be prepared for an abnormal or emergency situation, learn the first steps of each checklist, if not all the steps. If time permits, refer to the checklist to make sure all items have been covered. For more information on emergencies, refer to Chapter 11, Helicopter Emergencies and Hazards.

Manufacturers are encouraged to include an optional area titled Abnormal Procedures, which describes recommended procedures for handling malfunctions that are not considered to be emergencies. This information would most likely be found in larger helicopters.

Normal Procedures (Section 4)

The normal procedures section is the section most frequently used. It usually begins with a listing of airspeeds that may enhance the safety of normal operations. It is a good idea to learn the airspeeds that are used for normal flight operations. The next part of the section includes several checklists, which cover the preflight inspection, before-starting procedure, how to start the engine, rotor engagement, ground checks, takeoff, approach, landing, and shutdown. Some manufacturers also include the procedures for practice autorotations. To avoid skipping an important step, always use a checklist when one is available. More information on maneuvers can be found in Chapter 9, Basic Flight Maneuvers, and Chapter 10, Advanced Flight Maneuvers.

Performance (Section 5)

The performance section contains all the information required by the regulations and any additional performance information the manufacturer determines may enhance a pilot's ability to operate the helicopter safely. Although the performance section is not in the limitation section and is therefore not a limitation, operation outside or beyond the flight-tested and documented performance section can be expensive, slightly hazardous, or outright dangerous to life and property. If the helicopter is certificated under 14 CFR part 29, then the performance section may very well be a restrictive limitation. In any event, a pilot should determine the performance available and plan to stay within those parameters.

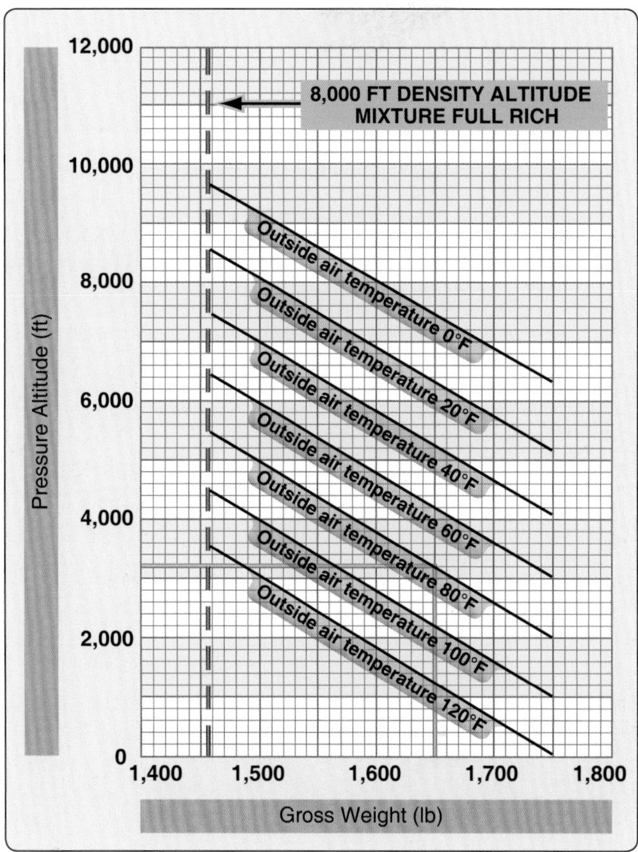

Figure 5-7. *One of the performance charts in the performance section is the In Ground Effect Hover Ceiling versus Gross Weight chart. This chart can be used to determine how much weight can be carried and still operate at a specific pressure altitude or, if carrying a specific weight, detrmine that specific altitude limitation.*

These charts, graphs, and tables vary in style, but all contain the same basic information. Some examples of the performance information that can be found in most flight manuals include a calibrated versus indicated airspeed conversion graph, hovering ceiling versus gross weight charts, and a height-velocity diagram. *[Figure 5-7]* For information on how to use the charts, graphs, and tables, refer to Chapter 7, Helicopter Performance.

Weight and Balance (Section 6)

The weight and balance section should contain all the information required by the FAA that is necessary to calculate weight and balance. To help compute the proper data, most manufacturers include sample problems. Weight and balance is detailed in Chapter 6, Weight and Balance.

Aircraft and Systems Description (Section 7)

The aircraft and systems description section is an excellent place to study all the systems found on an aircraft. The manufacturers should describe the systems in a manner that is understandable to most pilots. For larger, more complex helicopters, the manufacturer may assume a higher degree of knowledge. For more information on helicopter systems, refer to Chapter 4, Helicopter Components, Sections, and Systems.

Handling, Servicing, and Maintenance (Section 8)

The handling, servicing, and maintenance section describes the maintenance and inspections recommended by the manufacturer, as well as those required by the regulations, and airworthiness directive (AD) compliance procedures. There are also suggestions on how the pilot/operator can ensure that the work is done properly.

This section also describes preventative maintenance that may be accomplished by certificated pilots, as well as the manufacturer's recommended ground handling procedures, including considerations for hangaring, tie down, and general storage procedures for the helicopter.

Supplements (Section 9)

The supplements section describes pertinent information necessary to operate optional equipment installed on the helicopter that would not be installed on a standard aircraft. Some of this information may be supplied by the aircraft manufacturer, or by the maker of the optional equipment. The information is then inserted into the flight manual at the time the equipment is installed.

Since civilian manuals are not updated to the extent of military manuals, the pilot must learn to read the supplements after determining what equipment is installed and amend their daily use checklists to integrate the supplemental instructions and procedures. This is why air carriers must furnish checklists to their crews. Those checklists furnished to the crews must incorporate all procedures from any and all equipment actually installed in the aircraft and the approved company procedures.

Safety and Operational Tips (Section 10)

The safety and operational tips section is optional and contains a review of information that could enhance the safety of the operation. Manufacturers may include best operating practices and other recommended procedures for the enhancement of safety and reducing accidents. Some examples of the information that might be covered include physiological factors, general weather information, fuel conservation procedures, external load warnings, low rotor rpm considerations, and recommendations that if not adhered to, could lead to an emergency.

Chapter Summary

This chapter familiarized the reader with the RFM. It detailed each section and explained how to follow and better understand the flight manual to enhance safety of flight.

Chapter 6

Weight and Balance

Introduction

It is vital to comply with weight and balance limits established for helicopters. Operating above the maximum weight limitation compromises the structural integrity of the helicopter and adversely affects performance. Balance is also critical because, on some fully loaded helicopters, center of gravity (CG) deviations as small as three inches can dramatically change a helicopter's handling characteristics. Operating a helicopter that is not within the weight and balance limitations is unsafe. Refer to FAA-H-8083-1 (as revised), Aircraft Weight and Balance Handbook, for more detailed information.

Weight

When determining if a helicopter is within the weight limits, consider the weight of the basic helicopter, crew, passengers, cargo, and fuel. Although the effective weight (load factor) varies during maneuvering flight, this chapter primarily addresses the weight of the loaded helicopter while at rest.

It is critical to understand that the maximum allowable weight may change during the flight. When operations include out of ground effect (OGE) hovers and confined areas, planning must be done to ensure that the helicopter is capable of lifting the weight during all phases of flight. The weight may be acceptable during the early morning hours, but as the density altitude increases during the day, the maximum allowable weight may have to be reduced to keep the helicopter within its capability.

The following terms are used when computing a helicopter's weight:

- Basic Empty Weight
- Maximum Gross Weight
- Weight Limitations

Basic Empty Weight

The starting point for weight computations is the basic empty weight. This is the weight of the standard helicopter, optional equipment, unusable fuel, and all operating fluids including engine and transmission oil, and hydraulic fluid for those aircraft so equipped. Some helicopters might use the term "licensed empty weight," which is nearly the same as basic empty weight, except that it does not include full engine and transmission oil, just undrainable oil. If flying a helicopter that lists a licensed empty weight, be sure to add the weight of the oil to the computations.

Maximum Gross Weight

The maximum weight of the helicopter is referred to its maximum gross weight. Most helicopters have an internal maximum gross weight, which refers to the weight within the helicopter structure and an external maximum gross weight, which refers to the weight of the helicopter with an external load. The external maximum weight may vary depending on where it is attached to the helicopter. Some large cargo helicopters may have several attachment points for sling load or winch operations. These helicopters can carry a tremendous amount of weight when the attachment point is directly under the CG of the aircraft.

Weight Limitations

Weight limits are necessary to guarantee the structural integrity of the helicopter, enable pilots to predict helicopter performance and insure aircraft controllability. Although aircraft manufacturers build in safety factors, a pilot should never intentionally exceed the load limits for which a helicopter is certificated.

Operating below a minimum weight could adversely affect the handling characteristics of the helicopter. During single-pilot operations in some helicopters, a pilot needs to use a large amount of forward cyclic to maintain a hover. By adding ballast to the helicopter, the neutral cyclic position can be shifted toward the center of its range, thus giving a greater range of control outward from neutral in every direction. When operating at or below the minimum weight of the helicopter, additional weight also improves autorotational characteristics since the autorotational descent can be established sooner. In addition, operating below minimum weight could prevent achieving the desirable rotor revolutions per minute (rpm) during autorotations.

Operating above a maximum weight could result in structural deformation or failure during flight if encountering excessive load factors, strong wind gusts, or turbulence. Weight and maneuvering limitations also are factors in establishing fatigue life of components. Overweight, meaning overstressed, parts fail sooner than anticipated. Therefore, premature failure is a major consideration in determination of fatigue life and life cycles of parts.

Although a helicopter is certificated for a specified maximum gross weight, it is not safe to take off with this load under some conditions. Anything that adversely affects takeoff, climb, hovering, and landing performance may require off-loading of fuel, passengers, or baggage to some weight less than the published maximum. Factors that can affect performance include high altitude, high temperature, and high humidity conditions, which result in a high-density altitude. In-depth performance planning is critical when operating in these conditions.

Balance

Helicopter performance is not only affected by gross weight, but also by the position of that weight. It is essential to load the aircraft within the allowable CG range specified in the rotorcraft flight manual's (RFM) weight and balance limitations. Loading outside approved limits can result in insufficient control travel for safe operation.

Center of Gravity

The pilot should ensure that the helicopter is properly balanced and within its center of gravity limitations, so that minimal cyclic input is required during hovering flight, except for any wind corrections. Since the fuselage acts as a pendulum suspended from the rotor, changing the CG changes the angle at which the aircraft hangs from the rotor. When the CG is directly under the rotor mast, the helicopter hangs horizontally; if the CG is too far forward of the mast, the helicopter hangs with its nose tilted down; if the CG is too far aft of the mast, the nose tilts up. *[Figure 6-1]*

CG Forward of Forward Limit

A forward CG may occur when a heavy pilot and passenger take off without baggage or proper ballast located aft of the rotor mast. This situation becomes worse if the fuel tanks are located aft of the rotor mast because as fuel burns the CG continues to shift forward.

This condition is easily recognized when coming to a hover following a vertical takeoff. The helicopter has a nose-low attitude, and excessive rearward displacement of the cyclic control is needed to maintain a hover in a no-wind condition. Do not continue flight in this condition, since a pilot could rapidly lose rearward cyclic control as fuel is consumed. A pilot may also find it impossible to decelerate sufficiently to bring the helicopter to a stop. In the event of engine failure and the resulting autorotation, there may not be enough cyclic control to flare properly for the landing.

A forward CG is not as obvious when hovering into a strong wind, since less rearward cyclic displacement is required than when hovering with no wind. When determining whether a critical balance condition exists, it is essential to consider the wind velocity and its relation to the rearward displacement of the cyclic control.

CG Aft of Aft Limit

Without proper ballast in the cockpit, exceeding the aft CG may occur when:

- A lightweight pilot takes off solo with a full load of fuel located aft of the rotor mast.
- A lightweight pilot takes off with maximum baggage allowed in a baggage compartment located aft of the rotor mast.
- A lightweight pilot takes off with a combination of baggage and substantial fuel where both are aft of the rotor mast.

A pilot can recognize the aft CG condition when coming to a hover following a vertical takeoff. The helicopter will have a tail-low attitude and will need excessive forward displacement of cyclic control to maintain a hover in a no-wind condition. When facing upwind, even greater forward cyclic is needed.

If flight is continued in this condition, it may be impossible to fly in the upper allowable airspeed range due to inadequate forward cyclic authority to maintain a nose-low attitude. In addition, with an extreme aft CG, gusty or rough air could accelerate the helicopter to a speed faster than that produced with full forward cyclic control. In this case, dissymmetry of lift and blade flapping could cause the rotor disk to tilt aft. With full forward cyclic control already applied, a pilot might not be able to lower the rotor disk, resulting in possible loss of control, or the rotor blades striking the tailboom.

Lateral Balance

For smaller helicopters, it is generally unnecessary to determine the lateral CG for normal flight instruction and passenger flights. This is because helicopter cabins are relatively narrow and most optional equipment is located near the centerline. However, some helicopter manuals specify the seat from which a pilot must conduct solo flight. In addition, if there is an unusual situation that could affect the lateral CG, such as a heavy pilot and a full load of fuel on one side of the helicopter, its position should be checked against the CG envelope. If carrying external loads in a position that requires large lateral cyclic control displacement to maintain level flight, fore and aft cyclic effectiveness could be limited dramatically. Manufacturers generally account for known lateral CG displacements by locating external attachment points opposite the lateral imbalance. Examples are placement of hoist systems attached to the side, and wing stores commonly used on military aircraft for external fuel pods or armament systems.

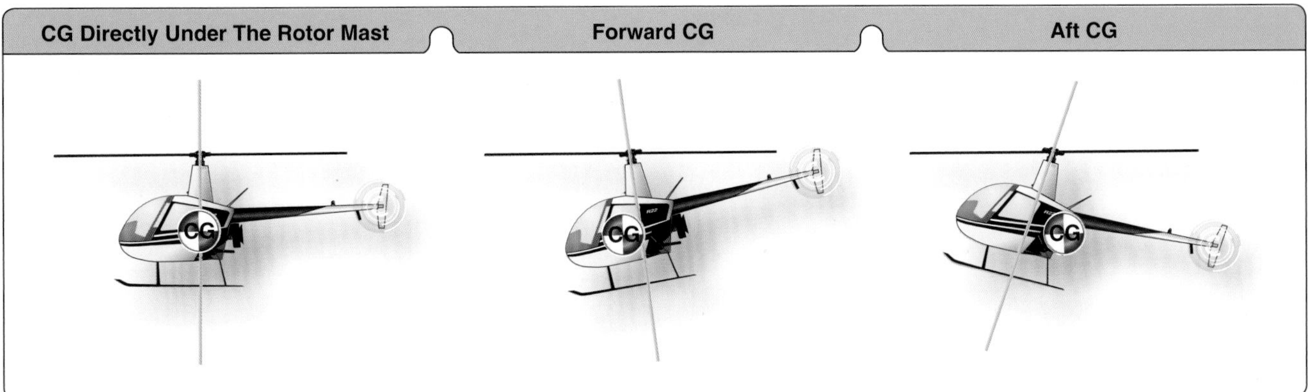

Figure 6-1. *The location of the CG strongly influences how the helicopter handles.*

Weight and Balance Calculations

When determining whether a helicopter is properly loaded, two questions must be answered:

1. Is the gross weight less than or equal to the maximum allowable gross weight?
2. Is the CG within the allowable CG range, and will it stay within the allowable range throughout the duration of flight including all loading configurations that may be encountered?

To answer the first question, just add the weight of the items comprising the useful load (pilot, passengers, fuel, oil [if applicable] cargo, and baggage) to the basic empty weight of the helicopter. Ensure that the total weight does not exceed the maximum allowable gross weight.

To answer the second question, use CG or moment information from loading charts, tables, or graphs in the RFM. It is important to note that any weight and balance computation is only as accurate as the information provided. Therefore, ask passengers what they weigh and add a few pounds to account for the additional weight of clothing, especially during the winter months. Baggage should be weighed on a scale, if practical. If a scale is not available, compute personal loading values according to each individual estimate. *Figure 6-2* indicates the standard weights for specific operating fluids. These values are used when computing a helicopter's balance.

Reference Datum

Balance is determined by the location of the CG, which is usually described as a given number of inches from the reference datum. The horizontal reference datum is an imaginary vertical plane or point, arbitrarily fixed somewhere along the longitudinal axis of the helicopter, from which all horizontal distances are measured for weight and balance purposes. There is no fixed rule for its location. It may be located at the rotor mast, the nose of the helicopter, or even at a point in space ahead of the helicopter. *[Figure 6-3]*

Fluid	Weight
Aviation Gasoline (AVGAS)	6 lb/gal
Jet Fuel (JP-4)	6.5 lb/gal
Jet Fuel (JP-5)	6.8 lb/gal
Reciprocating Engine Oil	7.5 lb/gal*
Turbine Engine Oil	Varies between 6 and 8 lb/gal*
Water	8.35 lb/gal

*Oil weight is given in pounds per gallon while oil capacity is usually given in quarts; therefore, convert the amount of oil to gallons before calculating its weight. Remember, four quarts equal one gallon.

Figure 6-2. *When making weight and balance computations, always use actual weights if they are available, especially if the helicopter is loaded near the weight and balance limits.*

The lateral reference datum is usually located at the center of the helicopter. The location of the reference datum is established by the manufacturer and is defined in the RFM. *[Figure 6-4]*

Chapter Summary

This chapter discusses the importance of computing the weight and balance of the helicopter. The chapter also discusses the common terms and meanings associate with weight and balance.

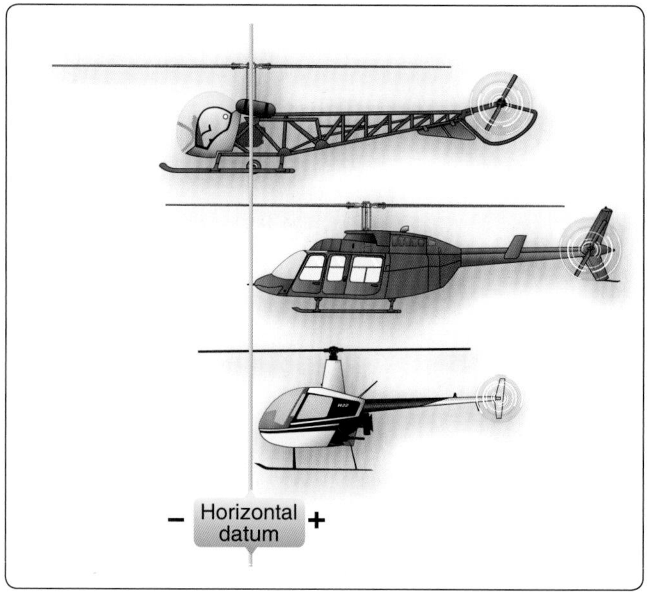

Figure 6-3. *While the horizontal reference datum can be anywhere the manufacturer chooses, some manufacturers choose the datum line at or ahead of the most forward structural point on the helicopter, in which case all moments are positive. This aids in simplifying calculations. Other manufacturers choose the datum line at some point in the middle of the helicopter, in which case moments produced by weight in front of the datum are negative and moments produced by weight aft of the datum are positive.*

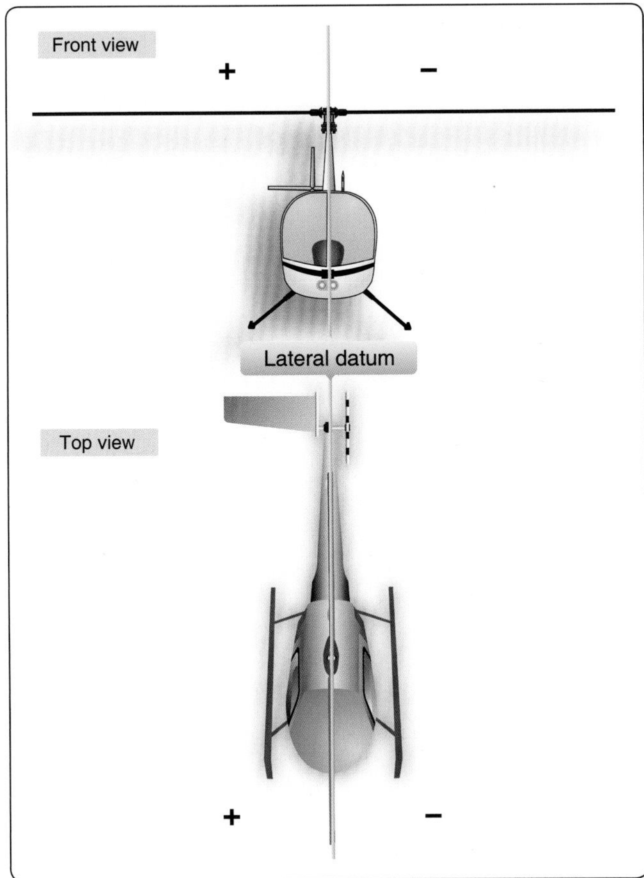

Figure 6-4. *The lateral reference datum is located longitudinally through the center of the helicopter; therefore, there are positive and negative values.*

Chapter 7

Helicopter Performance

Introduction

A pilot's ability to predict the performance of a helicopter is extremely important. It helps to determine how much weight the helicopter can carry before takeoff, if the helicopter can safely hover at a specific altitude and temperature, the distance required to climb above obstacles, and what the maximum climb rate will be.

Factors Affecting Performance

A helicopter's performance is dependent on the power output of the engine and the lift produced by the rotors, whether it is the main rotor(s) or tail rotor. Any factor that affects engine and rotor efficiency affects performance. The three major factors that affect performance are density altitude, weight, and wind. The Pilot's Handbook of Aeronautical Knowledge, FAA-H-8083-25 (as revised), discusses these factors in great detail.

Moisture (Humidity)

Humidity alone is usually not considered an important factor in calculating density altitude and helicopter performance; however, it does contribute. There are no rules of thumb used to compute the effects of humidity on density altitude, but some manufacturers include charts with 80 percent relative humidity columns as additional information. There appears to be an approximately 3–4 percent reduction in performance compared to dry air at the same altitude and temperature, so expect a decrease in hovering and takeoff performance in high humidity conditions. Although 3–4 percent seems insignificant, it can be the cause of a mishap when already operating at the limits of the helicopter.

Weight

Weight is one of the most important factors because the pilot can control it. Most performance charts include weight as one of the variables. By reducing the weight of the helicopter, a pilot may be able to take off or land safely at a location that otherwise would be impossible. However, if ever in doubt about whether a takeoff or landing can be performed safely, delay your takeoff until more favorable density altitude conditions exist. If airborne, try to land at a location that has more favorable conditions, or one where a landing can be made that does not require a hover.

In addition, at higher gross weights, the increased power required to hover produces more torque, which means more antitorque thrust is required. In some helicopters during high altitude operations, the maximum antitorque produced by the tail rotor during a hover may not be sufficient to overcome torque even if the gross weight is within limits.

Winds

Wind direction and velocity also affect hovering, takeoff, and climb performance. Translational lift occurs any time there is relative airflow over the rotor disk. This occurs whether the relative airflow is caused by helicopter movement or by the wind. Assuming a headwind, as wind speed increases, translational lift increases, resulting in less power required to hover.

The wind direction is also an important consideration. Headwinds are the most desirable as they contribute to the greatest increase in performance. Strong crosswinds and tailwinds may require the use of more tail rotor thrust to maintain directional control. This increased tail rotor thrust absorbs power from the engine, which means there is less power available to the main rotor for the production of lift. Some helicopters even have a critical wind azimuth or maximum safe relative wind chart. Operating the helicopter beyond these limits could cause loss of tail rotor effectiveness.

Takeoff and climb performance is greatly affected by wind. When taking off into a headwind, effective translational lift is achieved earlier, resulting in more lift and a steeper climb angle. When taking off with a tailwind, more distance is required to accelerate through translation lift.

Performance Charts

In developing performance charts, aircraft manufacturers make certain assumptions about the condition of the helicopter and the ability of the pilot. It is assumed that the helicopter is in good operating condition, calm wind, and the engine is developing its rated power. The pilot is assumed to be following normal operating procedures and to have average flying abilities. Average means a pilot capable of doing each of the required tasks correctly and at the appropriate times.

Using these assumptions, the manufacturer develops performance data for the helicopter based on actual flight tests. However, they do not test the helicopter under each and every condition shown on a performance chart. Instead, they evaluate specific data and mathematically derive the remaining data.

Height/Velocity Diagram

The height/velocity (H/V) diagram shows the combinations of airspeed and height above the ground, which will allow an average pilot to successfully complete a landing after an engine failure. By carefully studying the height/velocity diagram, a pilot is able to avoid the combinations of altitude and airspeed that may not allow sufficient time or altitude to enter a stabilized autorotative descent. Refer to *Figure 7-1* during the remainder of the discussion on the height/velocity diagram.

In the simplest explanation, the H/V diagram is a diagram in which the shaded areas should be avoided, as the pilot may be unable to complete an autorotation landing without damage. The H/V diagram usually contains a takeoff profile, where the diagram can be traversed from zero height and zero speed to

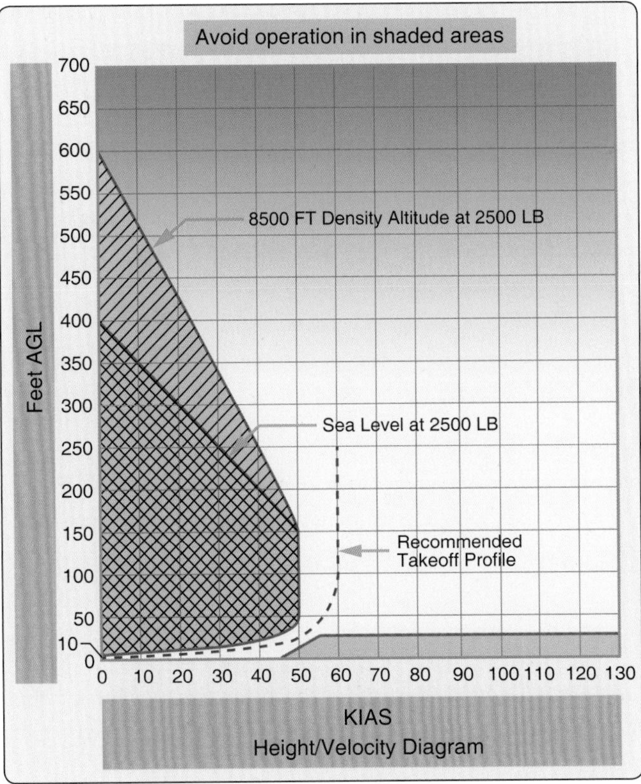

Figure 7-1. *Sample height/velocity diagram for a Robinson Model R44 II.*

cruise, without entering the shaded areas or with minimum exposure to shaded areas.

The grey portion on the left side of the diagram marks a flight profile that probably does not allow the pilot to complete an autorotation successfully, primarily due to having insufficient airspeed to enter an autorotative profile in time to avoid a crash. The shaded area on the lower right is dangerous due to the airspeed and proximity to the ground resulting in dramatically reduced reaction time for the pilot in the case of mechanical failure, or other in-flight emergencies. This shaded area at the lower right is not portrayed in H/V diagrams for multiengine helicopters capable of safely hovering and flying with a single engine failure.

The following examples further illustrate the relevance of the H/V diagram to a single-engine helicopter.

At low heights with low airspeed, such as a hover taxi, the pilot can simply use the kinetic energy from the rotor disk to cushion the landing with collective, converting rotational inertia to lift. The aircraft is in a safe part of the H/V diagram. At the extreme end of the scale (e.g., a three-foot hover taxi at walking pace) even a complete failure to recognize the power loss resulting in an uncushioned landing would probably be survivable.

As the airspeed increases without an increase in height, there comes a point at which the pilot's reaction time would be insufficient to react with a flare in time to prevent a high speed, and thus probably fatal, ground impact. Another thing to consider is the length of the tailboom and the response time of the helicopter flight controls at slow airspeeds and low altitudes. Even small increases in height give the pilot much greater time to react; therefore, the bottom right part of the H/V diagram is usually a shallow gradient. If airspeed is above ideal autorotation speed, the pilot's instinct is usually to flare to convert speed to height and increase rotor rpm through coning, which also immediately gets them out of the dead man's curve.

Conversely, an increase in height without a corresponding increase in airspeed puts the aircraft above a survivable uncushioned impact height, and eventually above a height where rotor inertia can be converted to sufficient lift to enable a survivable landing. This occurs abruptly with airspeeds much below the ideal autorotative speed (typically 40–80 knots). The pilot must have enough time to accelerate to autorotation speed in order to autorotate successfully; this directly relates to a requirement for height. Above a certain height the pilot can achieve autorotation speed even from a zero knot start, thus putting high OGE hovers outside the curve.

The typical safe takeoff profile involves initiation of forward flight from a 2–3 feet landing gear height, only gaining altitude as the helicopter accelerates through translational lift, as airspeed approaches a safe autorotative speed. At this point, some of the increased thrust available may be used to attain safe climb airspeed, which will keep the helicopter out of the shaded or hatched areas of the H/V diagram. Although helicopters are not restricted from conducting maneuvers that will place them in the shaded area of the H/V diagram, it is important for pilots to understand that operation in those shaded areas exposes pilot, aircraft, and passengers to a certain hazard should the engine or driveline malfunction. The pilot should always evaluate the risk of the maneuver versus the operational value.

The Effect of Weight Versus Density Altitude

The height/velocity diagram *[Figure 7-1]* depicts altitude and airspeed situations from which a successful autorotation can be made. The time required, and therefore, altitude necessary to attain a steady state autorotative descent, is dependent on the weight of the helicopter and the density altitude. For this reason, the H/V diagram is valid only when the helicopter is operated in accordance with the gross weight versus density altitude chart. If published, this chart is found in the RFM for the particular helicopter. *[Figure 7-2]* The gross weight

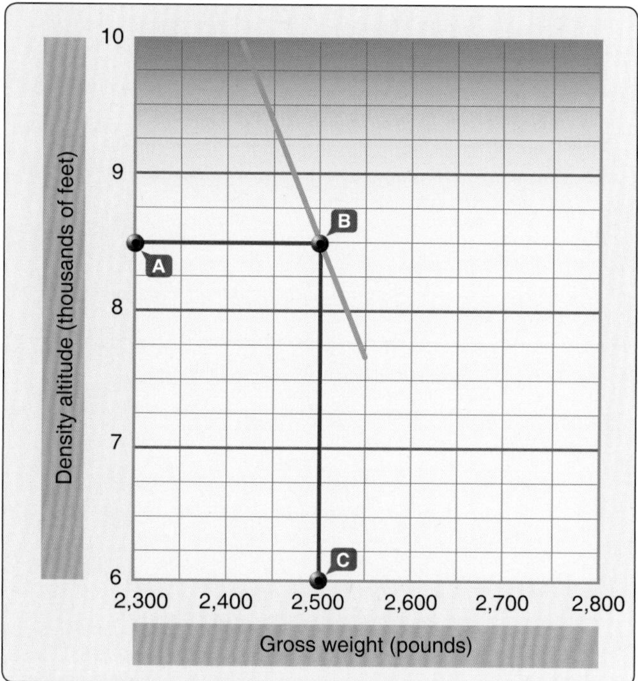

Figure 7-2. *Gross weight versus density altitude.*

versus density altitude chart is not intended to provide a restriction to gross weight, but to be an advisory of the autorotative capability of the helicopter during takeoff and climb. A pilot must realize, however, that at gross weights above those recommended by the gross weight versus density altitude chart, the values are unknown.

Assuming a density altitude of 8,500 feet, the height/velocity diagram in *Figure 7-1* would be valid up to a gross weight of approximately 2,500 pounds. This is found by entering the graph in *Figure 7-2* at a density altitude of 8,500 feet (point A), then moving horizontally to the solid line (point B). Moving vertically to the bottom of the graph (point C), with the existing density altitude, the maximum gross weight under which the height/velocity diagram is applicable is 2,500 pounds.

The production of performance charts and diagrams for helicopters are regulatory as set out in Title 14 of the Code of Federal Regulations (14 CFR) Part 27, Airworthiness Standards. These charts establish safer parameters for operation. Although not regulatory, the pilot should carry out a full risk assessment to carefully consider the higher risk before operating within the shaded areas of the height/velocity diagram.

Autorotational Performance

Most autorotational performance charts state that autorotational descent performance is a function of indicated airspeed (IAS) and is essentially unaffected by density altitude and gross weight. Keep in mind that, at some point, the potential energy expended during the autorotation is converted into kinetic energy for the flare and touchdown phase of the maneuver. It is at that point that increased density altitudes and heavier gross weights have a great impact on the successful completion of the autorotation. The rotor disk must be able to overcome the downward momentum of the helicopter and provide enough lift to cushion the landing. With increased density altitudes and gross weights, the lift potential is reduced and a higher collective pitch angle (angle of incidence) is required.

During autorotation gravity provides the source of energy powering the rotor by causing upflow up through the rotor during descent. This is the same as saying that potential energy is being traded for kinetic energy to turn the rotor as the aircraft descends.

In *Figure 7-3*, the S-300 curve shows the various combinations of horizontal and vertical speeds that supply the required energy to keep the rotor turning at a constant 471 rpm. For example, an airspeed of 54 mph with a corresponding vertical speed of 1,600 feet per minute (fpm) will provide enough kinetic energy to maintain the rotor at a 471 rpm. The rotor does not care if the air is coming from the front or the bottom so long as the total is sufficient to maintain the rpm. Any point on the curve will maintain rotor speed. However, the pilot does care because if he or she, for example, glides at 30 knots, the corresponding rate of descent will be over 2,200 fpm. Since there is little airspeed for a deceleration (or "flare") to reduce the rate of decent before touchdown, the collective pitch application (increasing blade pitch and giving a final temporary increase in lift before the blades slow down) may be insufficient to arrest the rate of descent.

Students who fully comprehend this relationship understand why training autorotations are usually limited to airspeeds between the minimum rate of descent airspeed and the maximum range airspeed (usually about 25 percent faster than the minimum rate of descent airspeed).

Referencing a curve similar to the one shown in *Figure 7-3* is useful to understand the consequences of not maintaining the target airspeed when executing an autorotation. Simply put, the pilot should know why airspeed is the most significant factor affecting the rate of descent.

Hovering Performance

Helicopter performance revolves around whether or not the helicopter can be hovered. More power is required during the hover than in any other flight regime. Obstructions aside, if a hover can be maintained, a takeoff can be made, especially with the additional benefit of translational lift. Hover charts are provided for in ground effect (IGE) hover and out of ground effect (OGE) hover under various conditions of

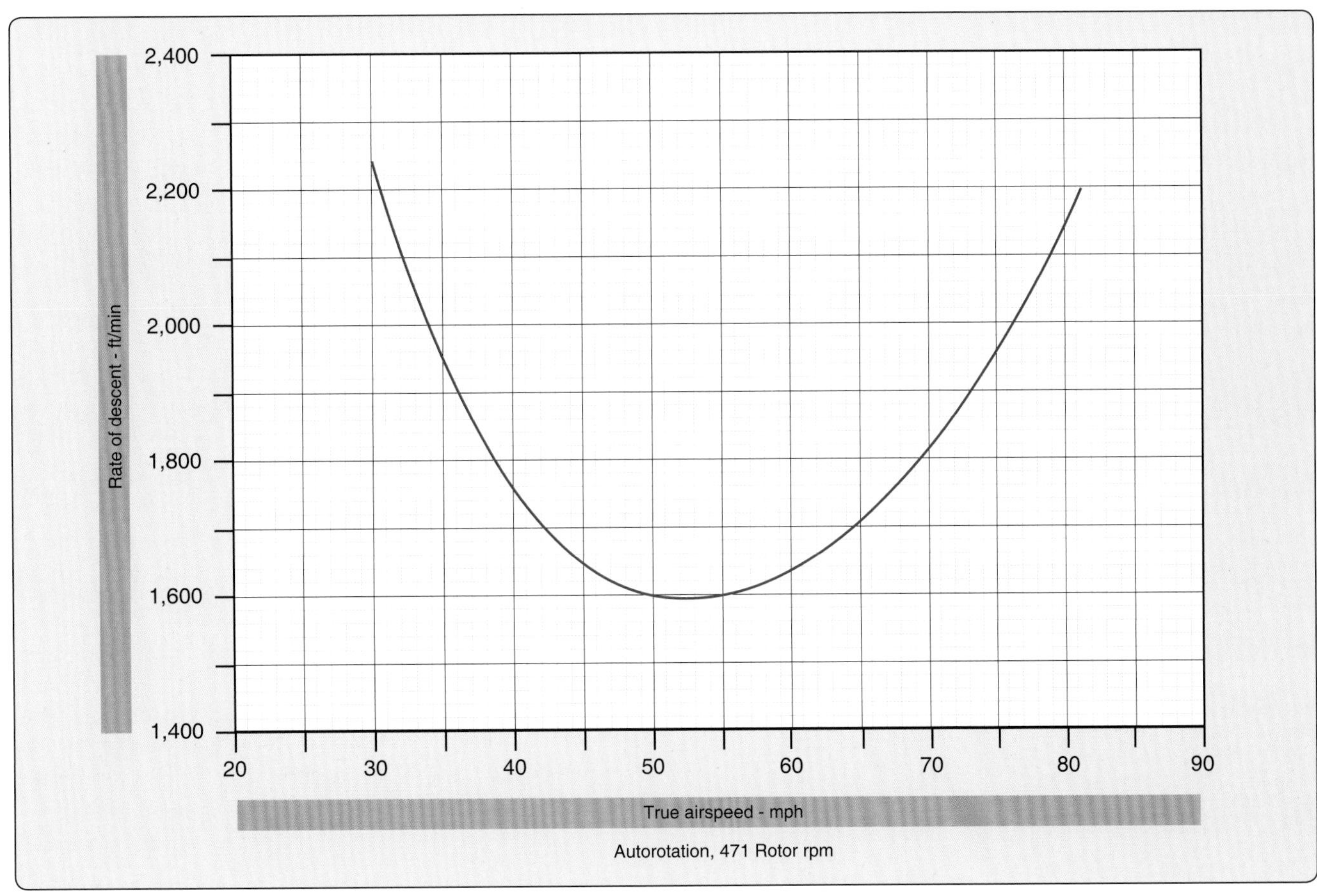

Figure 7-3. *An autorotation curve for the S-300 shows the various combinations of horizontal and vertical speeds that supply the required energy to keep the rotor turning at a constant 471 rpm.*

gross weight, altitude, temperature, and power. The IGE hover ceiling is usually higher than the OGE hover ceiling because of the added lift benefit produced by ground effect. See Chapter 2, Aerodynamics of Flight, for more details on IGE and OGE hover. A pilot should always plan an OGE hover when landing in an area that is uncertain or unverified. As density altitude increases, more power is required to hover. At some point, the power required is equal to the power available. This establishes the hovering ceiling under the existing conditions. Any adjustment to the gross weight by varying fuel, payload, or both, affects the hovering ceiling. The heavier the gross weight, the lower the hovering ceiling. As gross weight is decreased, the hover ceiling increases.

Sample Hover Problem 1

You are to fly a photographer to a remote location to take pictures of the local wildlife. Using *Figure 7-4*, can you safely hover in ground effect at your departure point with the following conditions?

A. Pressure Altitude................................8,000 feet

B. Temperature...+15 °C

C. Takeoff Gross Weight.........................1,250 lb

rpm..104 percent

First enter the chart at 8,000 feet pressure altitude (point A), then move right until reaching a point midway between the +10 °C and +20 °C lines (point B). From that point, proceed down to find the maximum gross weight where a 2 foot hover can be achieved. In this case, it is approximately 1,280 pounds (point C).

Since the gross weight of your helicopter is less than this, you can safely hover with these conditions.

Sample Hover Problem 2

Once you reach the remote location in the previous problem, you will need to hover OGE for some of the pictures. The pressure altitude at the remote site is 9,000 feet, and you will use 50 pounds of fuel getting there. (The new gross weight is now 1,200 pounds.) The temperature will remain at +15 °C. Using *Figure 7-5*, can you accomplish the mission?

Enter the chart at 9,000 feet (point A) and proceed to point B (+15 °C). From there, determine that the maximum gross

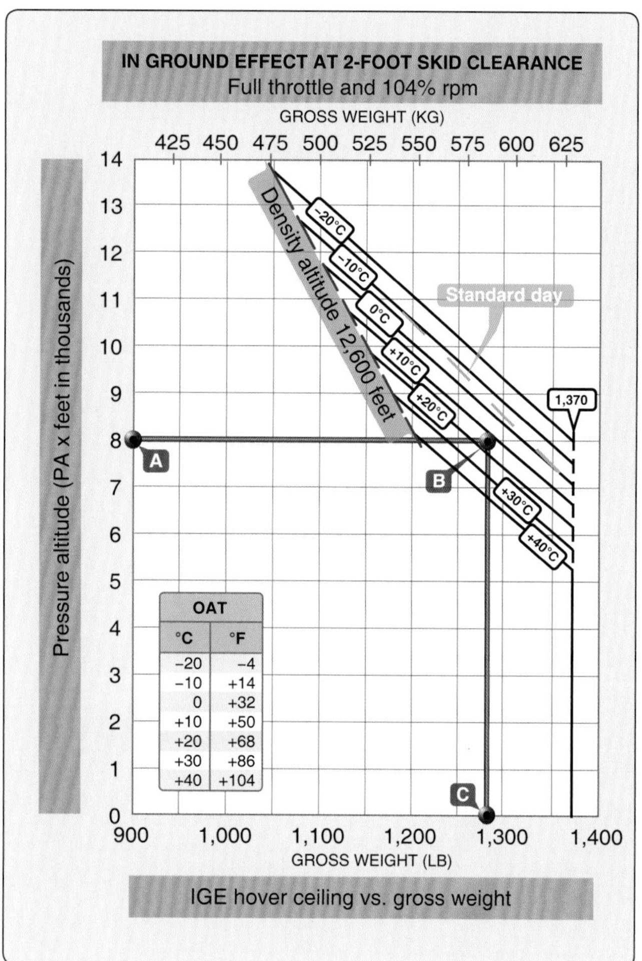

Figure 7-4. *In ground effect hovering ceiling versus gross weight chart.*

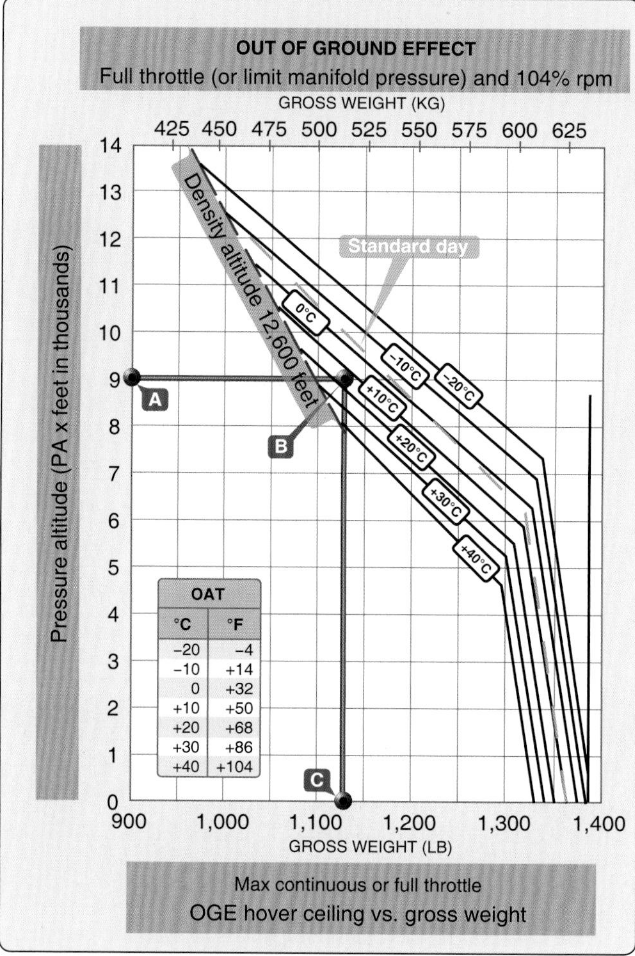

Figure 7-5. *Out of ground effect hover ceiling versus gross weight chart.*

weight to hover OGE is approximately 1,130 pounds (point C). Since your gross weight is higher than this value, you will not be able to hover in these conditions. To accomplish the mission, you will need to remove approximately 70 pounds before you begin the flight.

These two sample problems emphasize the importance of determining the gross weight and hover ceiling throughout the entire flight operation. Being able to hover at the takeoff location with a specific gross weight does not ensure the same performance at the landing point. If the destination point is at a higher density altitude because of higher elevation, temperature, and/or relative humidity, more power is required to hover there. You should be able to predict whether hovering power will be available at the destination by knowing the temperature and wind conditions, using the performance charts in the helicopter flight manual, and making certain power checks during hover and in flight prior to commencing the approach and landing.

For helicopters with dual engines, performance charts provide torque amounts for both engines.

Sample Hover Problem 3

Using *Figure 7-6*, determine what torque is required to hover. Use the following conditions:

A. Pressure Altitude....................9,500 feet
B. Outside Air Temperature..................0 °C
C. Gross Weight........................4,250 lb
D. Desired Skid Height.....................5 feet

First, enter the chart at 9,500 feet pressure altitude, then move right to outside air temperature, 0 °C. From that point, move down to 4,250 pounds gross weight and then move left to 5-foot skid height. Drop down to read 66 percent torque required to hover.

Climb Performance

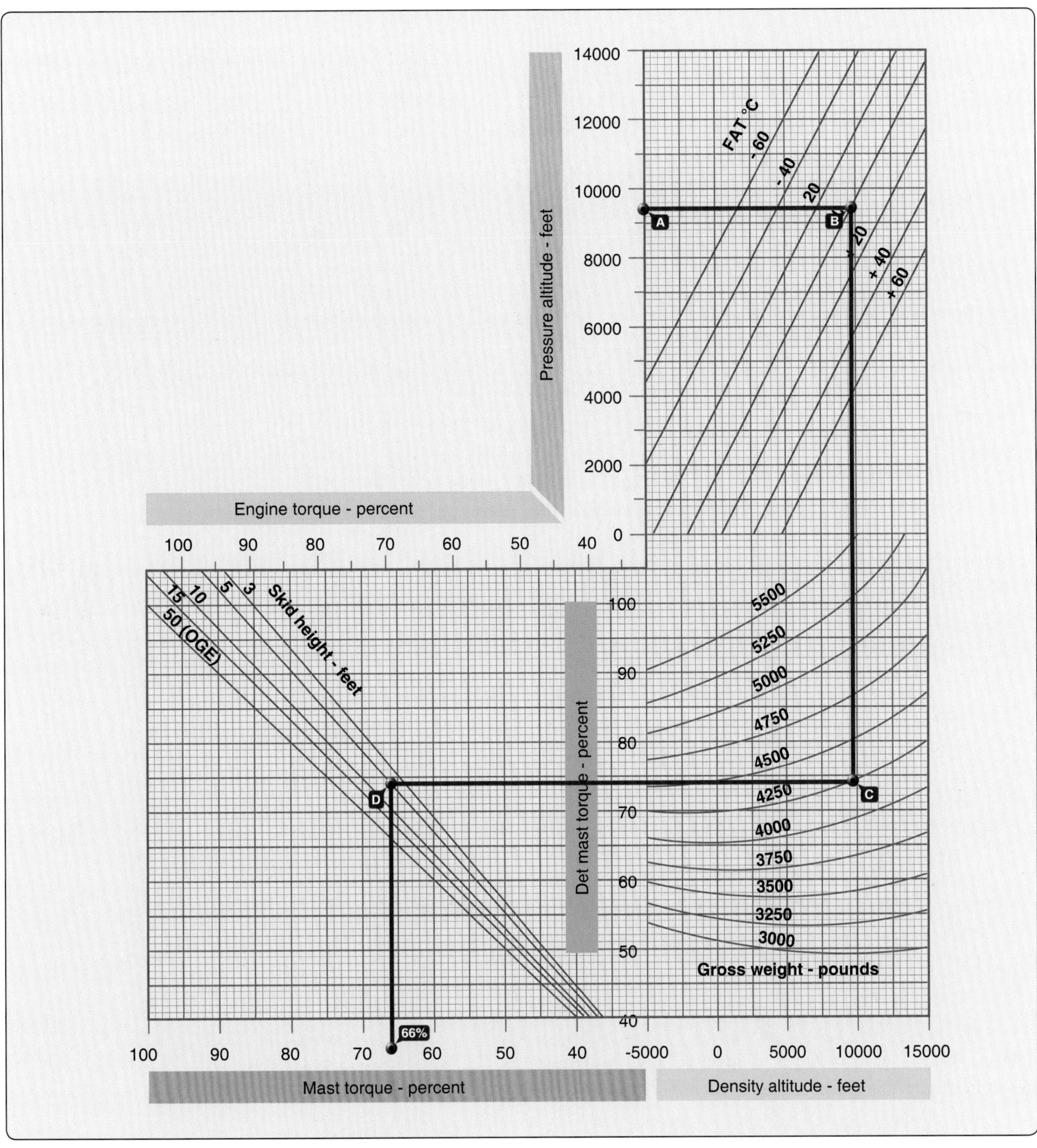

Figure 7-6. *Torque required for cruise or level flight.*

Most of the factors affecting hover and takeoff performance also affect climb performance. In addition, turbulent air, pilot techniques, and overall condition of the helicopter can cause climb performance to vary.

A helicopter flown at the best rate-of-climb speed (V_Y) obtains the greatest gain in altitude over a given period of time. This speed is normally used during the climb after all obstacles have been cleared and is usually maintained until reaching cruise altitude. Rate of climb must not be confused with angle of climb. Angle of climb is a function of altitude gained over a given distance. The V_Y results in the highest climb rate, but not the steepest climb angle, and may not be sufficient to clear obstructions. The best angle of climb speed (V_X) depends upon the power available. If there is a surplus of power available, the helicopter can climb vertically, so V_X is zero.

Wind direction and speed have an effect on climb performance, but it is often misunderstood. Airspeed is the speed at which the helicopter is moving through the atmosphere and is unaffected by wind. Atmospheric wind affects only the groundspeed, or speed at which the helicopter is moving over the Earth's surface. Thus, the only climb performance affected by atmospheric wind is the angle of climb and not the rate of climb.

When planning for climb performance, it is first important to plan for torque settings at level flight. Climb performance charts show the change in torque, above or below torque, required for level flight under the same gross weight and atmospheric conditions to obtain a given rate of climb or descent.

Sample Cruise or Level Flight Problem

Determine torque setting for cruise or level flight using *Figure 7-7.* Use the following conditions:

Pressure Altitude... 8,000 feet

Outside Air Temperature.................................... +15 °C

A. Indicated Airspeed.......................................80 knots

B. Maximum Gross Weight...............................5,000 lb

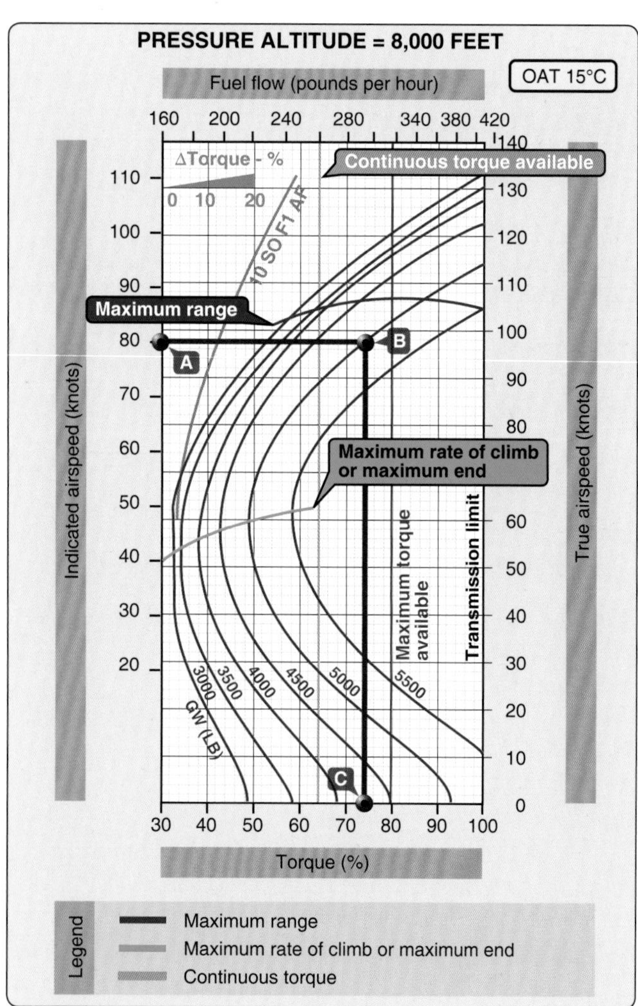

Figure 7-7. *Maximum rate-of-climb chart.*

With this chart, first confirm that it is for a pressure altitude of 8,000 feet with an OAT of 15°. Begin on the left side at 80 knots indicated airspeed (point A) and move right to maximum gross weight of 5,000 lb (point B). From that point, proceed down to the torque reading for level flight, which is 74 percent torque (point C). This torque setting is used in the next problem to add or subtract cruise/descent torque percentage from cruise flight.

Sample Climb Problem

Determine climb/descent torque percentage using *Figure 7-8.* Use the following conditions:

A. Rate of Climb or Descent. . . . . 500 fpm

B. Maximum Gross Weight . . . . 5,000 lb

With this chart, first locate a 500-fpm rate of climb or descent (point A), and then move to the right to a maximum gross

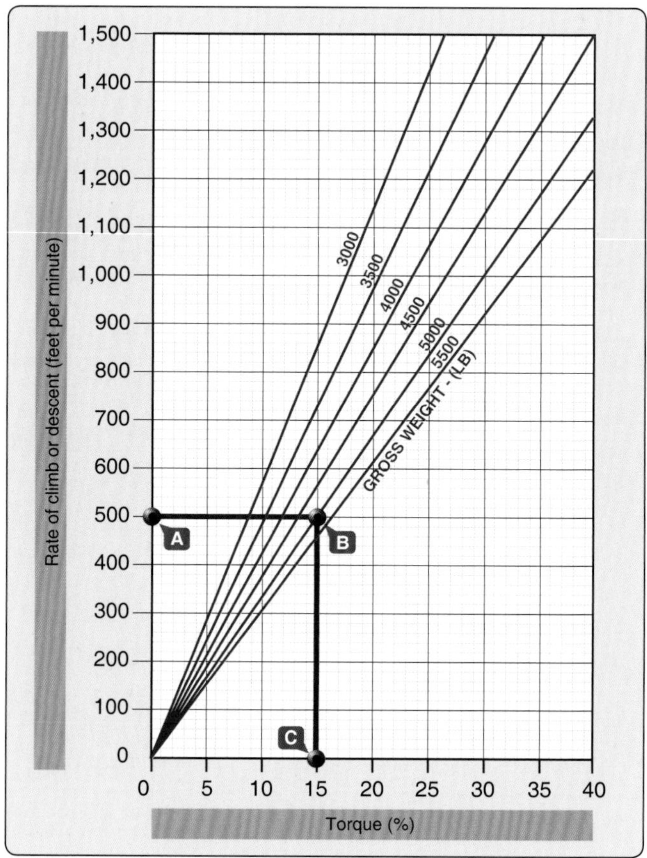

Figure 7-8. *Climb/descent torque percentage chart.*

weight of 5,000 lb (point B). From that point, proceed down to the torque percentage, which is 15 percent torque (point C). For climb or descent, 15 percent torque should be added/ subtracted from the 74 percent torque needed for level flight. For example, if the numbers were to be used for a climb torque, the pilot would adjust torque settings to 89 percent for optimal climb performance.

Chapter Summary

This chapter discussed the factors affecting performance: density altitude, weight, and wind. Five sample problems were also given with performance charts to calculate different flight conditions and determine the performance of the helicopter.

Chapter 8

Ground Procedures and Flight Preparations

Introduction

Once a pilot takes off, it is up to him or her to make sound, safe decisions throughout the flight. It is equally important for the pilot to use the same diligence when conducting a preflight inspection, making maintenance decisions, refueling, and conducting ground operations. This chapter discusses the responsibility of the pilot regarding ground safety in and around the helicopter and when preparing to fly.

Preflight

Before any flight, ensure the helicopter is airworthy by inspecting it according to the rotorcraft flight manual (RFM), pilot's operating handbook (POH), or other information supplied either by the operator or the manufacturer. Remember that it is the responsibility of the pilot in command (PIC) to ensure the aircraft is in an airworthy condition.

In preparation for flight, the use of a checklist is important so that no item is overlooked. *[Figure 8-1]* Follow the manufacturer's suggested outline for both the inside and outside inspection. This ensures that all the items the manufacturer feels are important are checked. If supplemental equipment has been added to the helicopter, these procedures should be included on the checklist as well.

Minimum Equipment Lists (MELs) and Operations with Inoperative Equipment

Title 14 of the Code of Federal Regulations (14 CFR) requires that all aircraft instruments and installed equipment be operative prior to each departure. However, when the Federal Aviation Administration (FAA) adopted the minimum equipment list (MEL) concept for 14 CFR part 91 operations, flights were allowed with inoperative items, as long as the inoperative items were determined to be nonessential for safe flight. At the same time, it allowed part 91 operators, without an MEL, to defer repairs on nonessential equipment within the guidelines of part 91.

Figure 8-1. *The pilot in command is responsible for the airworthy condition of the aircraft and using checklists to ensure proper inspection of the helicopter prior to flight.*

There are two primary methods of deferring maintenance on rotorcraft operating under part 91. They are the deferral provision of 14 CFR part 91, section 91.213(d) and an FAA-approved MEL.

The deferral provision of 14 CFR section 91.213(d) is widely used by most pilot/operators. Its popularity is due to simplicity and minimal paperwork. When inoperative equipment is found during preflight or prior to departure, the decision should be to cancel the flight, obtain maintenance prior to flight, determine if the flight can be made under the limitations imposed by the defective equipment, or to defer the item or equipment.

Maintenance deferrals are not used for in-flight discrepancies. The manufacturer's RFM/POH procedures are to be used in those situations. The discussion that follows is an example of a pilot who wishes to defer maintenance that would ordinarily be required prior to flight.

If able to use the deferral provision of 14 CFR section 91.213(d), the pilot determines whether the inoperative equipment is required by type design or 14 CFR. If the inoperative item is not required, and the helicopter can be safely operated without it, the deferral may be made. The inoperative item shall be deactivated or removed and an INOPERATIVE placard placed near the appropriate switch, control, or indicator. If deactivation or removal involves maintenance (removal always does), it must be accomplished by certificated maintenance personnel.

For example, if the position lights (installed equipment) were discovered to be inoperative prior to a daytime flight, the pilot would follow the requirements of 14 CFR section 91.213(d). The pilot must then decide if the flight can be accomplished prior to night, when the lights will be needed.

The deactivation may be a process as simple as the pilot positioning a circuit breaker to the *off* position, or as complex as rendering instruments or equipment totally inoperable. Complex maintenance tasks require a certificated and appropriately rated maintenance person to perform the deactivation. In all cases, the item or equipment must be placarded INOPERATIVE.

When an operator requests an MEL, and a Letter of Authorization (LOA) is issued by the FAA, then the use of the MEL becomes mandatory for that helicopter. All maintenance deferrals must be accomplished in accordance with the terms and conditions of the MEL and the operator-generated procedures document.

The use of an MEL for rotorcraft operated under part 91 also allows for the deferral of inoperative items or equipment. The primary guidance becomes the FAA-approved MEL issued to that specific operator and N-numbered helicopter.

The FAA has developed master minimum equipment lists (MMELs) for rotorcraft in current use. Upon written request by a rotorcraft operator, the local FAA Flight Standards District Office (FSDO) may issue the appropriate make and model MMEL, along with an LOA, and the preamble. The operator then develops operations and maintenance (O&M) procedures from the MMEL. This MMEL with O&M procedures now becomes the operator's MEL. The MEL, LOA, preamble, and procedures document developed by the operator must be on board the helicopter when it is operated.

The FAA considers an approved MEL to be a supplemental type certificate (STC) issued to an aircraft by serial number and registration number. It therefore becomes the authority to operate that aircraft in a condition other than originally type certificated.

With an approved MEL, if the position lights were discovered inoperative prior to a daytime flight, the pilot would make an entry in the maintenance record or discrepancy record provided for that purpose. The item is then either repaired or deferred in accordance with the MEL. Upon confirming that daytime flight with inoperative position lights is acceptable in accordance with the provisions of the MEL, the pilot would leave the position lights switch *off*, open the circuit breaker (or whatever action is called for in the procedures document), and placard the position light switch as INOPERATIVE.

There are exceptions to the use of the MEL for deferral. For example, should a component fail that is not listed in the MEL as deferrable (the rotor tachometer, engine tachometer, or cyclic trim, for example), then repairs are required to be performed prior to departure. If maintenance or parts are not readily available at that location, a special flight permit can be obtained from the nearest FSDO. This permit allows the helicopter to be flown to another location for maintenance. This allows an aircraft that may not currently meet applicable airworthiness requirements, but is capable of safe flight, to be operated under the restrictive special terms and conditions attached to the special flight permit.

Deferral of maintenance is not to be taken lightly, and due consideration should be given to the effect an inoperative component may have on the operation of a helicopter, particularly if other items are inoperative. Further information regarding MELs and operations with inoperative equipment can be found in AC 9 1-67, Minimum Equipment Requirements for General Aviation Operations Under FAR Part 91.

Engine Start and Rotor Engagement

During the engine start, rotor engagement, and systems ground check, use the manufacturer's checklists. If a problem arises, have it checked before continuing. Prior to performing these tasks, however, make sure the area around and above the helicopter is clear of personnel and equipment. Position the rotor blades so that they are not aligned with the fuselage. This may prevent the engine from being started with the blades still fastened. For a two-bladed rotor system, position the blades so that they are perpendicular to the fuselage and easily seen from the cockpit. Helicopters are safe and efficient flying machines as long as they are operated within the parameters established by the manufacturer.

Rotor Safety Considerations

The exposed nature of the main and tail rotors deserves special caution. Exercise extreme care when taxiing near hangars or obstructions since the distance between the rotor blade tips and obstructions is very difficult to judge. *[Figure 8-2]* In addition, the tail rotor of some helicopters cannot be seen from the cabin. Therefore, when hovering backward or turning in those helicopters, allow plenty of room for tail rotor clearance. It is a good practice to glance over your shoulder to maintain this clearance

Another rotor safety consideration is the thrust a helicopter generates. The main rotor system is capable of blowing sand, dust, snow, ice, and water at high velocities for a significant distance causing injury to nearby people and damage to buildings, automobiles, and other aircraft. Loose snow, sand, or soil can severely reduce visibility and obscure outside visual references. There is also the possibility of sand and snow being ingested into the engine intake, which can overwhelm filters and cutoff air to the engine or allow unfiltered air into the engine, leading to premature failure. Any airborne debris near the helicopter can be ingested into the engine air intake or struck by the main and tail rotor blades.

Figure 8-2. *Exercise extreme caution when hovering near buildings or other aircraft.*

Aircraft Servicing

The helicopter rotor blades are usually stopped, and both the aircraft and the refueling unit properly grounded prior to any refueling operation. The pilot should ensure that the proper grade of fuel and the proper additives, when required, are being dispensed.

Refueling of a turbine aircraft while the blades are turning, known as "hot refueling," may be practical for certain types of operation. However, this can be hazardous if not properly conducted. Pilots should remain at the flight controls; and refueling personnel should be knowledgeable about the proper refueling procedures and properly briefed for specific helicopter makes and models.

The pilot may need to train the refueling personnel on proper hot refueling procedures for that specific helicopter. The pilot should explain communication signs or calls, normal servicing procedures, and emergency procedures as a minimum. At all times during the refueling process, the pilot should remain vigilant and ready to immediately shut down the engine(s) and egress the aircraft. Several accidents have occurred due to hot refueling performed by improperly trained personnel.

Refueling units should be positioned to ensure adequate rotor blade clearance. Persons not involved with the refueling operation should keep clear of the area. Smoking must be prohibited in and around the aircraft during all refueling operations.

If operations dictate that the pilot must leave the helicopter during refueling operations, the throttle should be rolled back to flight idle and flight control friction firmly applied to prevent uncommanded control movements. The pilot should be thoroughly trained on setting the controls and egressing/ingressing the helicopter.

Safety in and Around Helicopters

People have been injured, some fatally, in helicopter accidents that would not have occurred had they been informed of the proper method of boarding or deplaning. *[Figure 8-3]* A properly briefed passenger should never be endangered by a spinning rotor. The simplest method of avoiding accidents of this sort is to stop the rotors before passengers are boarded or allowed to depart. Because this action is not always practicable, and to realize the vast and unique capabilities of the helicopter, it is often necessary to take on passengers or have them exit the helicopter while the engine and rotors are turning. To avoid accidents, it is essential that all persons associated with helicopter operations, including passengers, be made aware of all possible hazards and instructed how those hazards can be avoided.

Ramp Attendants and Aircraft Servicing Personnel

These personnel should be instructed as to their specific duties and the proper method of fulfilling them. In addition, the ramp attendant should be taught to:

1. Keep passengers and unauthorized persons out of the helicopter landing and takeoff area.
2. Brief passengers on the best way to approach and board a helicopter with its rotors turning.

Persons directly involved with boarding or deplaning passengers, aircraft servicing, rigging, or hooking up external loads, etc., should be instructed as to their duties. It would be difficult, if not impossible, to cover each and every type of operation related to helicopters. A few of the more obvious and common ones are covered below.

Passengers

Passengers increase the responsibility, workload, and risk for the pilot. The workload and distractions seem magnified to inexperienced pilots while they are developing confidence and ability to operate in the aviation environment. Inexperienced pilots should consider building up their passenger carrying experience while remaining in good flying conditions and in a familiar area.

All persons boarding a helicopter while its rotors are turning should be briefed on the safest means of doing so. The pilot in command (PIC) should always brief the passengers prior to engine start to ensure complete understanding of all procedures. The exact procedures may vary slightly from one helicopter model to another, but the following should suffice as a generic guide.

When boarding—

1. Stay away from the rear of the helicopter.
2. Approach or leave the helicopter in a crouching manner.
3. Approach from the side of the helicopter but never out of the pilot's line of vision. Certain rotor system designs allow for rotor blades to pass closer to the ground towards the front of the helicopter. For that reason, it is generally accepted for personnel to approach from the side of the helicopter. Helicopters designed to be loaded from the rear require personnel to exercise extreme caution due to tailrotor hazards.
4. Carry tools horizontally, below waist level—never upright or over the shoulder.
5. Hold firmly onto hats and loose articles.
6. Never reach up or dart after a hat or other object that might be blown off or away.
7. Protect eyes by shielding them with a hand or by

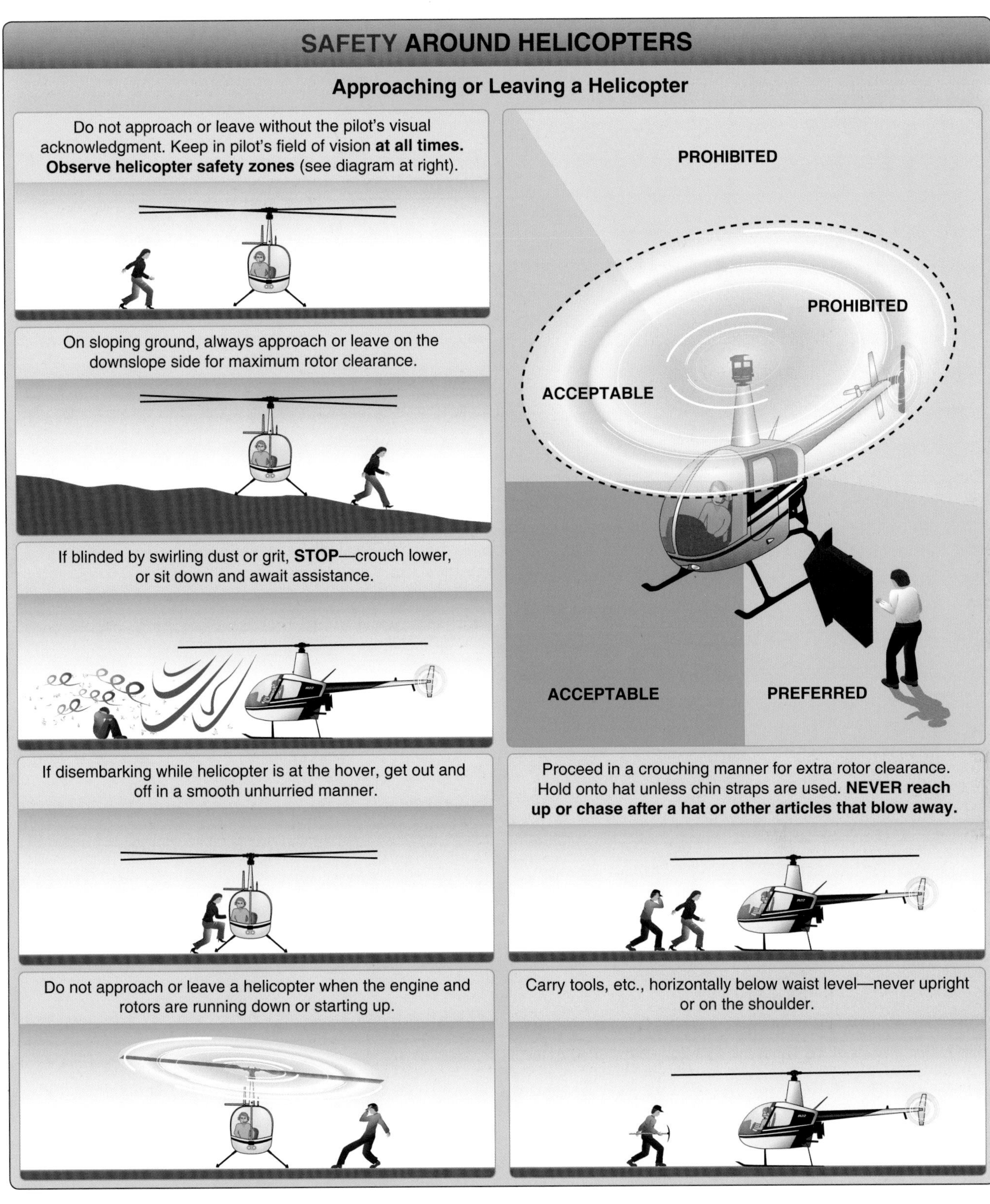

Figure 8-3. *Safety procedures for approaching or leaving a helicopter.*

squinting.

8. If suddenly blinded by dust or a blowing object, stop and crouch lower; better yet, sit down and wait for help.
9. Never grope or feel your way toward or away from the helicopter.
10. Protect hearing by wearing earplugs or earmuffs.

Since few helicopters carry cabin attendants, the pilot must conduct the pretakeoff and prelanding briefings, usually before takeoff due to noise and cockpit layout. The type of operation dictates what sort of briefing is necessary. All briefings should include the following:

1. Passengers should be briefed and understand the proper use of seatbelts, doors, and headsets/intercom system.
2. The safe entry and exit paths (away from the tail rotor and within the pilot's view).
3. If possible, remove front passenger flight controls and ensure all passenger personal items, such as cameras and mobile phones are secure.
4. For over water flights, the location and use of flotation gear and other survival equipment that are onboard. Pilot instructions should include how and when to exit the helicopter should ditching or a water landing occur.
5. For flights over rough or isolated terrain, the pilot should brief all occupants regarding the location of maps and survival equipment.
6. Passengers should be informed as to what actions and precautions to take in the event of an emergency, such as the body position for best spinal protection against a high vertical impact (erect with back firmly against the seat back); and when and how to exit. Ensure that passengers are aware of the location of the fire extinguisher, survival equipment and, if equipped, how to use and locate the Emergency Position Indicator Radio Beacon (EPIRB).

When passengers are approaching or leaving a helicopter that is sitting on a slope with the rotors turning, they should approach and depart downhill. This affords the greatest distance between the rotor blades and the ground. If this involves walking around the helicopter, they should always go around the front—never the rear.

Pilot at the Flight Controls

Many helicopter operators have been lured into a "quick turnaround" ground operation to avoid delays at airport terminals and to minimize stop/start cycles of the engine. As part of this quick turn-around, the pilot might leave the cockpit with the engine and rotors turning. Such an operation can be extremely hazardous if a gust of wind disturbs the rotor disk, or the collective flight control moves causing lift to be generated by the rotor system. Either occurrence may cause the helicopter to roll or pitch, resulting in a rotor blade striking the tail boom or the ground. Good operating procedures dictate that, generally, pilots remain at the flight controls whenever the engine is running, and the rotors are turning.

If operations require the pilot to leave the cockpit to refuel, the throttle should be rolled back to flight idle and all controls firmly frictioned to prevent uncommanded control movements. The pilot should be well trained on setting controls and exiting the cockpit without disturbing the flight or power controls.

After Landing and Securing

When the flight is terminated, park the helicopter where it does not interfere with other aircraft and is not a hazard to people during shutdown. For many helicopters, it is advantageous to land with the wind coming from the right over the tail boom (counterrotating blades). This tends to lift the blades over the tail boom but lowers the blades in front of the helicopter. This action decreases the likelihood of a main rotor strike to the tail boom due to gusty winds. Rotor downwash can cause damage to other aircraft in close proximity, and spectators may not realize the danger or see the rotors turning. Passengers should remain in the helicopter with their seats belts secured until the rotors have stopped turning. During the shutdown and postflight inspection, follow the manufacturer's checklist. Any discrepancies should be noted and, if necessary, reported to maintenance personnel.

Chapter Summary

This chapter explained the importance of preflight and safety when conducting helicopter ground operations. Proper procedures for engine run-up, refueling, and ground safety were detailed and the responsibilities of the pilot when maintenance issues occur before flight.

Chapter 9

Basic Flight Maneuvers

Introduction

From the previous chapters, it should be apparent that no two helicopters perform the same way. Even when flying the same model of helicopter, wind, temperature, humidity, weight, and equipment make it difficult to predict just how the helicopter will perform. Therefore, this chapter presents the basic flight maneuvers in a way that would apply to the majority of helicopters. In most cases, the techniques described apply to small training helicopters with:

- A single, main rotor rotating in a counterclockwise direction (looking downward on the rotor).
- An antitorque system.

Where a technique differs, it is noted. For example, a power increase on a helicopter with a clockwise rotor system requires right antitorque pedal pressure instead of left pedal pressure. In many cases, the terminology "apply proper pedal pressure" is used to indicate both types of rotor systems. However, when discussing throttle coordination to maintain proper rotations per minute (rpm), there is no differentiation between those helicopters with a governor and those without. In a sense, the governor is doing the work for you. In addition, instead of using the terms "collective pitch control" and "cyclic pitch control" throughout the chapter, these controls are referred to as just "collective" and "cyclic."

Because helicopter performance varies with weather conditions and aircraft loading, specific nose attitudes and power settings are not detailed in this handbook. In addition, this chapter does not detail every attitude of a helicopter in the various flight maneuvers, nor every move that must be made in order to perform a given maneuver.

When a maneuver is presented, there is a brief description, followed by the technique to accomplish the maneuver. In most cases, there is a list of common errors at the end of the discussion.

The Four Fundamentals

There are four fundamentals of flight upon which all maneuvers are based: straight-and-level flight, turns, climbs, and descents. All controlled flight maneuvers consist of one or more of these four fundamentals of flight. If a student pilot is able to perform these maneuvers well, and the student's proficiency is based on accurate "feel" and control analysis rather than mechanical movements, the ability to perform any assigned maneuver is only a matter of obtaining a clear visual and mental conception of it. The flight instructor must impart a good knowledge of these basic elements to the student and must combine them and plan their practice so that proper performance of each is instinctive without conscious effort. The importance of this to the success of flight training cannot be overemphasized. As the student progresses to more complex maneuvers, discounting any difficulties in visualizing the maneuvers, most student difficulties are caused by a lack of training, practice, or understanding of the principles of one or more of these fundamentals.

Guidelines

Good practices to follow during maneuvering flight include:

1. Move the cyclic only as fast as trim, torque, and rotor speed can be maintained. When entering a maneuver and the trim, rotor, or torque reacts quicker than anticipated, pilot limitations have been exceeded. If continued, an aircraft limitation will be exceeded. Perform the maneuver with less intensity until all aspects of the machine can be controlled. The pilot must be aware of the sensitivity of the flight controls due to the high speed of the main rotor.
2. Anticipate changes in aircraft performance due to loading or environmental condition. The normal collective increase to check rotor speed at sea level standard (SLS) may not be sufficient at 4,000 feet pressure altitude (PA) and 95 °F.
3. The following flight characteristics may be expected during maneuvering flight and will be discussed and demonstrated by your Flight Instructor:
    - Left turns, torque increases (more antitorque). This applies to most helicopters, but not all.
    - Right turns, torque decreases (less antitorque). This applies to most helicopters, but not all.
    - Application of aft cyclic, torque decreases and rotor speed increases.
    - Application of forward cyclic (especially when immediately following aft cyclic application), torque increases and rotor speed decreases.
    - Always leave a way out.
    - Know where the winds are.
    - Engine failures can occur during power changes and cruise flight. One possible cause of engine failure during cruise flight can be attributed to the pilot ignoring carburetor air temperatures, which could lead to carburetor icing and, subsequently, engine failure.
    - Crew coordination is critical. Everyone needs to be fully aware of what is going on, and each crewmember has a specific duty.
    - In steep turns, the nose drops. In most cases, energy (airspeed) must be traded to maintain altitude as the required excess engine power may not be available (to maintain airspeed in a 2G/60° turn, rotor thrust/engine power must increase by 100 percent). Failure to anticipate this at low altitude endangers the crew and passengers. The rate of pitch change is proportional to gross weight and density altitude.
    - Normal helicopter landings usually require high power settings, with terminations to a hover requiring the highest power setting.
    - The cyclic position relative to the horizon determines the helicopter's travel and attitude.

Straight-and-Level Flight

Straight-and-level flight is flight in which constant altitude and heading are maintained. The attitude of the rotor disk relative to the horizon determines the airspeed. The horizontal stabilizer design determines the helicopter's attitude when stabilized at an airspeed and altitude. Altitude is primarily controlled by use of the collective.

Technique

To maintain forward flight, the rotor tip-path plane must be tilted forward to obtain the necessary horizontal thrust component from the main rotor. By doing this, it causes the nose of the helicopter to lower which in turn will cause the airspeed to increase. In order to counteract this, the pilot must find the correct power setting to maintain level flight by adjusting the collective. *[Figure 9-1]* The horizontal stabilizer aids in trimming the helicopter about its transverse, horizontal axis, and reduces the amount of nose tuck that would occur. On several helicopters, it is designed as a negative lift airfoil, which produces a lifting force in a downward direction.

When in straight-and-level flight, any increase in the collective, while holding airspeed constant, causes the helicopter to climb. A decrease in the collective, while holding airspeed constant, causes the helicopter to descend. A change in the collective requires a coordinated change of the throttle to maintain a constant rpm. Additionally, the antitorque pedals need to keep the helicopter in trim around the vertical axis.

To increase airspeed in straight-and-level flight, apply forward pressure on the cyclic and raise the collective as necessary to maintain altitude. To decrease airspeed, apply rearward pressure on the cyclic and lower the collective, as necessary, to maintain altitude.

Figure 9-1. *Maintain straight-and-level flight by adjusting the rotor tip-path plane forward but adjusting the collective as necessary to maintain a constant airspeed and altitude. The natural horizon line can be used as an aid in maintaining straight-and-level flight. If the horizon line begins to rise, slight power may be required or the nose of the helicopter may be too low. If the horizon line is slowly dropping, some power may need to be taken out or the nose of the helicopter may be too high, requiring a cyclic adjustment.*

Although the cyclic is sensitive, there is a slight delay in control reaction, and it is necessary to anticipate actual movement of the helicopter. When making cyclic inputs to control the altitude or airspeed of a helicopter, take care not to overcontrol. If the nose of the helicopter rises above the level-flight attitude, apply forward pressure to the cyclic to bring the nose down. If this correction is held too long, the nose drops too low. Since the helicopter continues to change attitude momentarily after the controls reach neutral, return the cyclic to neutral slightly before the desired attitude is reached. This principle holds true for any cyclic input.

Since helicopters are not very stable, but are inherently very controllable, if a gust or turbulence causes the nose to drop, the nose tends to continue to drop instead of returning to a straight-and-level attitude as it would on a fixed-wing aircraft. Therefore, a pilot must remain alert and fly the helicopter at all times.

Common Errors

1. Failure to trim the helicopter properly, tending to hold antitorque pedal pressure and opposite cyclic. This is commonly called cross-controlling.
2. Failure to maintain desired airspeed.
3. Failure to hold proper control position to maintain desired ground track.
4. Failure to allow helicopter to stabilize at new airspeed.

Turns

A turn is a maneuver used to change the heading of the helicopter. The aerodynamics of a turn were previously discussed in Chapter 2, Aerodynamics of Flight.

Technique

Before beginning any turn, the area in the direction of the turn must be cleared not only at the helicopter's altitude, but also above and below. To enter a turn from straight-and-level flight, apply sideward pressure on the cyclic in the direction the turn is to be made. This is the only control movement needed to start the turn. Do not use the pedals to assist the turn. Use the pedals only to compensate for torque to keep the helicopter in trim around the vertical axis. *[Figure 9-2]* Keeping the fuselage in the correct streamlined position around the vertical axis facilitates the helicopter flying forward with the least drag. Trim is indicated by a yaw string in the center, or a centered ball on a turn and slip indicator. A yaw string (also referred to as a slip string) is a tool used to indicate slip or skid during flight. It is simply a string attached to the nose or canopy of an aircraft so that it is visible to the pilot during flight. The string measures sideslip and offers a visual cue to the pilot in order to make yaw corrections.

Figure 9-2. *During a level, coordinated turn, the rate of turn is commensurate with the angle of bank used, and inertia and horizontal component of lift (HCL) are equal.*

How fast the helicopter banks depends on how much lateral cyclic pressure is applied. How far the helicopter banks (the steepness of the bank) depends on how long the cyclic is displaced. After establishing the proper bank angle, return the cyclic toward the neutral position. When the bank is established, returning the cyclic to neutral (or holding it inclined relative to the horizon) will maintain the helicopter at that bank angle. Increase the collective and throttle to maintain altitude and rpm. As the torque increases, increase the proper antitorque pedal pressure to maintain longitudinal trim. Depending on the degree of bank, additional forward cyclic pressure may be required to maintain airspeed.

Rolling out of the turn to straight-and-level flight is the same as the entry into the turn, except that pressure on the cyclic is applied in the opposite direction. Since the helicopter continues to turn as long as there is any bank, start the rollout before reaching the desired heading.

The discussion on level turns is equally applicable to making turns while climbing or descending. The only difference is that the helicopter is in a climbing or descending attitude rather than that of level flight. If a so-called simultaneous entry (entering a turn while, at the same time, climbing or descending) is desired, merely combine the techniques of both maneuvers—climb or descent entry and turn entry. When recovering from a climbing or descending turn, the desired heading and altitude are rarely reached at the same time. If the heading is reached first, stop the turn and maintain the climb or descent until reaching the desired altitude. On the other hand, if the altitude is reached first, establish the level flight attitude and continue the turn to the desired heading.

Slips

A slip occurs when the helicopter slides sideways toward the center of the turn. *[Figure 9-3]* It is caused by an insufficient amount of antitorque pedal in the direction of the turn, or too much in the direction opposite the turn, in relation to the amount of power used. In other words, if you hold improper antitorque pedal pressure, which keeps the nose from following the turn, the helicopter slips sideways toward the center of the turn.

Skids

A skid occurs when the helicopter slides sideways away from the center of the turn. *[Figure 9-4]* It is caused by too much antitorque pedal pressure in the direction of the turn, or by too little in the direction opposite the turn in relation to the amount of power used. If the helicopter is forced to turn faster with increased pedal pressure instead of by increasing the degree of the bank, it skids sideways away from the center of the turn instead of flying in its normal curved path.

In summary, a skid occurs when the rate of turn is too great for the amount of bank being used, and a slip occurs when the rate of turn is too low for the amount of bank being used. *[Figure 9-5]*

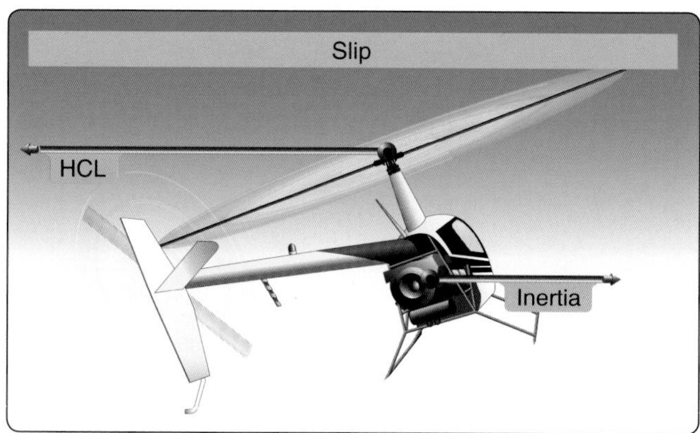

Figure 9-3. *During a slip, the rate of turn is too low for the angle of bank used, and the horizontal component of lift (HCL) exceeds inertia.*

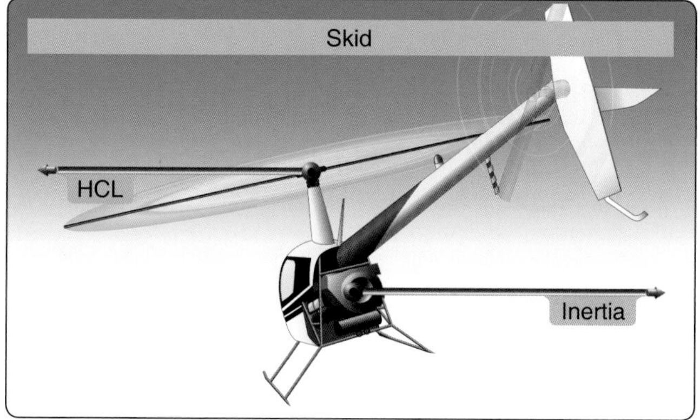

Figure 9-4. *During a skid, the rate of turn is too great for the angle of bank used, and inertia exceeds the horizontal component of lift (HCL).*

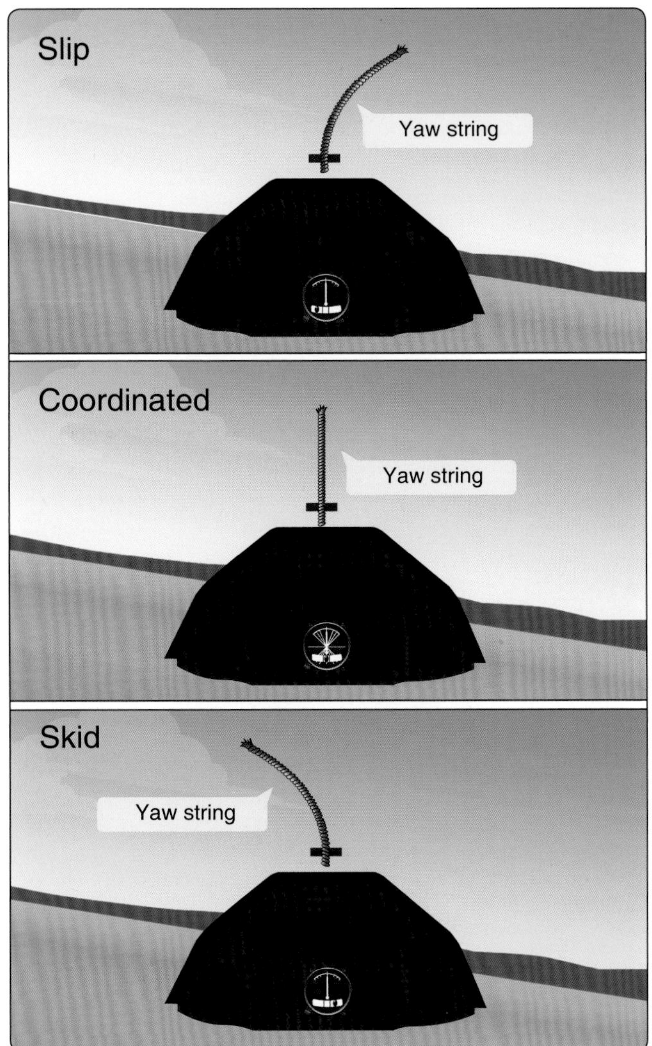

Figure 9-5. *Cockpit view of a slip and skid.*

Normal Climb

The entry into a climb from a hover has already been described in the Normal Takeoff from a Hover subsection; therefore, this discussion is limited to a climb entry from cruising flight.

Technique

To enter a climb in a helicopter while maintaining airspeed, the first actions are increasing the collective and throttle, and adjusting the pedals as necessary to maintain a centered ball in the slip/skid indicator. Moving the collective up requires a slight aft movement of the cyclic to direct all of the increased power into lift and maintain the airspeed. Remember, a helicopter can climb with the nose down and descend with the nose up. Helicopter attitude changes mainly reflect acceleration or deceleration, not climb or descent. Therefore, the climb attitude is approximately the same as level flight in a stable climb, depending on the aircraft's horizontal stabilizer design.

If the pilot wishes to climb faster, with a decreased airspeed, then the climb can be initiated with aft cyclic. Depending on initial or entry airspeed for the climb, the climb can be accomplished without increasing the collective, if a much slower airspeed is acceptable. However, as the airspeed decreases, the airflow over the vertical fin decreases necessitating more antitorque (left) pedal application.

To level off from a climb, start adjusting the attitude to the level flight attitude a few feet prior to reaching the desired altitude. The amount of lead depends on the rate of climb at the time of level-off (the higher the rate of climb, the more the lead). Generally, the lead is 10 percent of the climb rate. For example, if the climb rate is 500 feet per minute (fpm), you should lead the level-off by 50 feet.

To begin the level-off, apply forward cyclic to adjust and maintain a level flight attitude, which can be slightly nose low. Maintain climb power until the airspeed approaches the desired cruising airspeed, then lower the collective to obtain cruising power and adjust the throttle to obtain and maintain cruising rpm. Throughout the level-off, maintain longitudinal trim with the antitorque pedals.

Common Errors

1. Failure to maintain proper power and airspeed.
2. Holding too much or too little antitorque pedal.
3. In the level-off, decreasing power before adjusting the nose to cruising attitude.

Normal Descent

A normal descent is a maneuver in which the helicopter loses altitude at a controlled rate in a controlled attitude.

Technique

To establish a normal descent from straight-and-level flight at cruising airspeed, lower the collective to obtain proper power, adjust the throttle to maintain rpm, and increase right antitorque pedal pressure to maintain heading in a counterclockwise rotor system (or left pedal pressure in a clockwise system). If cruising airspeed is the same as or slightly above descending airspeed, simultaneously apply the necessary cyclic pressure to obtain the approximate descending attitude. If the pilot wants to decelerate, the cyclic must be moved aft. If the pilot desires to descend with increased airspeed, then forward cyclic is all that is required if airspeed remains under the limit. As the helicopter stabilizes at any forward airspeed, the fuselage attitude will streamline due to the airflow over the horizontal stabilizer. As the airspeed changes, the airflow over the vertical stabilizer or fin changes, so the pedals must be adjusted for trim.

The pilot should always remember that the total lift and thrust vectoring is controlled by the cyclic. If a certain airspeed is desired, it will require a certain amount of cyclic and collective movement for level flight. If the cyclic is moved, the thrust-versus-lift ratio is changed. Aft cyclic directs more power to lift, and altitude increases. Forward cyclic directs more power to thrust, and airspeed increases. If the collective is not changed and there is a change only in cyclic, the total thrust to lift ratio does not change: aft cyclic results in a climb, and forward cyclic results in a descent with the corresponding airspeed changes.

To level off from the descent, lead the desired altitude by approximately 10 percent of the rate of descent. For example, a 500-fpm rate of descent would require a 50-foot lead. At this point, increase the collective to obtain cruising power, adjust the throttle to maintain rpm, and increase left antitorque pedal pressure to maintain heading (right pedal pressure in a clockwise rotor system). Adjust the cyclic to obtain cruising airspeed and a level flight attitude as the desired altitude is reached.

Common Errors

1. Failure to maintain constant angle of decent during training.
2. Failure to level-off the aircraft sufficiently, which results in recovery below the desired altitude.
3. Failure to adjust antitorque pedal pressures for changes in power.

Vertical Takeoff to a Hover

A vertical takeoff to a hover involves flying the helicopter from the ground vertically to a skid height of two to three feet, while maintaining a constant heading. Once the desired skid height is achieved, the helicopter should remain nearly motionless over a reference point at a constant altitude and on a constant heading. The maneuver requires a high degree of concentration and coordination.

Technique

The pilot on the controls needs to clear the area left, right, and above to perform a vertical takeoff to a hover. The pilot should remain focused outside the aircraft and obtain clearance to take off from the controlling tower. If necessary, the pilot who is not on the controls assists in clearing the aircraft and provides adequate warning of any obstacles and any unannounced or unusual drift/altitude changes.

Heading control, direction of turn, and rate of turn at hover are all controlled by using the pedals. Hover height, rate of ascent, and the rate of descent are controlled by using the collective. Helicopter position and the direction of travel are controlled by the cyclic.

After receiving the proper clearance and ensuring that the area is clear of obstacles and traffic, begin the maneuver with the collective in the down position and the cyclic in a neutral position, or slightly into the wind. Very slowly increase the collective until the helicopter becomes light on the skids or wheels. As collective and torque increases, antitorque must be adjusted as well. Therefore, as the aircraft begins to get light on the landing gear, apply appropriate antitorque pedal to maintain aircraft heading. Continue to apply pedals as necessary to maintain heading and coordinate the cyclic for a vertical ascent. As the helicopter slowly leaves the ground, check for proper attitude control response and helicopter center of gravity. A slow ascent will allow stopping if responses are outside the normal parameters indicating hung or entangled landing gear, center of gravity problems, or control issues. If a roll or tilt begin, decrease the collective and determine the cause of the roll or tilt. Upon reaching the desired hover altitude, adjust the flight controls as necessary to maintain position over the intended hover area. Student pilots should be reminded that while at a hover, the helicopter is rarely ever level. Helicopters usually hover left side low due to the tail rotor thrust being counteracted by the main rotor tilt. A nose low or high condition is generally caused by loading. Once stabilized, check the engine instruments and note the power required to hover.

Excessive movement of any flight control requires a change in the other flight controls. For example, if the helicopter drifts to one side while hovering, the pilot naturally moves the cyclic in the opposite direction. When this is done, part of the vertical thrust is diverted, resulting in a loss of altitude. To maintain altitude, increase the collective. This increases drag on the blades and tends to slow them down. To counteract the drag and maintain rpm, increase the throttle. Increased throttle means increased torque, so the pilot must add more pedal pressure to maintain the heading. This can easily lead to overcontrolling the helicopter. However, as level of proficiency increases, problems associated with overcontrolling decrease. Helicopter controls are usually more driven by pressure than by gross control movements.

Common Errors

1. Failing to ascend vertically as the helicopter becomes airborne.
2. Pulling excessive collective to become airborne, causing the helicopter to gain too much altitude.
3. Overcontrolling the antitorque pedals, which not only changes the heading of the helicopter, but also changes the rpm.

4. Reducing throttle rapidly in situations in which proper rpm has been exceeded, usually resulting in exaggerated heading changes and loss of lift, resulting in loss of altitude.
5. Failing to ascend slowly.

Hovering

A stationary hover is a maneuver in which the helicopter is maintained in nearly motionless flight over a reference point at a constant altitude and on a constant heading.

Technique

To maintain a hover over a point, use sideview and peripheral vision to look for small changes in the helicopter's attitude and altitude. When these changes are noted, make the necessary control inputs before the helicopter starts to move from the point. To detect small variations in altitude or position, the main area of visual attention needs to be some distance from the aircraft, using various points on the helicopter or the tip-path plane as a reference. Looking too closely or looking down leads to overcontrolling. Obviously, in order to remain over a certain point, know where the point is, but do not focus all attention there.

As with a takeoff, the pilot controls altitude with the collective and maintains a constant rpm with the throttle. The cyclic is used to maintain the helicopter's position; the pedals, to control heading. To maintain the helicopter in a stabilized hover, make small, smooth, coordinated corrections. As the desired effect occurs, remove the correction in order to stop the helicopter's movement. For example, if the helicopter begins to move rearward, apply a small amount of forward cyclic pressure. However, neutralize this pressure just before the helicopter comes to a stop, or it will begin to move forward.

After experience is gained, a pilot develops a certain "feel" for the helicopter. Small deviations can be felt and seen, so you can make the corrections before the helicopter actually moves. A certain relaxed looseness develops, and controlling the helicopter becomes second nature, rather than a mechanical response.

Common Errors

1. Tenseness and slow reactions to movements of the helicopter.
2. Failure to allow for lag in cyclic and collective pitch, which leads to overcontrolling. It is very common for a student to get ahead of the helicopter. Due to inertia, it requires some small time period for the helicopter to respond.
3. Confusing attitude changes for altitude changes, which results in improper use of the controls.
4. Hovering too high, creating a hazardous flight condition. The height velocity chart should be referenced to determine the maximum skid height to hover and safely recover the helicopter should a malfunction occur.
5. Hovering too low, resulting in occasional touchdown.
6. Becoming overly confident over prepared surfaces when taking off to a hover. Be aware that dynamic rollover accidents usually occur over a level surface.

Hovering Turn

A hovering turn is a maneuver performed at hovering height in which the nose of the helicopter is rotated either left or right while maintaining position over a reference point on the surface. Hovering turns can also be made around the mast or tail of the aircraft. The maneuver requires the coordination of all flight controls and demands precise control near the surface. A pilot should maintain a constant altitude, rate of turn, and rpm.

Technique

Initiate the turn in either direction by applying anti-torque pedal pressure toward the desired direction. It should be noted that during a turn to the left, more power is required because left pedal pressure increases the pitch angle of the tail rotor, which, in turn, requires additional power from the engine. A turn to the right requires less power. (On helicopters with a clockwise rotating main rotor, right pedal increases the pitch angle and, therefore, requires more power.)

As the turn begins, use the cyclic as necessary (usually into the wind) to keep the helicopter over the desired spot. To continue the turn, add more pedal pressure as the helicopter turns to the crosswind position. This is because the wind is striking the tail surface and tail rotor area, making it more difficult for the tail to turn into the wind. As pedal pressures increase due to crosswind forces, increase the cyclic pressure into the wind to maintain position. Use the collective with the throttle to maintain a constant altitude and rpm. *[Figure 9-6]*

After the 90° portion of the turn, decrease pedal pressure slightly to maintain the same rate of turn. Approaching the 180°, or downwind portion, anticipate opposite pedal pressure due to the tail moving from an upwind position to a downwind position. At this point, the rate of turn has a tendency to increase at a rapid rate due to the tendency of the tail surfaces to weathervane. Because of the tailwind condition,

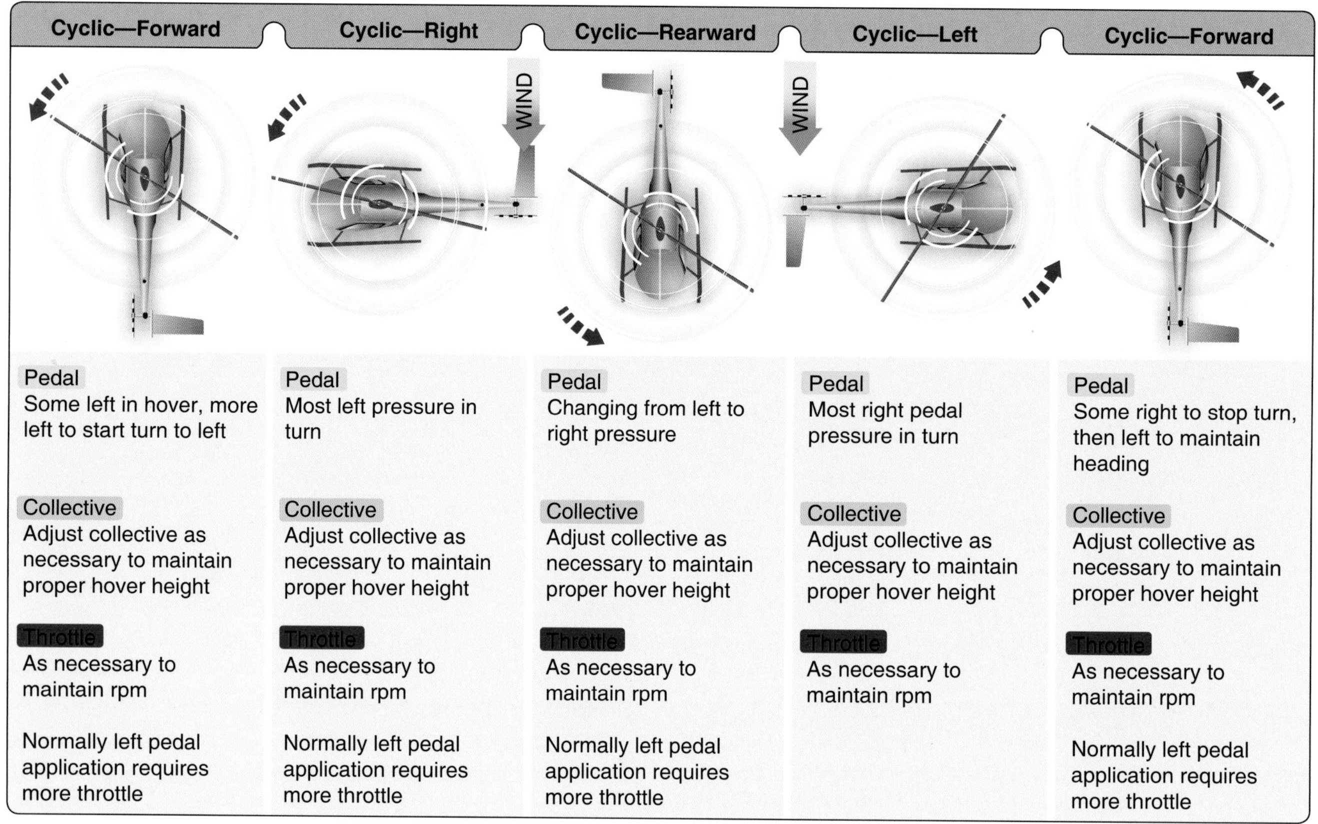

Figure 9-6. *Left turns in helicopters with a counterclockwise rotating main rotor are more difficult to execute because the tail rotor demands more power. This requires you to compensate with additional left pedal and increased throttle. Refer to this graphic throughout the remainder of the discussion on a hovering turn to the left.*

hold rearward cyclic pressure to keep the helicopter over the same spot.

The horizontal stabilizer has a tendency to lift the tail during a tailwind condition. This is the most difficult portion of the hovering turn. Horizontal and vertical stabilizers have several different designs and locations, including the canted stabilizers used on some Hughes and Schweizer helicopters. The primary purpose of the vertical stabilizer is to unload the work of the antitorque system and to aid in trimming the helicopter in flight should the antitorque system fail. The horizontal stabilizer provides for a more usable CG range and aids in trimming the helicopter longitudinally.

Because of the helicopter's tendency to weathervane, maintaining the same rate of turn from the 180° position actually requires some pedal pressure opposite the direction of turn. If a pilot does not apply opposite pedal pressure, the helicopter tends to turn at a faster rate. The amount of pedal pressure and cyclic deflection throughout the turn depends on the wind velocity. As the turn is finished on the upwind heading, apply opposite pedal pressure to stop the turn. Gradually apply forward cyclic pressure to keep the helicopter from drifting.

Control pressures and direction of application change continuously throughout the turn. The most dramatic change is the pedal pressure (and corresponding power requirement) necessary to control the rate of turn as the helicopter moves through the downwind portion of the maneuver.

Turns can be made in either direction; however, in a high wind condition, the tail rotor may not be able to produce enough thrust, which means the pilot cannot control a turn to the right in a counterclockwise rotor system. Therefore, if control is ever questionable, first attempt to make a 90° turn to the left. If sufficient tail rotor thrust exists to turn the helicopter crosswind in a left turn, a right turn can be successfully controlled. The opposite applies to helicopters with clockwise rotor systems. In this case, start the turn to the right. Hovering turns should be avoided in winds strong enough to preclude sufficient aft cyclic control to maintain the helicopter on the selected surface reference point when headed downwind. Check the flight manual for the manufacturer's recommendations for this limitation.

Common Errors

1. Failing to maintain a slow, constant rate of turn.
2. Failing to maintain position over the reference point.
3. Failing to maintain rpm within normal range.
4. Failing to maintain constant altitude.
5. Failing to use the antitorque pedals properly.

Hovering—Forward Flight

Forward hovering flight is normally used to move a helicopter to a specific location, and it may begin from a stationary hover. During the maneuver, constant groundspeed, altitude, and heading should be maintained.

Technique

Before starting, pick out two references directly in front and in line with the helicopter. These reference points should be kept in line throughout the maneuver. *[Figure 9-7]*

Begin the maneuver from a normal hovering height by applying forward pressure on the cyclic. As movement begins, return the cyclic toward the neutral position to maintain low groundspeed—no faster than a brisk walk. Throughout the maneuver, maintain a constant groundspeed and path over the ground with the cyclic, a constant heading with the antitorque pedals, altitude with the collective, and the proper rpm with the throttle.

To stop the forward movement, apply rearward cyclic pressure until the helicopter stops. As forward motion stops, return the cyclic to the neutral position to prevent rearward movement. Forward movement can also be stopped by simply applying rearward pressure to level the helicopter and allowing it to drift to a stop.

Common Errors

1. Exaggerated movement of the cyclic, resulting in erratic movement over the surface.
2. Failure to use proper antitorque pedal control, resulting in excessive heading change.
3. Failure to maintain desired hovering height.
4. Failure to maintain proper rpm.
5. Failure to maintain alignment with direction of travel.

Figure 9-7. *To maintain a straight ground track, use two reference points in line and at some distance in front of the helicopter.*

Hovering—Sideward Flight

Sideward hovering flight may be necessary to move the helicopter to a specific area when conditions make it impossible to use forward flight. During the maneuver, a constant groundspeed, altitude, and heading should be maintained.

Technique

Before starting sideward hovering flight, ensure the area for the hover is clear, especially at the tail rotor. Constantly monitor hover height and tail rotor clearance during all hovering maneuvers to prevent dynamic rollover or tail rotor strikes to the ground. Then, pick two points of in-line reference in the direction of sideward hovering flight to help maintain the proper ground track. These reference points should be kept in line throughout the maneuver. *[Figure 9-8]*

Begin the maneuver from a normal hovering height by applying cyclic toward the side in which the movement is desired. As the movement begins, return the cyclic toward the neutral position to maintain low groundspeed—no faster than a brisk walk. Throughout the maneuver, maintain a constant groundspeed and ground track with cyclic. Maintain heading,

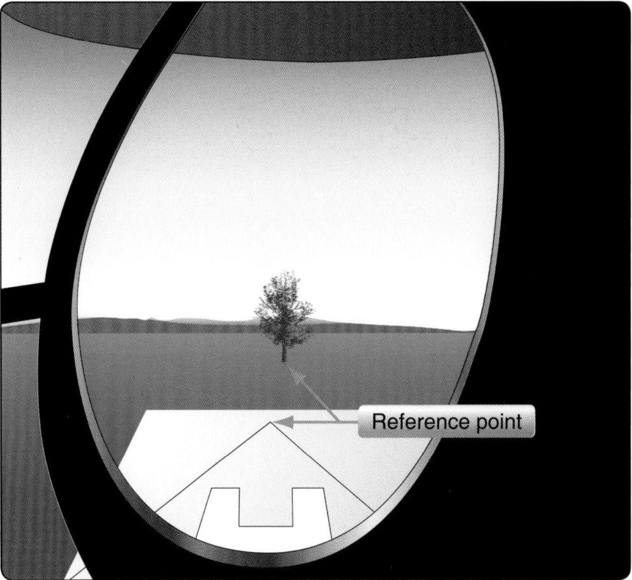

Figure 9-8. *The key to hovering sideward is establishing at least two reference points that help maintain a straight track over the ground while keeping a constant heading.*

which in this maneuver is perpendicular to the ground track, with the antitorque pedals, and a constant altitude with the collective. Use the throttle to maintain the proper operating rpm. Be aware that the nose tends to weathervane into the wind. Changes in the pedal position will change the rpm and must be corrected by collective and/or throttle changes to maintain altitude.

To stop the sideward movement, apply cyclic pressure in the direction opposite to that of movement and hold it until the helicopter stops. As motion stops, return the cyclic to the neutral position to prevent movement in the opposite direction. Applying sufficient opposite cyclic pressure to level the helicopter may also stop sideward movement. The helicopter then drifts to a stop.

Common Errors

1. Exaggerated movement of the cyclic, resulting in overcontrolling and erratic movement over the surface.
2. Failure to use proper antitorque pedal control, resulting in excessive heading change.
3. Failure to maintain desired hovering height.
4. Failure to maintain proper rpm.
5. Failure to make sure the area is clear prior to starting the maneuver.

Hovering—Rearward Flight

Rearward hovering flight may be necessary to move the helicopter to a specific area when the situation is such that forward or sideward hovering flight cannot be used. During the maneuver, maintain a constant groundspeed, altitude, and heading. Due to the limited visibility behind a helicopter, it is important that the area behind the helicopter be cleared before beginning the maneuver. Use of ground personnel is recommended.

Technique

Before starting rearward hovering flight, pick out two reference points in front of, and in line with the helicopter just like hovering forward. *[Figure 9-7]* The movement of the helicopter should be such that these points remain in line.

Begin the maneuver from a normal hovering height by applying rearward pressure on the cyclic. After the movement has begun, position the cyclic to maintain a slow groundspeed—no faster than a brisk walk. Throughout the maneuver, maintain constant groundspeed and ground track with the cyclic, a constant heading with the antitorque pedals, constant altitude with the collective, and the proper rpm with the throttle.

To stop the rearward movement, apply forward cyclic and hold it until the helicopter stops. As the motion stops, return the cyclic to the neutral position. Also, as in the case of forward and sideward hovering flight, opposite cyclic can be used to level the helicopter and let it drift to a stop. Tail rotor clearance must be maintained. Generally, a higher-than-normal hover altitude is preferred.

Common Errors

1. Exaggerated movement of the cyclic resulting in overcontrolling and an uneven movement over the surface.
2. Failure to use proper antitorque pedal control, resulting in excessive heading change.
3. Failure to maintain desired hovering height.
4. Failure to maintain proper rpm.
5. Failure to make sure the area is clear prior to starting the maneuver.

Taxiing

Taxiing refers to operations on or near the surface of taxiways or other prescribed routes. Helicopters utilize three different types of taxiing.

Hover Taxi

A hover taxi is used when operating below 25 feet above ground level (AGL). *[Figure 9-9]* Since hover taxi is just like forward, sideward, or rearward hovering flight, the technique to perform it is not presented here.

Air Taxi

An air taxi is preferred when movements require greater distances within an airport or heliport boundary. *[Figure 9-10]* In this case, fly to the new location; however, it is expected that the helicopter will remain below 100 feet AGL with an appropriate airspeed and will avoid over flight of other aircraft, vehicles, and personnel.

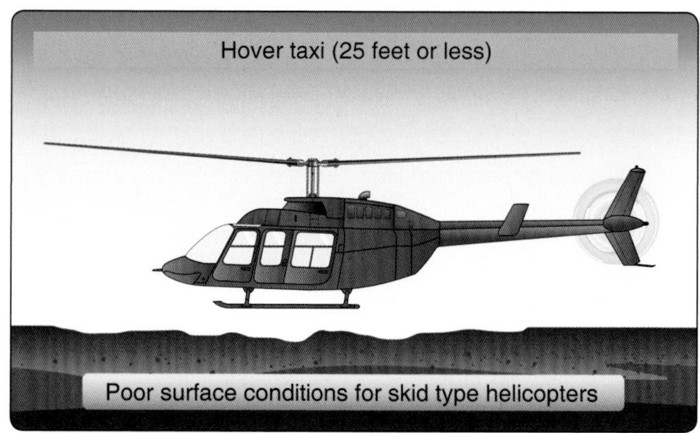

Figure 9-9. *Hover taxi.*

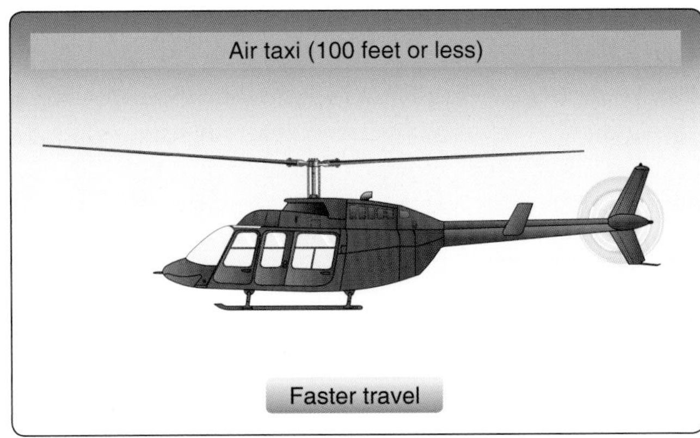

Figure 9-10. *Air taxi.*

Technique

Before starting, determine the appropriate airspeed and altitude combination to remain out of the cross-hatched or shaded areas of the height/velocity diagram (see *Figure 7-1*). Additionally, be aware of crosswind conditions that could lead to loss of tail rotor effectiveness. Pick out two references directly in front of the helicopter for the ground path desired. These reference points should be kept in line throughout the maneuver.

Begin the maneuver from a normal hovering height by applying forward pressure on the cyclic. As movement begins, attain the desired airspeed with the cyclic. Control the desired altitude with the collective and rpm with the throttle. Throughout the maneuver, maintain a desired groundspeed and ground track with the cyclic, a constant heading with antitorque pedals, the desired altitude with the collective, and proper operating rpm with the throttle.

To stop the forward movement, apply aft cyclic pressure to reduce forward speed. Simultaneously lower the collective to initiate a descent to hover altitude. As forward motion stops, return the cyclic to the neutral position to prevent rearward movement. As approaching the proper hover altitude, increase the collective as necessary to stop descent at hover altitude (much like a quick stop maneuver (see page 10-4)).

Common Errors

1. Erratic movement of the cyclic, resulting in improper airspeed control and erratic movement over the surface.
2. Failure to use proper antitorque pedal control, resulting in excessive heading change.
3. Failure to maintain desired altitude.
4. Failure to maintain proper rpm.
5. Overflying parked aircraft causing possible damage from rotor downwash.
6. Flying in the cross-hatched or shaded area of the height/velocity diagram.
7. Flying in a crosswind that could lead to loss of tail rotor effectiveness.
8. Excessive tail-low attitudes.
9. Excessive power used or required to stop.
10. Failure to maintain alignment with direction of travel.

Surface Taxi

A surface taxi is used to minimize the effects of rotor downwash in wheel-type helicopters. *[Figure 9-11]* Surface taxiing in skid type helicopters is generally not recommended due to the high risk of dynamic rollover; for more information, refer to Chapter 11, Helicopter Emergencies and Hazards.

Technique

The helicopter should be in a stationary position on the surface with the collective full down and the rpm the same as that used for a hover. This rpm should be maintained throughout the maneuver. Then, move the cyclic slightly forward and apply gradual upward pressure on the collective to move the helicopter forward along the surface. Use the antitorque pedals to maintain heading and the cyclic to maintain ground track. The collective controls starting, stopping, and speed while taxiing. The higher the collective pitch, the faster the taxi speed; however, do not taxi faster than a brisk walk. If the helicopter is equipped with brakes, use them to help slow down. Do not use the cyclic to control groundspeed.

During a crosswind taxi, hold the cyclic into the wind a sufficient amount to eliminate any drifting movement.

Common Errors

1. Improper use of cyclic.
2. Failure to use antitorque pedals for heading control.

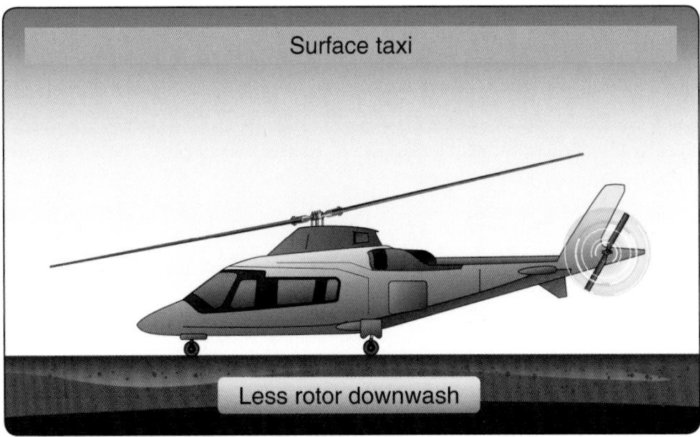

Figure 9-11. *Surface taxi.*

3. Improper use of the controls during crosswind operations.
4. Failure to maintain proper rpm.

Normal Takeoff from a Hover

A normal takeoff from a hover is an orderly transition to forward flight and is executed to increase altitude safely and expeditiously. Before initiating a takeoff, the pilot should ensure that the proper checklist has been completed and the helicopter systems are within normal limits. During the takeoff, fly a profile that avoids the cross-hatched or shaded areas of the height/velocity diagram.

Technique

Refer to *Figure 9-12* (position 1). Bring the helicopter to a hover and perform a hover and systems check, which includes power, balance, and flight controls prior to continuing flight. The power check should include an evaluation of the amount of excess power available; that is, the difference between the power being used to hover and the power available at the existing altitude and temperature conditions. The balance condition of the helicopter is indicated by the position of the cyclic when maintaining a stationary hover. Wind necessitates some cyclic deflection, but there should not be an extreme deviation from neutral. Flight controls must move freely, and the helicopter should respond normally. Then, visually clear the surrounding area.

Start the helicopter moving by smoothly and slowly easing the cyclic forward (position 2). As the helicopter starts to move forward, increase the collective, as necessary, to prevent the helicopter from sinking and adjust the throttle to maintain rpm. The increase in power requires an increase in the proper antitorque pedal to maintain heading. Maintain a straight takeoff path throughout the takeoff.

While accelerating through effective translational lift (position 3), the helicopter begins to climb, and the nose tends to rise due to increased lift. At this point, adjust the collective to obtain normal climb power and apply enough forward cyclic to overcome the tendency of the nose to rise. At position 4, hold an attitude that allows a smooth acceleration toward climbing airspeed and a commensurate gain in altitude so that the takeoff profile does not take the helicopter through any of the cross-hatched or shaded areas of the height/velocity diagram. As airspeed increases (position 5), place the aircraft in trim and allow a crab to take place to maintain ground track and a more favorable climb configuration. As the helicopter continues to climb and accelerate to best rate-of-climb, apply aft cyclic pressure to raise the nose smoothly to the normal climb attitude.

Common Errors

1. Failing to use sufficient collective pitch to prevent loss of altitude prior to attaining translational lift.
2. Adding power too rapidly at the beginning of the transition from hovering to forward flight without forward cyclic compensation, causing the helicopter to gain excessive altitude before acquiring airspeed.
3. Assuming an extreme nose-down attitude near the surface in the transition from hovering to forward flight.
4. Failing to maintain a straight flightpath over the surface (ground track).
5. Failing to maintain proper airspeed during the climb.
6. Failing to adjust the throttle to maintain proper rpm.
7. Failing to transition to a level crab to maintain ground track.

Figure 9-12. *The helicopter takes several positions during a normal takeoff from hover.*

Normal Takeoff from the Surface

Normal takeoff from the surface is used to move the helicopter from a position on the surface into effective translational lift and a normal climb using a minimum amount of power. If the surface is dusty or covered with loose snow, this technique provides the most favorable visibility conditions and reduces the possibility of debris being ingested by the engine.

Technique

Place the helicopter in a stationary position on the surface. Lower the collective to the full down position, and reduce the rpm below operating rpm. Visually clear the area and select terrain features or other objects to aid in maintaining the desired track during takeoff and climb out. Increase the throttle to the proper rpm, and raise the collective slowly until the helicopter is light on the skids. Hesitate momentarily and adjust the cyclic and antitorque pedals, as necessary, to prevent any surface movement. Continue to apply upward collective. As the helicopter leaves the ground, use the cyclic, as necessary, to begin forward movement as altitude is gained. Continue to accelerate. As effective translational lift is attained, the helicopter begins to climb. Adjust attitude and power, if necessary, to climb in the same manner as a takeoff from a hover. A second, less efficient, but acceptable, technique, is to attempt a vertical takeoff to evaluate if power or lift is sufficient to clear obstructions. This allows the helicopter to be returned to the takeoff position if required.

Common Errors

1. Departing the surface in an attitude that is too nose-low. This situation requires the use of excessive power to initiate a climb.
2. Using excessive power combined with a level attitude, which causes a vertical climb, unless needed for obstructions and landing considerations.
3. Application of the collective that is too abrupt when departing the surface, causing rpm and heading control errors.

Crosswind Considerations During Takeoffs

If the takeoff is made during crosswind conditions, the helicopter is flown in a slip during the early stages of the maneuver. *[Figure 9-13]* The cyclic is held into the wind a sufficient amount to maintain the desired ground track for the takeoff. The heading is maintained with the use of the antitorque pedals. In other words, the rotor is tilted into the wind so that the sideward movement of the helicopter is just enough to counteract the crosswind effect. To prevent the nose from turning in the direction of the rotor tilt, it is necessary to increase the antitorque pedal pressure on the side opposite the cyclic.

After approximately 50 feet of altitude is gained, make a coordinated turn into the wind to maintain the desired ground track. This is called crabbing into the wind. The stronger the crosswind, the more the helicopter has to be turned into the wind to maintain the desired ground track. *[Figure 9-14]*

Ground Reference Maneuvers

Ground reference maneuvers may be used as training exercises to help develop a division of attention between the flightpath and ground references, and while controlling

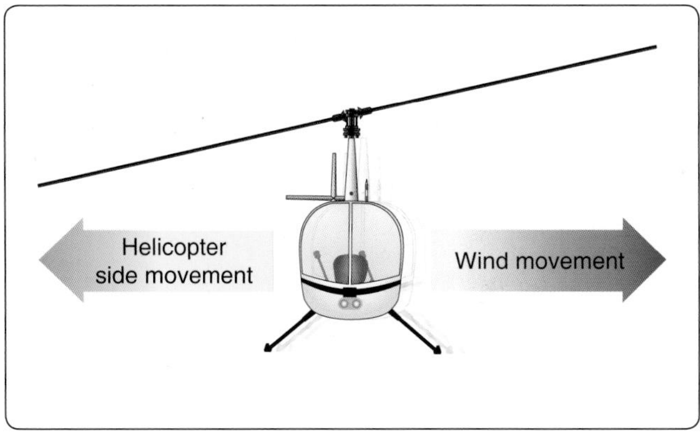

Figure 9-13. *During a slip, the rotor disk is tilted into the wind.*

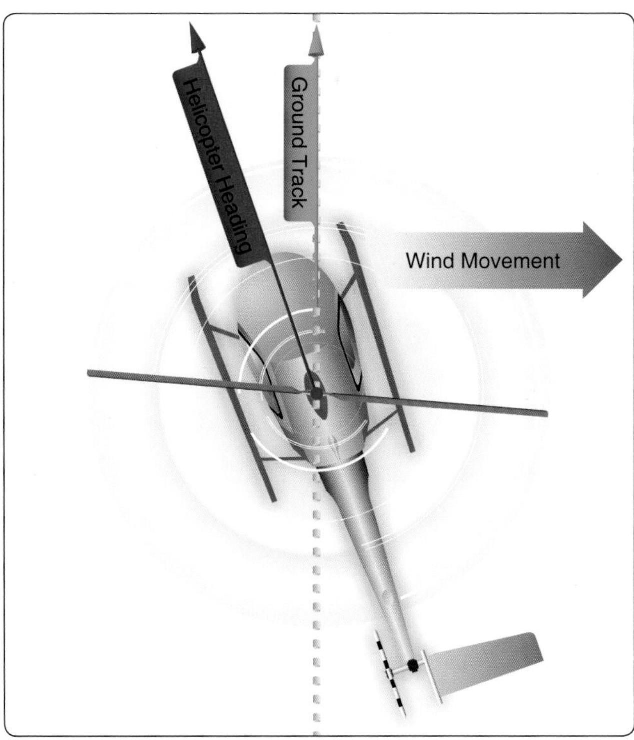

Figure 9-14. *To compensate for wind drift at altitude, crab the helicopter into the wind.*

the helicopter and watching for other aircraft in the vicinity. Other examples of ground reference maneuvers are flights for photographic or observation purposes, such as pipe line or power line checks. Prior to each maneuver, a clearing turn should be done to ensure the area is free of conflicting traffic.

Rectangular Course

The rectangular course is a training maneuver in which the ground track of the helicopter is kept equidistant from the sides of a selected rectangular area. While performing the maneuver, the altitude and airspeed should be held constant. The rectangular course helps develop recognition of a drift toward or away from a line parallel to the intended ground track. This is helpful in recognizing drift toward or from an airport runway during the various legs of the airport traffic pattern and is also useful in observation and photographic flights.

Technique

Maintaining ground track while trying to fly a straight line can be very difficult for new pilots to do. It is important to understand the effects of the wind and how to compensate for this. For this maneuver, pick a square or rectangular field, or an area bounded on four sides by section lines or roads, with sides approximately a mile in length. The area selected should be well away from other air traffic. Fly the maneuver approximately 500 to 1,000 feet above the ground as appropriate. If the student finds it difficult to maintain a proper ground track at that higher altitude, lower the altitude for better ground reference until they feel more comfortable and are able to grasp the concept better. Altitude can be raised up to 1,000 feet as proficiency improves.

Fly the helicopter parallel to and at a uniform distance, about one-fourth to one-half mile, from the field boundaries, and not directly above the boundaries. For best results, position flightpath outside the field boundaries just far enough away that they may be easily observed from either pilot seat by looking out the side of the helicopter. If an attempt is made to fly directly above the edges of the field, there will be no usable reference points to start and complete the turns. In addition, the closer the track of the helicopter is to the field boundaries, the steeper the bank necessary at the turning points. The edges of the selected field should be seen while seated in a normal position and looking out the side of the helicopter during either a left-hand or right-hand course. The distance of the ground track from the edges of the field should be the same regardless of whether the course is flown to the left or right. All turns should be started when the helicopter is abeam the corners of the field boundaries. The bank normally should not exceed 30°–45° in light winds. Strong winds may require more bank.

The pilot should understand that when trying to fly a straight line and maintain a specific heading, aircraft heading must be adjusted in order to compensate for the winds and stay on the proper ground track. Also, keep in mind that a constant scan of flight instruments and outside references aid in maintaining proper ground track.

Although the rectangular course may be entered from any direction, this discussion assumes entry on a downwind heading. *[Figure 9-15]* while approaching the field boundary on the downwind leg, begin planning for an upcoming turn. Since there is a tailwind on the downwind leg, the helicopter's groundspeed is increased (position 1). During the turn, the wind causes the helicopter to drift away from the field. To counteract this effect, the roll-in should be made at a fairly fast rate with a relatively steep bank (position 2). This is normally the steepest turn of the maneuver.

As the turn progresses, the tailwind component decreases, which decreases the groundspeed. Consequently, the bank angle and rate-of-turn must be reduced gradually to ensure that upon completion of the turn, the crosswind ground track continues to be the same distance from the edge of the field. Upon completion of the turn, the helicopter should be level and crabbed into the wind in order to maintain the proper ground track. Keep in mind that in order to maintain proper ground track the helicopter may have to be flown almost sideways depending on the amount of wind. The forward cyclic that is applied for airspeed will be in the direction of the intended flight path. For this example, it will be in the direction of the downwind corner of the field. However, since the wind is now pushing the helicopter away from the field, establish the proper drift correction by heading slightly into the wind. Therefore, the turn should be greater than a 90° change in heading (position 3). If the turn has been made properly, the field boundary again appears to be one-fourth to one-half mile away. While on the crosswind leg, the wind correction should be adjusted, as necessary, to maintain a uniform distance from the field boundary (position 4).

As the next field boundary is being approached (position 5), plan for the next turn. Since a wind correction angle is being held into the wind and toward the field, this next turn requires a turn of less than 90°. Since there is now a headwind, the groundspeed decreases during the turn, the bank initially must be medium and progressively decrease as the turn proceeds. To complete the turn, time the rollout so that the helicopter becomes level at a point aligned with the corner of the field just as the longitudinal axis of the helicopter again becomes parallel to the field boundary (position 6). The distance from the field boundary should be the same as on the other sides of the field.

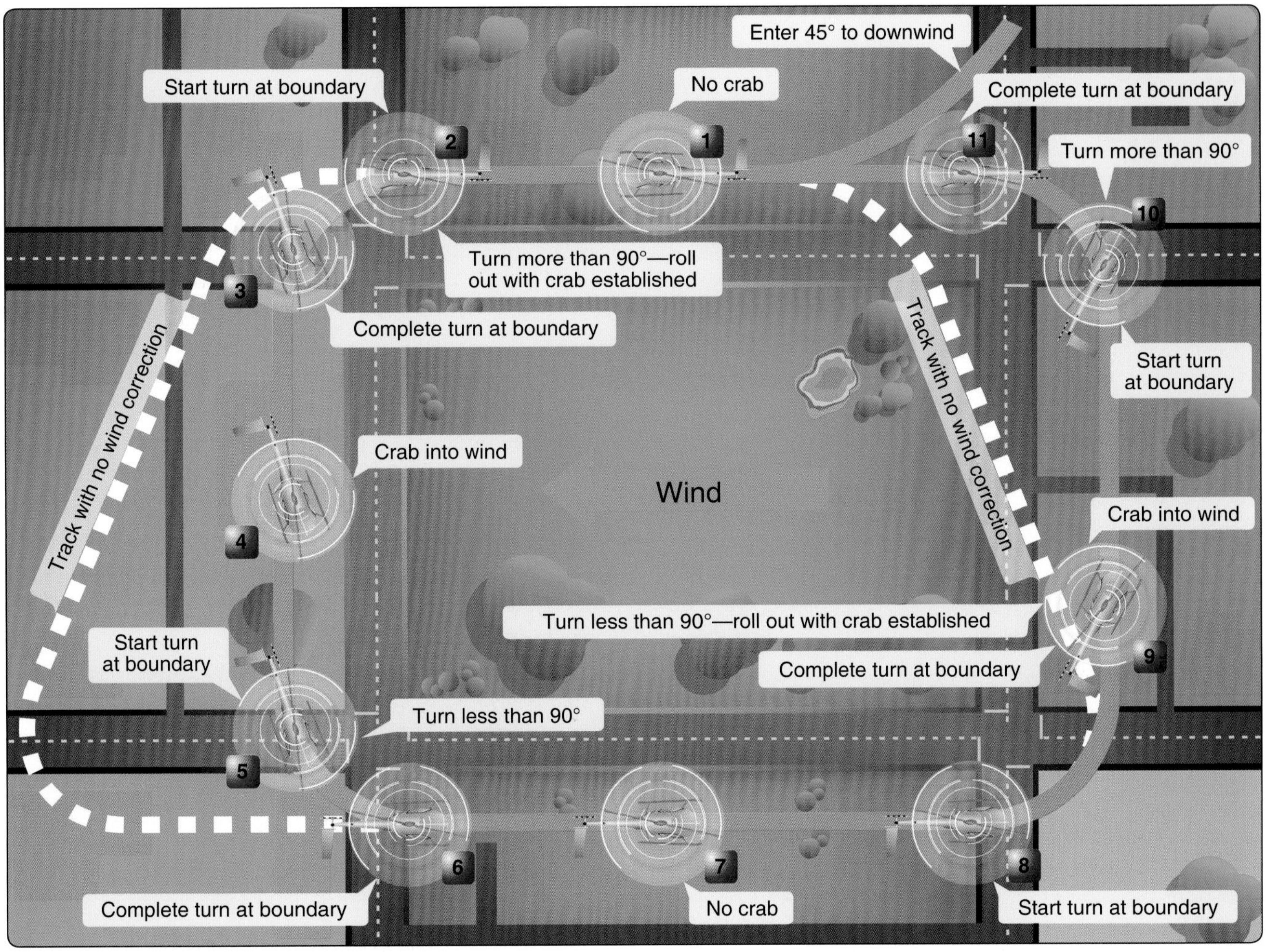

Figure 9-15. *Example of a rectangular course.*

Continue to evaluate each turn and determine the steepness or shallowness based on the winds. It is also important to remember that as the bank angles are adjusted in the turn, the pilot is subsequently forced to make changes with the flight controls.

Common Errors

1. Faulty entry technique.
2. Poor planning, orientation, and/or division of attention.
3. Uncoordinated flight control application.
4. Improper correction for wind drift.
5. Failure to maintain selected altitude and airspeed.
6. Selection of a ground reference with no suitable emergency landing area within gliding distance.
7. Not flying a course parallel to the intended area (e.g., traffic pattern or square field).

S-Turns

Another training maneuver to use is the S-turn, which helps correct for wind drift in turns. This maneuver requires turns to the left and right.

Technique

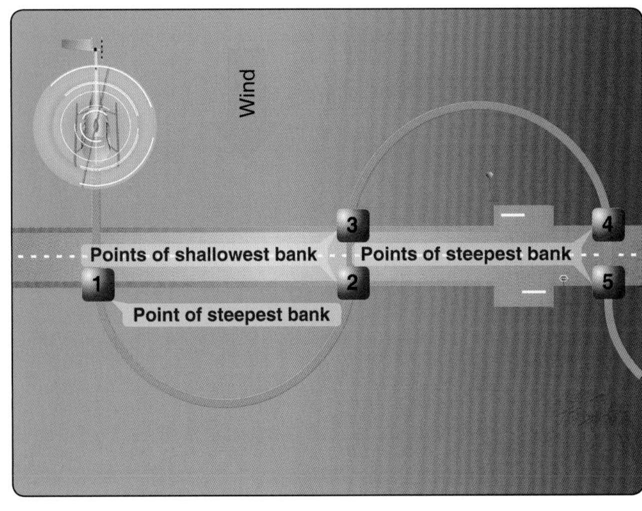

Figure 9-16. *S-turns across a road.*

The pilot can choose to use a road, a fence, or a railroad for a reference line. Regardless of what is used, it should be straight for a considerable distance and should extend as nearly perpendicular to the wind as possible. The object of S-turns is to fly a pattern of two half circles of equal size on opposite sides of the reference line. *[Figure 9-16]* The maneuver should be performed at a constant altitude between 500 and 800 feet above the terrain. As mentioned previously, if the student pilot is having a difficult time maintaining the proper altitude and airspeed, have him or her attempt the S-turn at a lower altitude, providing better ground reference. The discussion that follows is based on choosing a reference line perpendicular to the wind and starting the maneuver with the helicopter facing downwind.

As the helicopter crosses the reference line, immediately establish a bank. This initial bank is the steepest used throughout the maneuver since the helicopter is headed directly downwind and the groundspeed is greatest (position 1). Gradually reduce the bank, as necessary, to describe a ground track of a half circle. Time the turn so that, as the rollout is completed, the helicopter is crossing the reference line perpendicular to it and heading directly upwind (position 2). Immediately enter a bank in the opposite direction to begin the second half of the "S" (position 3). Since the helicopter is now on an upwind heading, this bank (and the one just completed before crossing the reference line) is the shallowest in the maneuver. Gradually increase the bank, as necessary, to describe a ground track that is a half circle identical in size to the one previously completed on the other side of the reference line (position 4). The steepest bank in this turn should be attained just prior to rollout when the helicopter is approaching the reference line nearest the downwind heading. Time the turn so that as the rollout is complete, the helicopter is perpendicular to the reference line and is again heading directly downwind (position 5).

In summary, the angle of bank required at any given point in the maneuver is dependent on the groundspeed. The faster the groundspeed is, the steeper the bank is; the slower the groundspeed is, the shallower the bank is. To express it another way, the more nearly the helicopter is to a downwind heading, the steeper the bank; the more nearly it is to an upwind heading, the shallower the bank. In addition to varying the angle of bank to correct for drift in order to maintain the proper radius of turn, the helicopter must also be flown with a drift correction angle (crab) in relation to its ground track; except, of course, when it is on direct upwind or downwind headings or there is no wind.

One would normally think of the fore and aft axis of the helicopter as being tangent to the ground track pattern at each point. However, this is not the case. During the turn on the upwind side of the reference line (side from which the wind is blowing), crab the nose of the helicopter toward the outside of the circle. During the turn on the downwind side of the reference line (side of the reference line opposite to the direction from which the wind is blowing), crab the nose of the helicopter toward the inside of the circle. In either case, it is obvious that the helicopter is being crabbed into the wind just as it is when trying to maintain a straight ground track. The amount of crab depends on the wind velocity and how close the helicopter is to a crosswind position. The stronger the wind is, the greater the crab angle is at any given position for a turn of a given radius. The more nearly the helicopter is to a crosswind position, the greater the crab angle. The maximum crab angle should be at the point of each half circle farthest from the reference line.

A standard radius for S-turns cannot be specified, since the radius depends on the airspeed of the helicopter, the velocity of the wind, and the initial bank chosen for entry. The only standard is crossing the ground reference line straight and level and having equal radius semi-circles on both sides.

Common Errors

1. Using antitorque pedal pressures to assist turns.
2. Slipping or skidding in the turn.
3. An unsymmetrical ground track during S-turns across a road.
4. Improper correction for wind drift.
5. Failure to maintain selected altitude or airspeed.
6. Excessive bank angles.

Turns Around a Point

This training maneuver requires flying constant radius turns around a preselected point on the ground using a bank angle of approximately 30°–45°, while maintaining both a constant altitude and the same distance from the point throughout the maneuver. *[Figure 9-17]* The objective, as in other ground reference maneuvers, is to develop the ability to subconsciously control the helicopter while dividing attention between flightpath, how the winds are affecting the turn and ground references and watching for other air traffic in the vicinity. This is also used in high reconnaissance, observation, and photography flight.

Technique

The factors and principles of drift correction that are involved in S-turns are also applicable to this maneuver. As in other ground track maneuvers, a constant radius around a point requires the pilot to change the angle of bank constantly and make numerous control changes to compensate for

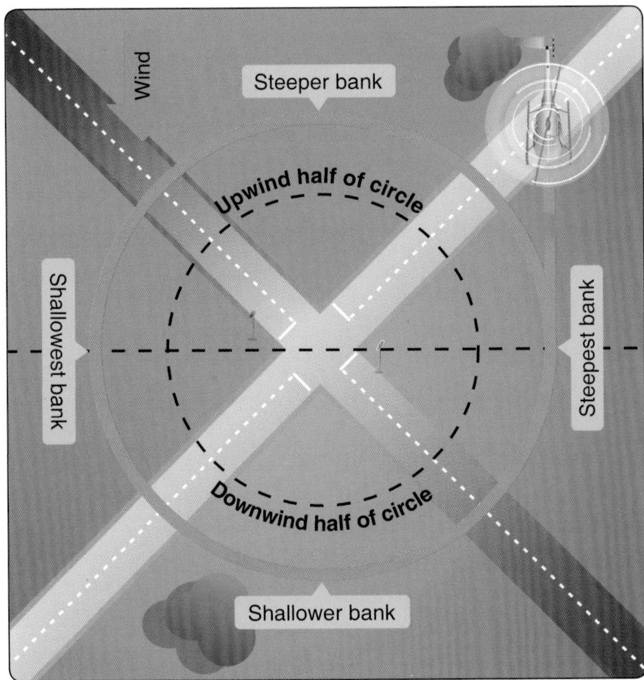

Figure 9-17. *Turns around a point.*

the wind. The closer the helicopter is to a direct downwind heading at which the groundspeed is greatest, the steeper the bank and the greater the rate of turn required to establish the proper wind correction angle. The closer the helicopter is to a direct upwind heading at which the groundspeed is least, the shallower the bank and the lower the rate of turn required to establish the proper wind correction angle. Therefore, throughout the maneuver, the bank and rate of turn must be varied gradually and in proportion to the groundspeed corrections made for the wind.

The point selected for turns should be prominent and easily distinguishable, yet small enough to present a precise reference. Isolated trees, crossroads, or other similar small landmarks are usually suitable. The point should be in an area away from communities, livestock, or groups of people on the ground to prevent possible annoyance or hazard to others. Additionally, the area should be clear and suitable for any emergency landings should they be required.

Just as S-turns require that the helicopter be turned into the wind in addition to varying the bank, so do turns around a point. During the downwind half of the circle, the helicopter's nose must be progressively turned toward the inside of the circle; during the upwind half, the nose must be progressively turned toward the outside. The downwind half of the turn around the point may be compared to the downwind side of the S-turn, while the upwind half of the turn around a point may be compared to the upwind side of the S-turn.

Upon gaining experience in performing turns around a point and developing a good understanding of the effects of wind drift and varying of the bank angle and wind correction angle as required, entry into the maneuver may be from any point. When entering this maneuver at any point, the radius of the turn must be carefully selected, taking into account the wind velocity and groundspeed so that an excessive bank is not required later to maintain the proper ground track.

Common Errors

1. Faulty entry technique.
2. Poor planning, orientation, or division of attention.
3. Uncoordinated flight control application.
4. Improper correction for wind drift.
5. Failure to maintain selected altitude or airspeed.
6. Failure to maintain an equal distance around the point.
7. Excessive bank angles.

Traffic Patterns

A traffic pattern promotes safety by establishing a common track to help pilots determine their landing order and provide common reference. A traffic pattern is also useful to control the flow of traffic, particularly at airports without operating control towers. It affords a measure of safety, separation, protection, and administrative control over arriving, departing, and circling aircraft. Due to specialized operating characteristics, airplanes and helicopters do not mix well in the same traffic environment. At multiple-use airports, regulation states that helicopters should always avoid the flow of fixed-wing traffic. To do this, be familiar with the patterns typically flown by airplanes. In addition, learn how to fly these patterns in case air traffic control (ATC) requests a fixed-wing traffic pattern be flown. Traffic patterns are initially taught during the day. Traffic patterns at night may need to be adjusted; for more information, refer to Chapter 12, Night Operations.

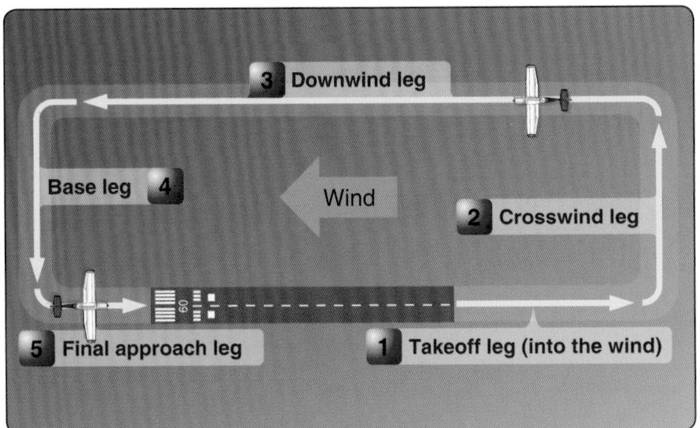

Figure 9-18. *A standard fixed-wing traffic pattern consists of left turns, has five designated legs, and is flown at 1,000' AGL.*

A normal airplane traffic pattern is rectangular, has five named legs, and a designated altitude, usually 1,000 feet AGL. While flying the traffic pattern, pilots should always keep in mind noise abatement rules and flying friendly to avoid dwellings and livestock. A pattern in which all turns are to the left is called a standard pattern. *[Figure 9-18]* The takeoff leg (item 1) normally consists of the aircraft's flightpath after takeoff. This leg is also called the departure leg. Turn to the crosswind leg (item 2) after passing the departure end of the runway when at a safe altitude. Fly the downwind leg (item 3) parallel to the runway at the designated traffic pattern altitude and distance from the runway. Begin the base leg (item 4) at a point selected according to other traffic and wind conditions. If the wind is very strong, begin the turn sooner than normal. If the wind is light, delay the turn to base. The final approach (item 5) is the path the aircraft flies immediately prior to touchdown.

Flying a fixed wing traffic pattern at 1,000 feet AGL upon the request of ATC should not be a problem for a helicopter unless conducting specific maneuvers that require specific altitudes. There are variations at different localities and at airports with operating control towers. For example, ATC may have airplanes in a left turn pattern (as airplane pilots are usually seated in the left), seat and a right turn pattern for helicopters (as those pilots are usually in the right seat). This arrangement affords the best view from each of the respective cockpits. Always consult the Airport/Facility Directory for the traffic pattern procedures at your airport/heliport.

When approaching an airport with an operating control tower in a helicopter, it is possible to expedite traffic by stating intentions. The communication consists of:

1. The helicopter's call sign, "Helicopter 8340J."
2. The helicopter's position, "10 miles west."
3. The "request for landing and hover to ..."

To avoid the flow of fixed-wing traffic, the tower often clears direct to an approach point or to a particular runway intersection nearest the destination point. At uncontrolled airports, if at all possible, adhere to standard practices and patterns.

Traffic pattern entry procedures at an airport with an operating control tower are specified by the controller. At uncontrolled airports, traffic pattern altitudes and entry procedures may vary according to established local procedures. Helicopter pilots should be aware of the standard airplane traffic pattern and avoid it. Generally, helicopters make a lower altitude pattern opposite from the fixed wing pattern and make their approaches to some point other than the runway in use by the fixed wing traffic. Chapter 7 of the Airplane flying Handbook, FAA-H-8083-3 discusses this in greater detail. For information concerning traffic pattern and landing direction, utilize airport advisory service or UNICOM, when available.

The standard departure procedure when using the fixed-wing traffic pattern is usually a straight-out, downwind, or right-hand departure. When a control tower is in operation, request the type of departure desired. In most cases, helicopter departures are made into the wind unless obstacles or traffic dictate otherwise. At airports without an operating control tower, comply with the departure procedures established for that airport, if any.

A helicopter traffic pattern is flown at 500-1,000 feet AGL depending on considerations such as terrain, obstacles, and other aircraft traffic. *[Figure 9-19]* This keeps the helicopter out of the flow of fixed-wing traffic. A helicopter may take off from a helipad into the wind with a turn to the right after 300 feet AGL or as needed to be in range of forced landing areas. When 500 feet AGL is attained, a right turn to parallel the takeoff path is made for the downwind. Then, as the intended landing point is about 45 degrees behind the abeam position of the helicopter, a right turn is made, and a descent is begun from downwind altitude to approximately 300 feet AGL for a base leg.

As the helicopter nears the final approach path, the turn to final should be made considering winds and obstructions. Depending on obstructions and forced landing areas, the final approach may need to be accomplished from as high as 500 feet AGL. The landing area should always be in sight and the angle of approach should never be too high (indicating that the base leg is too close) to the landing area or too low (indicating that the landing area is too far away).

Approaches

An approach is the transition from traffic pattern altitude to either a hover or to the surface. The approach should

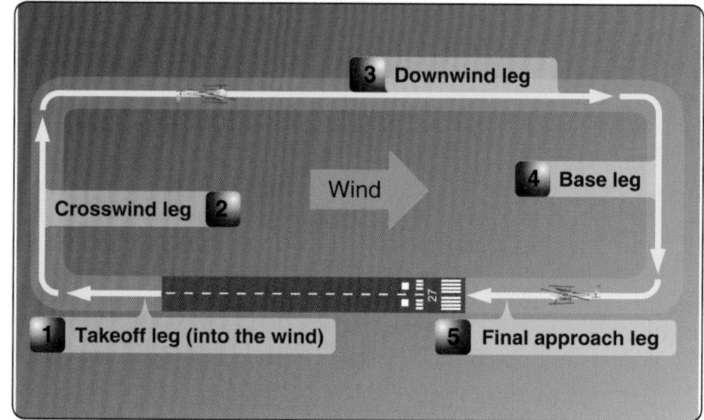

Figure 9-19. *A standard helicopter traffic pattern consists of right turns, has 5 designated legs, and is flown at 500' AGL.*

terminate at the hover altitude with the rate of descent and groundspeed reaching zero at the same time. Approaches are categorized according to the angle of descent as normal, steep, or shallow. In this chapter, concentration is on the normal approach. Steep and shallow approaches are discussed in the next chapter.

Use the type of approach best suited to the existing conditions. These conditions may include obstacles, size and surface of the landing area, density altitude, wind direction and speed, and weight. Regardless of the type of approach, it should always be made to a specific, predetermined landing spot.

Normal Approach to a Hover

A normal approach uses a descent angle of between 7° and 12°.

Technique

On final approach, at the recommended approach airspeed and at approximately 300 feet AGL, the helicopter should be on the correct ground track (or ground alignment) for the intended landing site, but the axis of the helicopter does not have to be aligned until about 50-100 feet AGL to facilitate a controlled approach. *[Figure 9-20]* Just prior to reaching the desired approach angle, begin the approach by lowering the collective sufficiently to get the helicopter decelerating and descending down the approach angle. With the decrease in the collective, the nose tends to pitch down, requiring aft cyclic to maintain the recommended approach airspeed attitude. Adjust antitorque pedals, as necessary, to maintain trim. Pilots should visualize the angle from the landing point to the middle of the skids or landing gear underneath them in the cockpit and maneuver the helicopter down that imaginary slope until the helicopter is at a hover centered over the landing point or touching down centered on the landing point. The most important standard for a normal approach is maintaining a consistent angle of approach to the termination point. The collective controls the angle of approach. Use the cyclic to control the rate of closure or how fast the helicopter is moving towards the touchdown point. Maintain entry airspeed until the apparent groundspeed and rate of closure appear to be increasing. At this point, slowly begin decelerating with slight aft cyclic, and smoothly lower the collective to maintain approach angle. Use the cyclic to maintain a rate of closure equivalent to a brisk walk.

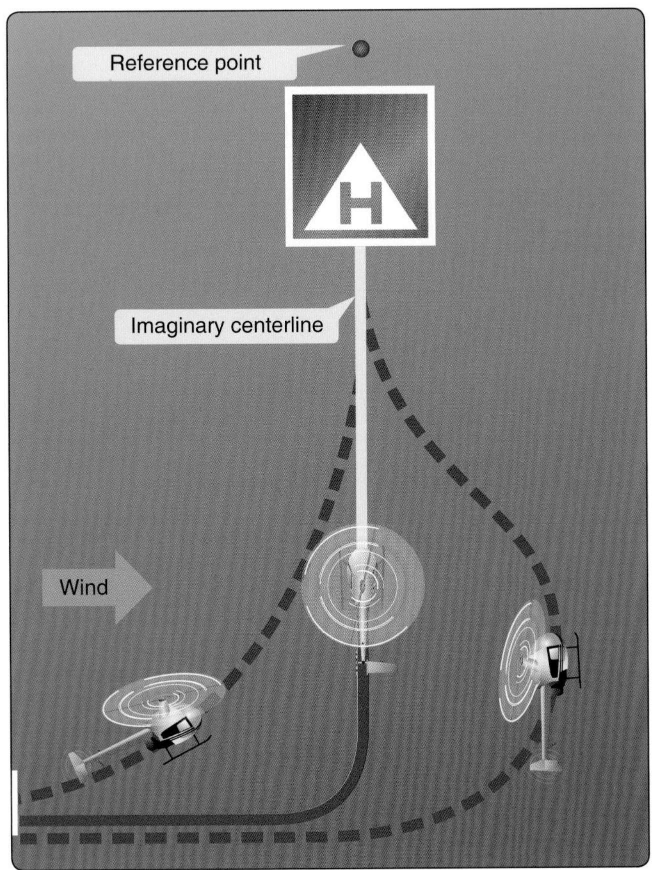

Figure 9-20. *Plan the turn to final so the helicopter rolls out on an imaginary extension of the centerline for the final approach path. This path should neither angle to the landing area, as shown by the helicopter on the left, nor require an S-turn, as shown by the helicopter on the right.*

At approximately 25 knots, depending on wind, the helicopter begins to lose effective translational lift. To compensate for loss of effective translational lift, increase the collective to maintain the approach angle, while maintaining the proper rpm. The increase of collective pitch tends to make the nose rise, requiring forward cyclic to maintain the proper rate of closure.

As the helicopter approaches the recommended hover altitude, increase the collective sufficiently to maintain the hover. Helicopters require near maximum power to land because the inertia of the helicopter in a descent must be overcome by lift in the rotor system. At the same time, apply aft cyclic to stop any forward movement while controlling the heading with antitorque pedals.

Common Errors

1. Failing to maintain proper rpm during the entire approach.
2. Improper use of the collective in controlling the angle of descent.
3. Failing to make antitorque pedal corrections to compensate for collective changes during the approach.
4. Maintaining a constant airspeed on final approach

instead of an apparent brisk walk.

5. Failing to simultaneously arrive at hovering height and attitude with zero groundspeed.
6. Low rpm in transition to the hover at the end of the approach.
7. Using too much aft cyclic close to the surface, which may result in tail rotor strikes.
8. Failure to crab above 100'AGL and slip below 100'AGL.

Normal Approach to the Surface

A normal approach to the surface or a no-hover landing is often used if loose snow or dusty surface conditions exist. These situations could cause severely restricted visibility, or the engine could possibly ingest debris when the helicopter comes to a hover. The approach is the same as the normal approach to a hover; however, instead of terminating at a hover, continue the approach to touchdown. Touchdown should occur with the skids level, zero groundspeed, and a rate of descent approaching zero.

Technique

As the helicopter nears the surface, increase the collective, as necessary, to cushion the landing on the surface, terminate in a skids-level attitude with no forward movement.

Common Errors

1. Terminating to a hover, and then making a vertical landing.
2. Touching down with forward movement.
3. Approaching too slow, requiring the use of excessive power during the termination.
4. Approaching too fast, causing a hard landing
5. Not maintaining skids aligned with direction of travel at touchdown. Any movement or misalignment of the skids or gear can induce dynamic rollover

Crosswind During Approaches

During a crosswind approach, crab into the wind. At approximately 50-100 feet of altitude, use a slip to align the fuselage with the ground track. The rotor is tilted into the wind with cyclic pressure so that the sideward movement of the helicopter and wind drift counteracts each other. Maintain the heading and ground track with the antitorque pedals. Under crosswind approaches, ground track is always controlled by the cyclic movement. The heading of the helicopter in hovering maneuvers is always controlled by the pedals. The collective controls power, which is altitude at a hover. This technique should be used on any type of crosswind approach, whether it is a shallow, normal, or steep approach.

Go-Around

A go-around is a procedure for remaining airborne after an intended landing is discontinued. A go-around may be necessary when:

- Instructed by the control tower.
- Traffic conflict occurs.
- The helicopter is in a position from which it is not safe to continue the approach. Any time an approach is uncomfortable, incorrect, or potentially dangerous, abandon the approach. The decision to make a go-around should be positive and initiated before a critical situation develops. When the decision is made, carry it out without hesitation. In most cases, when initiating the go-around, power is at a low setting. Therefore, the first response is to increase collective to takeoff power. This movement is coordinated with the throttle to maintain rpm, and with the proper antitorque pedal to control heading. Then, establish a climb attitude and maintain climb speed to go around for another approach.

Chapter Summary

This chapter introduced basic flight maneuvers and the techniques to perform each of them. Common errors and why they happen were also described to help the pilot achieve a better understanding of the maneuver.

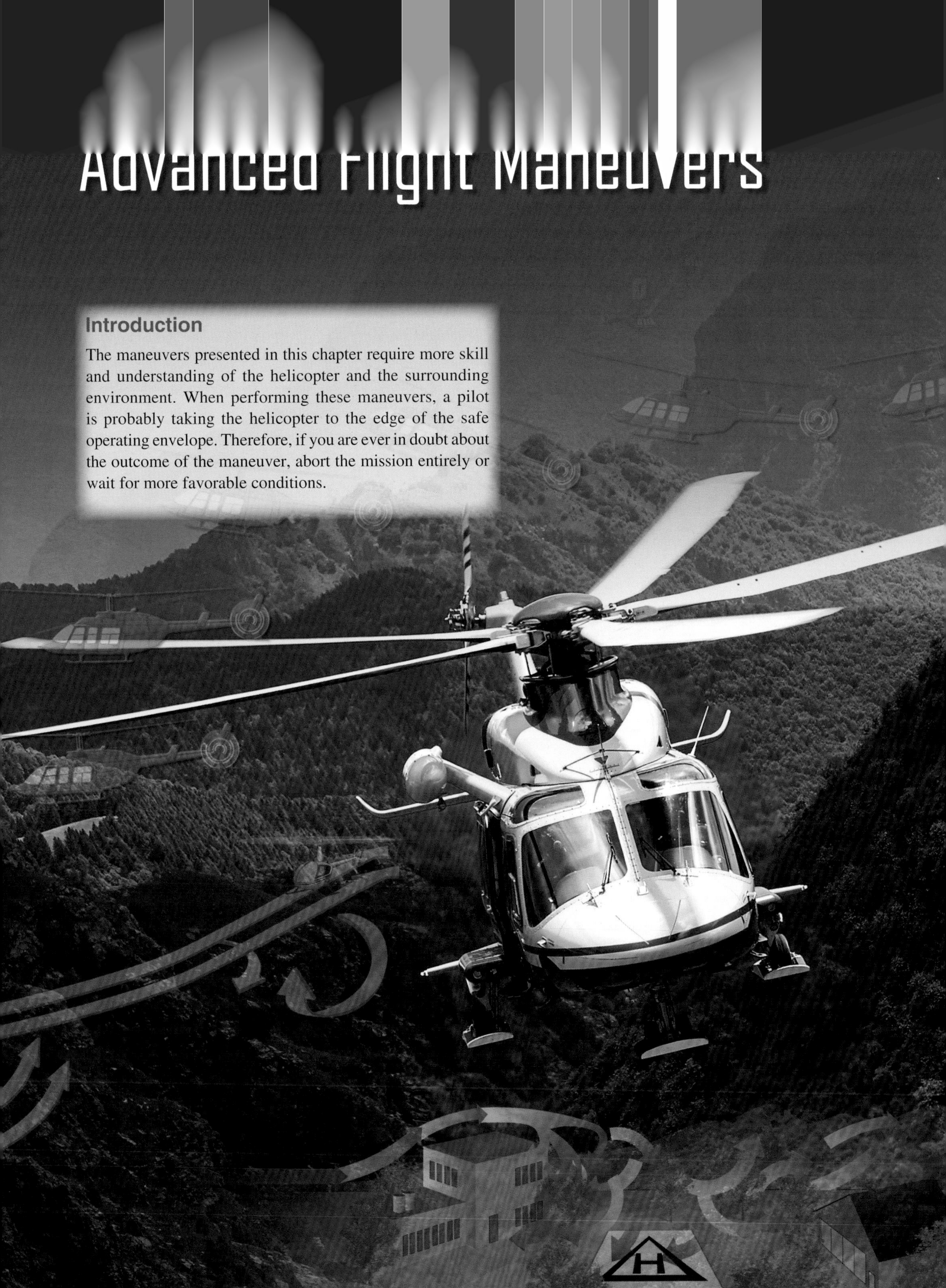

Advanced Flight Maneuvers

Introduction

The maneuvers presented in this chapter require more skill and understanding of the helicopter and the surrounding environment. When performing these maneuvers, a pilot is probably taking the helicopter to the edge of the safe operating envelope. Therefore, if you are ever in doubt about the outcome of the maneuver, abort the mission entirely or wait for more favorable conditions.

Reconnaissance Procedures

When planning to land or takeoff at an unfamiliar site, gather as much information as possible about the area. Reconnaissance techniques are ways of gathering this information.

High Reconnaissance

The purpose of conducting a high reconnaissance is to determine direction and speed of the wind, a touchdown point, suitability of the landing area, approach and departure axes, and obstacles for both the approach and departure. The pilot should also give particular consideration to forced landing areas in case of an emergency.

Altitude, airspeed, and flight pattern for a high reconnaissance are governed by wind and terrain features. It is important to strike a balance between a reconnaissance conducted too high and one too low. It should not be flown so low that a pilot must divide attention between studying the area and avoiding obstructions to flight. A high reconnaissance should be flown at an altitude of 300 to 500 feet above the surface. A general rule to follow is to ensure that sufficient altitude is available at all times to land into the wind in case of engine failure. In addition, a 45° angle of observation generally allows the best estimate of the height of barriers, the presence of obstacles, the size of the area, and the slope of the terrain. Always maintain safe altitudes and airspeeds and keep a forced landing area within reach whenever possible.

Low Reconnaissance

A low reconnaissance is accomplished during the approach to the landing area. When flying the approach, verify what was observed in the high reconnaissance, and check for anything new that may have been missed at a higher altitude, such as wires and their supporting structures (poles, towers, etc.), slopes, and small crevices. If the pilot determines that the area chosen is safe to land in, the approach can be continued. However, the decision to land or go around must be made prior to decelerating below effective translational lift (ETL), or before descending below the barriers surrounding the confined area.

If a decision is made to complete the approach, terminate the landing to a hover in order to check the landing point carefully before lowering the helicopter to the surface. Under certain conditions, it may be desirable to continue the approach to the surface. Once the helicopter is on the ground, maintain operating revolutions per minute (rpm) until the stability of the helicopter has been checked to be sure it is in a secure and safe position.

Ground Reconnaissance

Prior to departing an unfamiliar location, make a detailed analysis of the area. There are several factors to consider during this evaluation. Besides determining the best departure path and identifying all hazards in the area, select a route that gets the helicopter from its present position to the takeoff point while avoiding all hazards, especially to the tail rotor and landing gear.

Some things to consider while formulating a takeoff plan are the aircraft load, height of obstacles, the shape of the area, direction of the wind, and surface conditions. Surface conditions can consist of dust, sand and snow, as well as mud and rocks. Dust landings and snow landings can lead to a brownout or whiteout condition, which is the loss of the horizon reference. Disorientation may occur, leading to ground contact, often with fatal results. Taking off or landing on uneven terrain, mud, or rocks can cause the tail rotor to strike the surface or if the skids get caught can lead to dynamic rollover. If the helicopter is heavily loaded, determine if there is sufficient power to clear the obstacles. Sometimes it is better to pick a path over shorter obstacles than to take off directly into the wind. Also evaluate the shape of the area so that a path can be chosen that will provide you the most room to maneuver and abort the takeoff if necessary. Positioning the helicopter at the most downwind portion of the confined area gives the pilot the most distance to clear obstacles.

Wind analysis also helps determine the route of takeoff. The prevailing wind can be altered by obstructions on the departure path and can significantly affect aircraft performance. There are several ways to check the wind direction before taking off. One technique is to watch the tops of the trees; another is to look for any smoke in the area. If there is a body of water in the area, look to see which way the water is rippling. If wind direction is still in question revert to the last report that was received by either the Automatic Terminal Information Service (ATIS) or airport tower.

Maximum Performance Takeoff

A maximum performance takeoff is used to climb at a steep angle to clear barriers in the flightpath. It can be used when taking off from small areas surrounded by high obstacles. Allow for a vertical takeoff, although not preferred, if obstruction clearance could be in doubt. Before attempting a maximum performance takeoff, know thoroughly the capabilities and limitations of the equipment. Also consider the wind velocity, temperature, density altitude, gross weight, center of gravity (CG) location, and other factors affecting pilot technique and the performance of the helicopter.

To accomplish this type of takeoff safely, there must be enough power to hover out of ground effect (OGE) in order to prevent the helicopter from sinking back to the surface after becoming airborne. A hover power check can be used to determine if there is sufficient power available to accomplish this maneuver.

The angle of climb for a maximum performance takeoff depends on existing conditions. The more critical the conditions are, such as high-density altitudes, calm winds, and high gross weights, the shallower the angle of climb is. In light or no wind conditions, it might be necessary to operate in the crosshatched or shaded areas of the height/velocity diagram during the beginning of this maneuver. Therefore, be aware of the calculated risk when operating in these areas. An engine failure at a low altitude and airspeed could place the helicopter in a dangerous position, requiring a high degree of skill in making a safe autorotative landing.

Technique

Before attempting a maximum performance takeoff, reposition the helicopter to the most downwind area to allow a longer takeoff climb, then bring the helicopter to a hover, and determine the excess power available by noting the difference between the power available and that required to hover. Also, perform a balance and flight control check and note the position of the cyclic. If the takeoff path allows, position the helicopter into the wind and return the helicopter to the surface. Normally, this maneuver is initiated from the surface. After checking the area for obstacles and other aircraft, select reference points along the takeoff path to maintain ground track. Also consider alternate routes in case the maneuver is not possible. *[Figure 10-1]*

Begin the takeoff by getting the helicopter light on the skids (position 1). Pause and neutralize all aircraft movement. Slowly increase the collective and position the cyclic to lift off in a 40-knot attitude. This is approximately the same attitude as when the helicopter is light on the skids. Continue to increase the collective slowly until the maximum power available is reached (takeoff power is normally 10 percent above power required for hover). This large collective movement requires a substantial increase in pedal pressure to maintain heading (position 2). Use the cyclic, as necessary, to control movement toward the desired flightpath and, therefore, climb angle during the maneuver (position 3). Maintain rotor rpm at its maximum, and do not allow it to decrease since you would probably need to lower the collective to regain it. Maintain these inputs until the helicopter clears the obstacle, or until reaching 50 feet for demonstration purposes (position 4). Then, establish a normal climb attitude and power setting (position 5). As in any maximum performance maneuver, the techniques used affect the actual results. Smooth, coordinated inputs coupled with precise control allow the helicopter to attain its maximum performance.

Figure 10-1. *Maximum performance takeoff.*

An acceptable method when departing from an area that does not allow for a takeoff with forward airspeed is to perform a vertical takeoff. This technique allows the pilot to descend vertically back into the confined area if the helicopter does not have the performance to clear the surrounding obstacles. During this maneuver, the helicopter must climb vertically and not be allowed to accelerate forward until the surrounding obstacles have been cleared. If not, a situation may develop where the helicopter does not have sufficient climb performance to avoid obstructions and may not have power to descend back to the takeoff point. The vertical takeoff might not be as efficient as the climbing profile but is much easier to abort from a vertical position directly over the landing point. The vertical takeoff, however, places the helicopter in the *avoid* area of the height/velocity diagram for a longer time. This maneuver requires hover OGE power to accomplish.

Common Errors

1. Failure to consider performance data, including height-velocity diagram.
2. Nose too low initially causing horizontal flight rather than more vertical flight.
3. Failure to maintain maximum permissible rpm.
4. Abrupt control movements.
5. Failure to resume normal climb power and airspeed after clearing the obstacle.

Running/Rolling Takeoff

A running takeoff in helicopter with fixed landing gear, such as skids, skis or floats, or a rolling takeoff in a wheeled helicopter is sometimes used when conditions of load and/or density altitude prevent a sustained hover at normal hovering height. For wheeled helicopters, a rolling takeoff is sometimes used to minimize the downwash created during a takeoff from a hover. Avoid a running/rolling maneuver if there is not sufficient power to hover, at least momentarily. If the helicopter cannot be hovered, its performance is unpredictable. If the helicopter cannot be raised off the surface at all, sufficient power might not be available to accomplish the maneuver safely. If a pilot cannot momentarily hover the helicopter, wait for conditions to improve or off-load some of the weight.

To accomplish a safe running or rolling takeoff, the surface area must be of sufficient length and smoothness, and there cannot be any barriers in the flightpath to interfere with a shallow climb.

Technique

Refer to *Figure 10-2*. To begin the maneuver, first align the helicopter to the takeoff path. Next, increase the throttle to obtain takeoff rpm, and increase the collective smoothly until the helicopter becomes light on the skids or landing gear (position 1). If taking off from the water, ensure that the floats are mostly out of the water. Then, move the cyclic slightly forward of the neutral hovering position, and apply additional collective to start the forward movement (position 2). To simulate a reduced power condition during practice, use one to two inches less manifold pressure, or three to five percent less torque than that required to hover. The landing gear must stay aligned with the takeoff direction until the helicopter leaves the surface to avoid dynamic rollover.

Maintain a straight ground track with lateral cyclic and heading with antitorque pedals until a climb is established. As effective translational lift is gained, the helicopter becomes airborne in a fairly level attitude with little or no pitching (position 3). Maintain an altitude to take advantage of ground effect, and allow the airspeed to increase toward normal climb speed. Then, follow a climb profile that takes the helicopter through the clear area of the height-velocity diagram (position 4). During practice maneuvers, after having climbed to an altitude of 50 feet, establish the normal climb power setting and attitude.

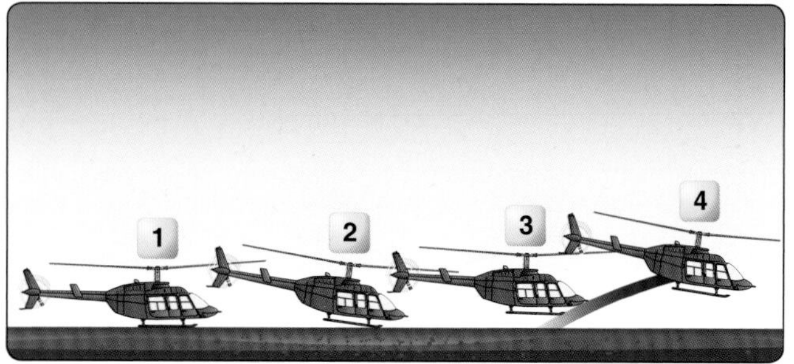

Figure 10-2. *Running/rolling takeoff.*

NOTE: It should be remembered that if a running takeoff is necessary for most modern helicopters, the helicopter is very close to, or has exceeded the maximum operating weight for the conditions (i.e., temperature and altitude).

The height/velocity parameters should be respected at all times. The helicopter should be flown to a suitable altitude to allow a safe acceleration in accordance with the height-velocity diagram.

Common Errors

1. Failing to align heading and ground track to keep surface friction to a minimum.
2. Attempting to become airborne before obtaining effective translational lift.
3. Using too much forward cyclic during the surface run.
4. Lowering the nose too much after becoming airborne, resulting in the helicopter settling back to the surface.
5. Failing to remain below the recommended altitude until airspeed approaches normal climb speed.

Rapid Deceleration or Quick Stop

This maneuver is used to decelerate from forward flight to a hover. It is often used to abort takeoffs, to stop if something blocks the helicopter flightpath, or simply to terminate an air taxi maneuver, as mentioned in the Aeronautical Information Manual (AIM). A quick stop is usually practiced on a runway, taxiway, or over a large grassy area away from other traffic or obstacles.

Technique

The maneuver requires a high degree of coordination of all controls. It is practiced at a height that permits a safe clearance between the tail rotor and the surface throughout the maneuver, especially at the point where the pitch attitude is highest. The height at completion should be no higher than the maximum safe hovering height prescribed by that particular helicopter's manufacturer. In selecting a height at which to begin the maneuver, take into account the overall length of the helicopter and its height/velocity diagram. Even though the maneuver is called a rapid deceleration or quick stop, it is performed slowly and smoothly with the primary emphasis on coordination.

During training, always perform this maneuver into the wind *[Figure 10-3, position 1]*. After leveling off at an altitude

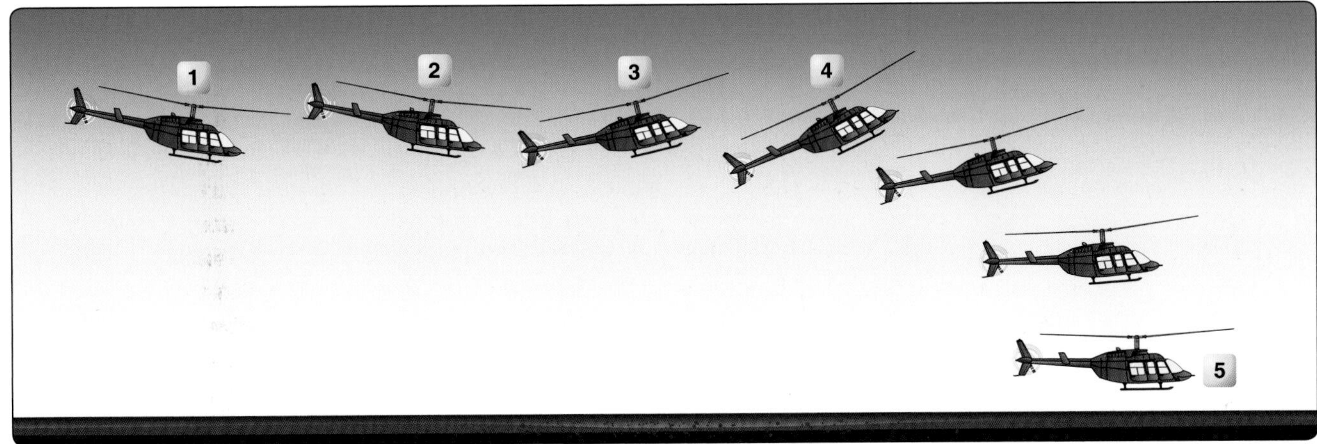

Figure 10-3. *Rapid deceleration or quick stop.*

between 25 and 40 feet, depending upon the manufacturer's recommendations, accelerate to the desired entry speed, which is approximately 45 knots for most training helicopters (position 2). The altitude chosen should be high enough to avoid danger to the tail rotor during the flare, but low enough to stay out of the hazardous areas of that helicopter's height-velocity diagram throughout the maneuver. In addition, this altitude should be low enough that the helicopter can be brought to a hover during the recovery.

At position 3, initiate the deceleration by applying aft cyclic to reduce forward groundspeed. Simultaneously, lower the collective, as necessary, to counteract any climbing tendency. The timing must be exact. If too little collective is taken out for the amount of aft cyclic applied, the helicopter climbs. If too much downward collective is applied, the helicopter will descend. A rapid application of aft cyclic requires an equally rapid application of down collective. As collective is lowered, apply proper antitorque pedal pressure to maintain heading, and adjust the throttle to maintain rpm. The G loading on the rotor system depends on the pitch-up attitude. If the attitude is too high, the rotor system may stall and cause the helicopter to impact the surface.

After attaining the desired speed (position 4), initiate the recovery by lowering the nose and allowing the helicopter to descend to a normal hovering height in level flight and zero groundspeed (position 5). During the recovery, increase collective pitch, as necessary, to stop the helicopter at normal hovering height, adjust the throttle to maintain rpm, and apply proper antitorque pedal pressure, as necessary, to maintain heading. During the maneuver, visualize rotating about the tail rotor's horizontal axis until a normal hovering height is reached.

Common Errors

1. Initiating the maneuver by lowering the collective without aft cyclic pressure to maintain altitude.
2. Initially applying aft cyclic stick too rapidly, causing the helicopter to balloon (climb).
3. Failing to effectively control the rate of deceleration to accomplish the desired results.
4. Allowing the helicopter to stop forward motion in a tail-low attitude.
5. Failing to maintain proper rotor rpm.
6. Waiting too long to apply collective pitch (power) during the recovery, resulting in an overtorque situation when collective pitch is applied rapidly.
7. Failing to maintain a safe clearance over the terrain.
8. Using antitorque pedals improperly, resulting in erratic heading changes.
9. Using an excessively nose-high attitude.

Steep Approach

A steep approach is used primarily when there are obstacles in the approach path that are too high to allow a normal approach. A steep approach permits entry into most confined areas and is sometimes used to avoid areas of turbulence around a pinnacle. An approach angle of approximately 13° to 15° is considered a steep approach. *[Figure 10-4]* Caution must be exercised to avoid the parameters for vortex ring state (20–100 percent of available power applied, airspeed of less than 10 knots, and a rate of descent greater than 300 feet per minute (fpm)). For additional information on vortex ring state (formerly referenced as settling-with-power), refer to Chapter 11, Helicopter Emergencies and Hazards.

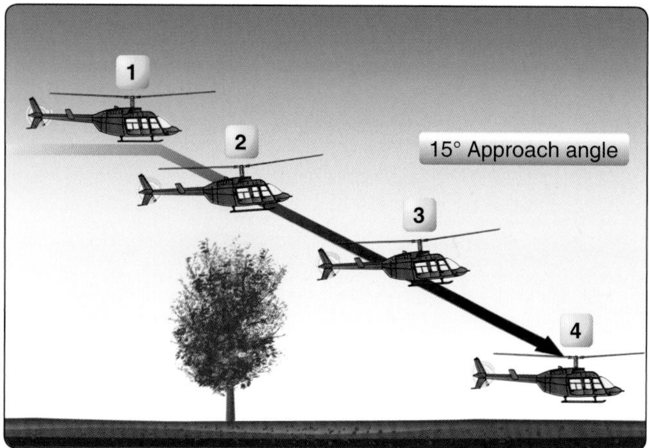

Figure 10-4. *Steep approach to a hover.*

Technique

On final approach, maintain track with the intended touchdown point and into the wind as much as possible at the recommended approach airspeed *[Figure 10-4, position 1]*. When intercepting an approach angle of 13° to 15°, begin the approach by lowering the collective sufficiently to start the helicopter descending down the approach path and decelerating (position 2). Use the proper antitorque pedal for trim. Since this angle is steeper than a normal approach angle, reduce the collective more than that required for a normal approach. Continue to decelerate with slight aft cyclic and smoothly lower the collective to maintain the approach angle.

The intended touchdown point may not always be visible throughout the approach, especially when landing to a hover. Pilots must learn to cue in to other references that are parallel to the intended landing area that will help them maintain ground track and position.

Constant management of approach angle and airspeed is essential to any approach. Aft cyclic is required to decelerate sooner than with a normal approach, and the rate of closure becomes apparent at a higher altitude. Maintain the approach angle and rate of descent with the collective, rate of closure with the cyclic, and trim with antitorque pedals.

The helicopter should be kept in trim just prior to loss of effective translational lift (approximately 25 knots). Below 100 feet above ground level (AGL), the antitorque pedals should be adjusted to align the helicopter with the intended touchdown point. Visualize the location of the tail rotor behind the helicopter and fly the landing gear to 3 feet above the intended landing point. In small confined areas, the pilot must precisely position the helicopter over the intended landing area. Therefore, the approach must stop at that point. Loss of effective translational lift occurs higher in a steep approach (position 3), requiring an increase in the collective to prevent settling, and more forward cyclic to achieve the proper rate of closure. Once the intended landing area is reached, terminate the approach to a hover with zero groundspeed (position 4). If the approach has been executed properly, the helicopter will come to a halt at a hover altitude of 3 feet over the intended landing point with very little additional power required to hold the hover.

The pilot must remain aware that any wind effect is lost once the aircraft has descended below the barriers surrounding a confined area, causing the aircraft to settle more quickly. Additional power may be needed on a strong wind condition as the helicopter descends below the barriers.

Common Errors

1. Failing to maintain proper rpm during the entire approach.
2. Using collective improperly in maintaining the selected angle of descent.
3. Failing to make antitorque pedal corrections to compensate for collective pitch changes during the approach.
4. Slowing airspeed excessively in order to remain on the proper angle of descent.
5. Failing to determine when effective translational lift is being lost.
6. Failing to arrive at hovering height and attitude, and zero groundspeed almost simultaneously.
7. Utilizing low rpm in transition to the hover at the end of the approach.
8. Using too much aft cyclic close to the surface, which may result in the tail rotor striking the surface.
9. Failure to align landing gear with direction of travel no later than beginning of loss of translational lift.

Shallow Approach and Running/Roll-On Landing

Use a shallow approach and running landing when a high-density altitude, a high gross weight condition, or some combination thereof, is such that a normal or steep approach cannot be made because of insufficient power to hover. *[Figure 10-5]* To compensate for this lack of power, a shallow approach and running landing makes use of translational lift until surface contact is made. If flying a wheeled helicopter, a roll-on landing can be used

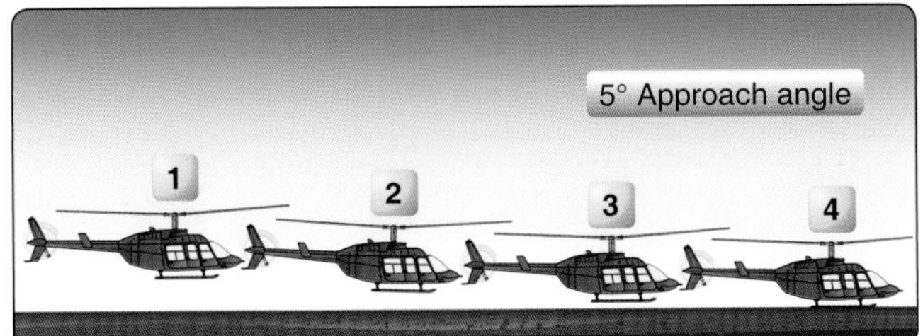

Figure 10-5. *Shallow approach and running landing.*

to minimize the effect of downwash. The glide angle for a shallow approach is approximately 3° to 5°. This angle is similar to the angle used on an instrument landing system (ILS) approach. Since the helicopter is sliding or rolling to a stop during this maneuver, the landing area should be smooth, and the landing gear must be aligned with the direction of travel to prevent dynamic rollover and must be long enough to accomplish this task. After landing, ensure that the pitch of the rotor blades is not too far aft as the main rotor blades could contact the tailboom.

Technique

A shallow approach is initiated in the same manner as the normal approach except that a shallower angle of descent is maintained. The power reduction to initiate the desired angle of descent is less than that for a normal approach since the angle of descent is less (position 1).

As the collective is lowered, maintain heading with proper antitorque pedal pressure and rpm with the throttle. Maintain approach airspeed until the apparent rate of closure appears to be increasing. Then, begin to slow the helicopter with aft cyclic (position 2).

As in normal and steep approaches, the primary control for the angle and rate of descent is the collective, while the cyclic primarily controls the groundspeed. However, there must be a coordination of all the controls for the maneuver to be accomplished successfully. The helicopter should arrive at the point of touchdown at or slightly above effective translational lift. Since translational lift diminishes rapidly at slow airspeeds, the deceleration must be coordinated smoothly, at the same time keeping enough lift to prevent the helicopter from settling abruptly.

Just prior to touchdown, place the helicopter in a level attitude with the cyclic, and maintain heading with the antitorque pedals. Use the cyclic to keep the direction of travel and ground track identical (position 3). Allow the helicopter to descend gently to the surface in a straight-and-level attitude, cushioning the landing with the collective. After surface contact, move the cyclic slightly forward to ensure clearance between the tail boom and the rotor disk. Use the cyclic to maintain the surface track (position 4). A pilot normally holds the collective stationary until the helicopter stops; however, to get more braking action, lower the collective slightly.

Keep in mind that, due to the increased ground friction when the collective is lowered or if the landing is being executed to a rough or irregular surface, the helicopter may come to an abrupt stop and the nose might pitch forward. Exercise caution not to correct this pitching movement with aft cyclic, which could result in the rotor making contact with the tail boom. An abrupt stop may also cause excessive transmission movement resulting in the transmission contacting its mount. During the landing, maintain normal rpm with the throttle and directional control with the antitorque pedals.

For wheeled helicopters, use the same technique except after landing, lower the collective, neutralize the controls, and apply the brakes, as necessary, to slow the helicopter. Do not use aft cyclic when bringing the helicopter to a stop.

Common Errors

1. Assuming excessive nose-high attitude to slow the helicopter near the surface.
2. Utilizing insufficient collective and throttle to cushion a landing.
3. Failure to maintain heading resulting in a turning or pivoting motion.
4. Failure to add proper antitorque pedal as collective is added to cushion landing, resulting in a touchdown while the helicopter is moving sideward.
5. Failure to maintain a speed that takes advantage of effective translational lift.
6. Touching down at an excessive groundspeed for the existing conditions. (Some helicopters have maximum touchdown groundspeeds.)
7. Failure to touch down in the appropriate attitude necessary for a safe landing. Appropriate attitude is based on the type of helicopter and the landing gear installed.
8. Failure to maintain proper rpm during and after touchdown.
9. Maintaining poor alignment with direction of travel during touchdown.

Slope Operations

Prior to conducting any slope operations, be thoroughly familiar with the characteristics of dynamic rollover and mast bumping, which are discussed in Chapter 11, Helicopter

Emergencies and Hazards. The approach to a slope is similar to the approach to any other landing area. During slope operations, make allowances for wind, barriers, and forced landing sites in case of engine failure. Since the slope may constitute an obstruction to wind passage, anticipate turbulence and downdrafts.

Slope Landing

A pilot usually lands a helicopter across the slope rather than with the slope. Landing with the helicopter facing down the slope or downhill is not recommended because of the possibility of striking the tail rotor on the surface.

Technique

Refer to *Figure 10-6.* At the termination of the approach, if necessary, move the helicopter slowly toward the slope, being careful not to turn the tail upslope. Position the helicopter across the slope at a stabilized hover headed into the wind over the intended landing spot (frame 1). Downward pressure on the collective starts the helicopter descending. As the upslope skid touches the ground, hesitate momentarily in a level attitude, then apply slight lateral cyclic in the direction of the slope (frame 2). This holds the skid against the slope while the pilot continues lowering the downslope skid with the collective. As the collective is lowered, continue to move the cyclic toward the slope to maintain a fixed position (frame 3) The slope must be shallow enough to hold the helicopter against it with the cyclic during the entire landing. A slope of 5° is recommended maximum for training in most helicopters. However, additional training to the manufacturer's limitations may be required. Consult the Rotorcraft Flight Manual (RFM) or Pilot's Operating Handbook (POH) for the specific limitations of the helicopter being flown.

Be aware of any abnormal vibration or mast bumping that signals maximum cyclic deflection. If helicopter mast moment or slope limits are reached before the helicopter is firmly on the ground, return the helicopter to a hover. Select a new area with a lesser degree of slope. In most helicopters with a counterclockwise rotor system, landings can be made on steeper slopes when holding the cyclic to the right. When landing on slopes using left cyclic, some cyclic input must be used to overcome the translating tendency. If wind is not a factor, consider the drifting tendency when determining landing direction.

After the downslope skid is on the surface, reduce the collective to full down, and neutralize the cyclic and pedals (frame 4). Normal operating rpm should be maintained until the full weight of the helicopter is on the landing gear.

This ensures adequate rpm for immediate takeoff in case the helicopter starts sliding down the slope. Use antitorque pedals as necessary throughout the landing for heading control. Before reducing the rpm, move the cyclic control as necessary to check that the helicopter is firmly on the ground.

Common Errors

1. Failing to consider wind effects during the approach and landing.
2. Failing to maintain proper rpm throughout the entire maneuver.
3. Failure to maintain heading resulting in a turning or pivoting motion.
4. Turning the tail of the helicopter into the slope.
5. Lowering the downslope skid or wheel too rapidly.
6. Applying excessive cyclic control into the slope, causing mast bumping.

Slope Takeoff

A slope takeoff is basically the reverse of a slope landing. *[Figure 10-7]* Conditions that may be associated with the slope, such as turbulence and obstacles, must be considered during the takeoff. Planning should include suitable forced landing areas.

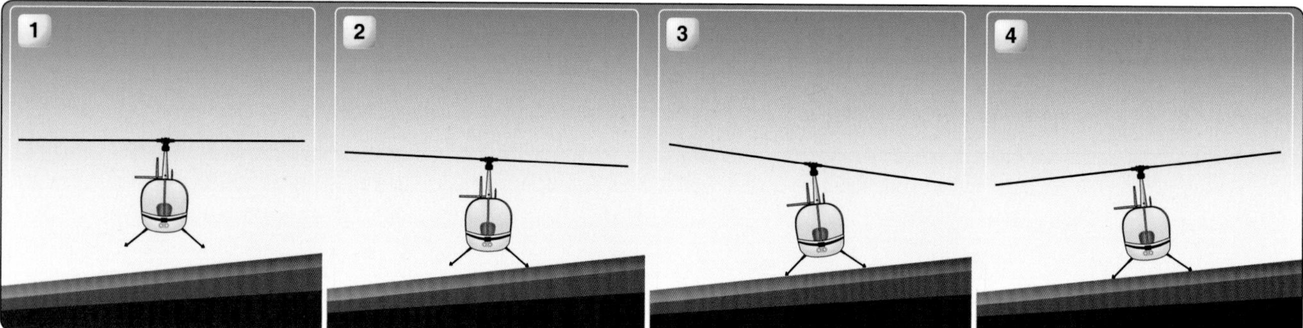

Figure 10-6. *Slope landing.*

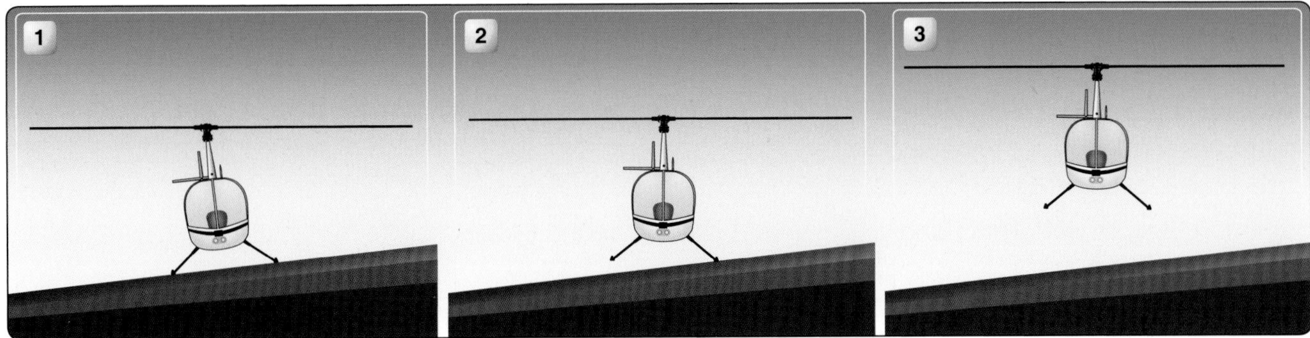

Figure 10-7. *Slope takeoff.*

Technique

Begin the takeoff by increasing rpm to the normal range with the collective full down. Then, move the cyclic toward the slope (frame 1). Holding the cyclic toward the direction of the slope causes the downslope skid to rise as the pilot slowly raises the collective (frame 2). As the skid comes up, move the cyclic as necessary to maintain a level attitude in relation to the horizon. If properly coordinated, the helicopter should attain a level attitude as the cyclic reaches the neutral position. At the same time, use antitorque pedal pressure to maintain heading and throttle to maintain rpm. With the helicopter level and the cyclic centered, pause momentarily to verify everything is correct, and then gradually raise the collective to complete the liftoff (frame 3). After reaching a hover, avoid hitting the ground with the tail rotor by not turning the helicopter tail upslope and gaining enough altitude to ensure the tail rotor is clear. If an upslope wind exists, execute a crosswind takeoff and then make a turn into the wind after clearing the ground with the tail rotor.

Common Errors

1. Failing to adjust cyclic control to keep the helicopter from sliding down slope.
2. Failing to maintain proper rpm.
3. Holding excessive cyclic into the slope as the down slope skid is raised.
4. Failure to maintain heading, resulting in a turning or pivoting motion.
5. Turning the tail of the helicopter into the slope during takeoff.

Confined Area Operations

A confined area is an area where the flight of the helicopter is limited in some direction by terrain or the presence of obstructions, natural or manmade. For example, a clearing in the woods, a city street, a road, a building roof, etc., can each be regarded as a confined area. The helicopter pilot has added responsibilities when conducting operations from a confined area that airplanes pilots do not. He or she assumes the additional roles of the surveyor, engineer, and manager when selecting an area to conduct operations. While airplane pilots generally operate from known pre-surveyed and improved landing areas, helicopter pilots fly into areas never used before for helicopter operations. Generally, takeoffs and landings should be made into the wind to obtain maximum airspeed with minimum groundspeed. The pilot should begin with as nearly accurate an altimeter setting as possible to determine the altitude.

There are several things to consider when operating in confined areas. One of the most important is maintaining a clearance between the rotors and obstacles forming the confined area. The tail rotor deserves special consideration because, in some helicopters, it is not always visible from the cabin. This not only applies while making the approach, but also while hovering. Another consideration is that wires are especially difficult to see; however, their supporting devices, such as poles or towers, serve as an indication of their presence and approximate height. If any wind is present, expect some turbulence. *[Figure 10-8]*

Something else to consider is the availability of forced landing areas during the planned approach. Think about the possibility of flying from one alternate landing area to another throughout the approach, while avoiding unfavorable areas. Always leave a way out in case the landing cannot be completed, or a go-around is necessary.

During the high reconnaissance, the pilot needs to formulate a takeoff plan as well. The heights of obstacles need to be determined. It is not good practice to land in an area and then determine that insufficient power exists to depart. Generally, more power is required to take off than to land so the takeoff criteria is most crucial. Fixing the departure azimuth or heading on the compass is a good technique to use. This ensures that the pilot is able to take off over the preselected departure path when it is not visible while sitting in the confined area.

Figure 10-8. *If the wind velocity is 10 knots or greater, expect updrafts on the windward side and downdrafts on the lee side of obstacles. Plan the approach with these factors in mind, but be ready to alter plans if the wind speed or direction changes.*

Approach

A high reconnaissance should be completed before initiating the confined area approach. Start the approach phase using the wind and speed to the best possible advantage. Keep in mind areas suitable for forced landing. It may be necessary to choose a crosswind approach that is over an open area, then one directly into the wind that is over trees. If these conditions exist, consider the possibility of making the initial phase of the approach crosswind over the open area and then turning into the wind for the final portion of the approach.

Always operate the helicopter as close to its normal capabilities as possible, taking into consideration the situation at hand. In all confined area operations, with the exception of a pinnacle operation (see next section, Takeoff), the angle of descent should be no steeper than necessary to clear any barrier with the tail rotor in the approach path and still land on the selected spot. The angle of climb on takeoff should be normal, or not steeper than necessary to clear any barrier. Clearing a barrier by a few feet and maintaining normal operating rpm, with perhaps a reserve of power, is better than clearing a barrier by a wide margin but with a dangerously low rpm and no power reserve.

Always make the landing to a specific point and not to some general area. This point should be located well forward, away from the approach end of the area. The more confined the area is, the more essential it is that the helicopter land precisely at a definite point. Keep this point in sight during the entire final approach.

When flying a helicopter near obstacles, always consider the tail rotor. A safe angle of descent over barriers must be established to ensure tail rotor clearance of all obstructions. After coming to a hover, avoid turning the tail into obstructions.

Takeoff

A confined area takeoff is considered an altitude over airspeed maneuver where altitude gain is more important to airspeed gain. Before takeoff, make a reconnaissance from the ground or cockpit to determine the type of takeoff to be performed, to determine the point from which the takeoff should be initiated to ensure the maximum amount of available area, and finally, how to maneuver the helicopter best from the landing point to the proposed takeoff position.

If wind conditions and available area permit, the helicopter should be brought to a hover, turned around, and hovered forward from the landing position to the takeoff position. Under certain conditions, sideward flight to the takeoff position may be preferred, but rearward flight may be necessary, stopping often while moving to check on the location of obstacles relative to the tail rotor.

When planning the takeoff, consider the direction of the wind, obstructions, and forced landing areas. To help fly up and over an obstacle, form an imaginary line from a point on the leading edge of the helicopter to the highest obstacle to be cleared. Fly this line of ascent with enough power to clear the obstacle by a safe distance. After clearing the obstacle, maintain the power setting and accelerate to the normal climb speed. Then, reduce power to the normal climb power setting.

Common Errors

1. Failure to perform, or improper performance of, a high or low reconnaissance.
2. Approach angle that is too steep or too shallow for the existing conditions.
3. Failing to maintain proper rpm.

4. Failure to consider emergency landing areas.
5. Failure to select a specific landing spot.
6. Failure to consider how wind and turbulence could affect the approach.
7. Improper takeoff and climb technique for existing conditions.
8. Failure to maintain safe clearance distance from obstructions.

Pinnacle and Ridgeline Operations

A pinnacle is an area from which the surface drops away steeply on all sides. A ridgeline is a long area from which the surface drops away steeply on one or two sides, such as a bluff or precipice. The absence of obstacles does not necessarily decrease the difficulty of pinnacle or ridgeline operations. Updrafts, downdrafts, and turbulence, together with unsuitable terrain in which to make a forced landing, may still present extreme hazards.

Approach and Landing

If there is a need to climb to a pinnacle or ridgeline, do it on the upwind side, when practicable, to take advantage of any updrafts. The approach flightpath should be parallel to the ridgeline and into the wind as much as possible. *[Figure 10-9]*

Load, altitude, wind conditions, and terrain features determine the angle to use in the final part of an approach. As a general rule, the greater the winds are, the steeper the approach needs to be to avoid turbulent air and downdrafts.

Figure 10-9. *When flying an approach to a pinnacle or ridgeline, avoid the areas where downdrafts are present, especially when excess power is limited. If downdrafts are encountered, it may become necessary to make an immediate turn away from the pinnacle to avoid being forced into the rising terrain.*

Groundspeed during a pinnacle approach is more difficult to judge because visual references are farther away than during approaches over trees or flat terrain. Pilots must continually perceive the apparent rate of closure by observing the apparent change in size of the landing zone features. Avoid the misperception of an increasing rate of closure to the landing site. The apparent rate of closure should be that of a brisk walk. If a crosswind exists, remain clear of down-drafts on the leeward or downwind side of the ridgeline. If the wind velocity makes the crosswind landing hazardous, it may be possible to make a low, coordinated turn into the wind just prior to terminating the approach. When making an approach to a pinnacle, avoid leeward turbulence and keep the helicopter within reach of a forced landing area as long as possible.

On landing, take advantage of the long axis of the area when wind conditions permit. Touchdown should be made in the forward portion of the area. When approaching to land on pinnacles, especially manmade areas such as rooftop pads, the pilot should determine the personnel access pathway to the helipad and ensure that the tail rotor is not allowed to intrude into that walkway or zone. Parking or landing with the tail rotor off the platform ensures personnel safety. Always perform a stability check prior to reducing rpm to ensure the landing gear is on firm terrain that can safely support the weight of the helicopter. Accomplish this by slowly moving the cyclic and pedals while lowering the collective. If movement is detected, reposition the aircraft.

Takeoff

A pinnacle takeoff is considered an airspeed over altitude maneuver which can be made from the ground or from a hover. Since pinnacles and ridgelines are generally higher than the immediate surrounding terrain, gaining airspeed on the takeoff is more important than gaining altitude. As airspeed increases, the departure from the pinnacle becomes more rapid, and helicopter time in the *avoid* area of the height/velocity area decreases. *[Figure 11-3]* In addition to covering unfavorable terrain rapidly, a higher airspeed affords a more favorable glide angle and thus contributes to the chances of reaching a safe area in the event of a forced landing. If a suitable forced landing area is not available, a higher airspeed also permits a more effective flare prior to making an autorotative landing.

On takeoff, as the helicopter moves out of ground effect, maintain altitude and accelerate to normal climb airspeed. When normal climb speed is attained, establish a normal climb attitude. Never dive the helicopter down the slope after clearing the pinnacle.

Common Errors

1. Failing to perform, or improper performance of, a high or low reconnaissance.
2. Flying the approach angle too steep or too shallow for the existing conditions.
3. Failing to maintain proper rpm.
4. Failing to consider emergency landing areas.
5. Failing to consider how wind and turbulence could affect the approach and takeoff.
6. Failure to maintain pinnacle elevation after takeoff.
7. Failure to maintain proper approach rate of closure.
8. Failure to achieve climb airspeed in timely manner.

Chapter Summary

This chapter described advanced flight maneuvers such as slope landings, confined area landings, and running takeoffs. The correlation between helicopter performance requirements, the environmental factors associated with different flight techniques, and safety considerations were also explained to familiarize the pilot with the measures that can be taken when performing these maneuvers to mitigate risks. Hazards associated with helicopter flight and certain aerodynamic considerations were also discussed.

Helicopter Emergencies and Hazards

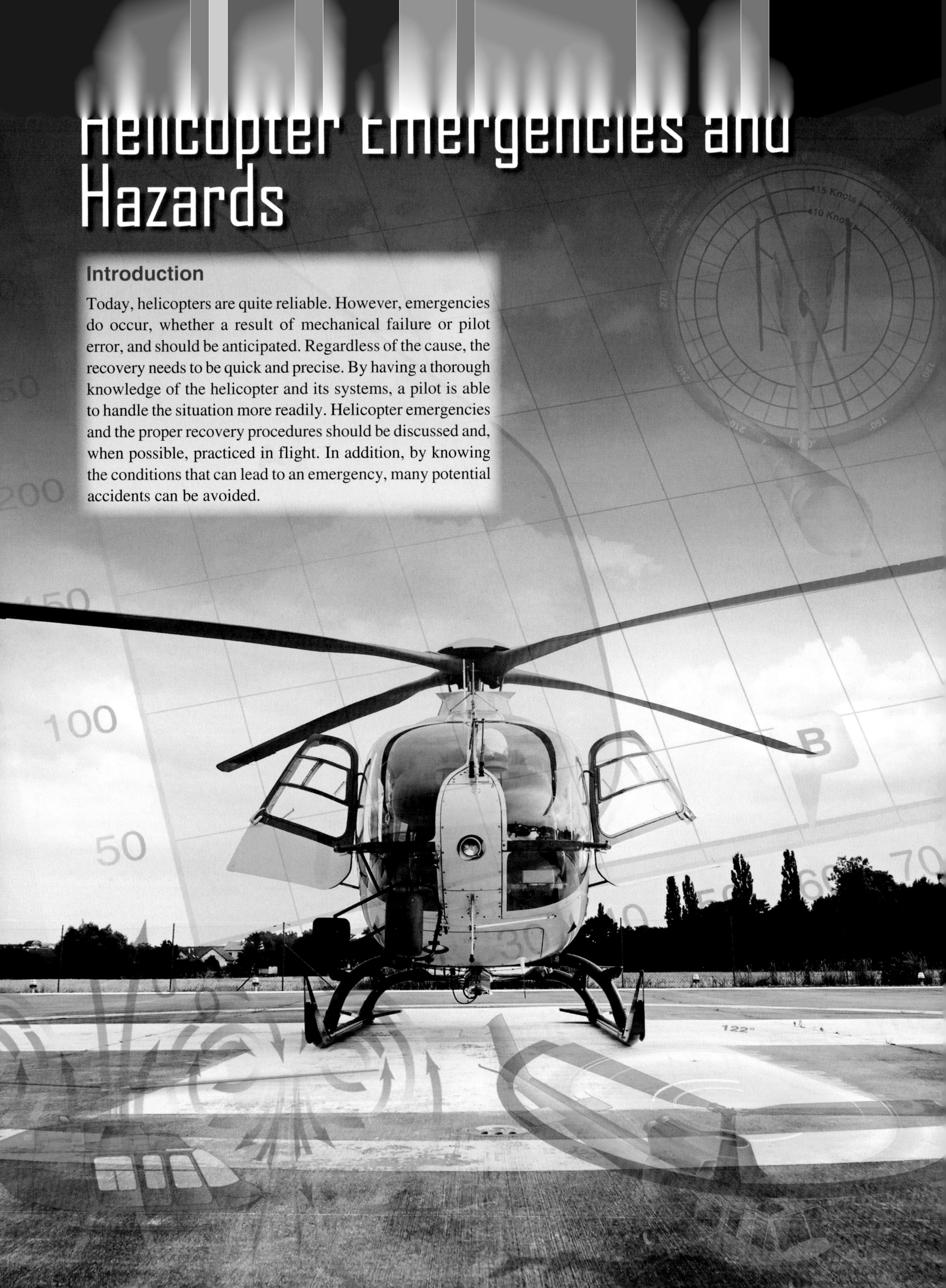

Introduction

Today, helicopters are quite reliable. However, emergencies do occur, whether a result of mechanical failure or pilot error, and should be anticipated. Regardless of the cause, the recovery needs to be quick and precise. By having a thorough knowledge of the helicopter and its systems, a pilot is able to handle the situation more readily. Helicopter emergencies and the proper recovery procedures should be discussed and, when possible, practiced in flight. In addition, by knowing the conditions that can lead to an emergency, many potential accidents can be avoided.

Autorotation

In a helicopter, an autorotative descent is a power-off maneuver in which the engine is disengaged from the main rotor disk and the rotor blades are driven solely by the upward flow of air through the rotor. *[Figure 11-1]* In other words, the engine is no longer supplying power to the main rotor.

The most common reason for an autorotation is failure of the engine or drive line, but autorotation may also be performed in the event of a complete tail rotor failure, since there is virtually no torque produced in an autorotation. In both cases, maintenance has often been a contributing factor to the failure. Engine failures are also caused by fuel contamination or exhaustion as well resulting in a forced autorotation.

If the engine fails, the freewheeling unit automatically disengages the engine from the main rotor, allowing it to rotate freely. Essentially, the freewheeling unit disengages anytime the engine revolutions per minute (rpm) is less than the rotor rpm.

At the instant of engine failure, the main rotor blades are producing lift and thrust from their angle of attack (AOA) and velocity. By lowering the collective (which must be done immediately in case of an engine failure), lift and drag are reduced, and the helicopter begins an immediate descent, thus producing an upward flow of air through the rotor disk. This upward flow of air through the rotor disk provides sufficient thrust to maintain rotor rpm throughout the descent. Since the tail rotor is driven by the main rotor transmission during autorotation, heading control is maintained with the antitorque pedals as in normal flight.

Several factors affect the rate of descent in autorotation: bank angle, density altitude, gross weight, rotor rpm, trim condition, and airspeed. The primary ways to control the rate of descent are with airspeed and rotor rpm. Higher or lower airspeed is obtained with the cyclic pitch control just as in normal powered flight. In theory, a pilot has a choice in the angle of descent, varying, from straight vertical to maximum horizontal range (which is the minimum angle of descent). Rate of descent is high at zero airspeed and decreases to a minimum at approximately 50–60 knots, depending upon the particular helicopter and the factors just mentioned. As the airspeed increases beyond that which gives minimum rate of descent, the rate of descent increases again.

When landing from an autorotation, the only energy available to arrest the descent rate and ensure a soft landing is the kinetic energy stored in the rotor blades. Tip weights can greatly increase this stored energy. A greater amount of rotor energy is required to stop a helicopter with a high rate of descent than is required to stop a helicopter that is descending more slowly. Therefore, autorotative descents at very low or very high airspeeds are more critical than those performed at the minimum rate of descent airspeed. Refer to the height/velocity diagram discussion in Chapter 7, Helicopter Performance.

Each type of helicopter has a specific airspeed and rotor rpm at which a power-off glide is most efficient. The specific airspeed is somewhat different for each type of helicopter, but certain factors affect all configurations in the same manner. In general, rotor rpm maintained in the low green area (see *Figure 5-3*) gives more distance in an autorotation. Heavier helicopter weights may require more collective to control rotor rpm. Some helicopters need slight adjustments to minimum rotor rpm settings for winter versus summer

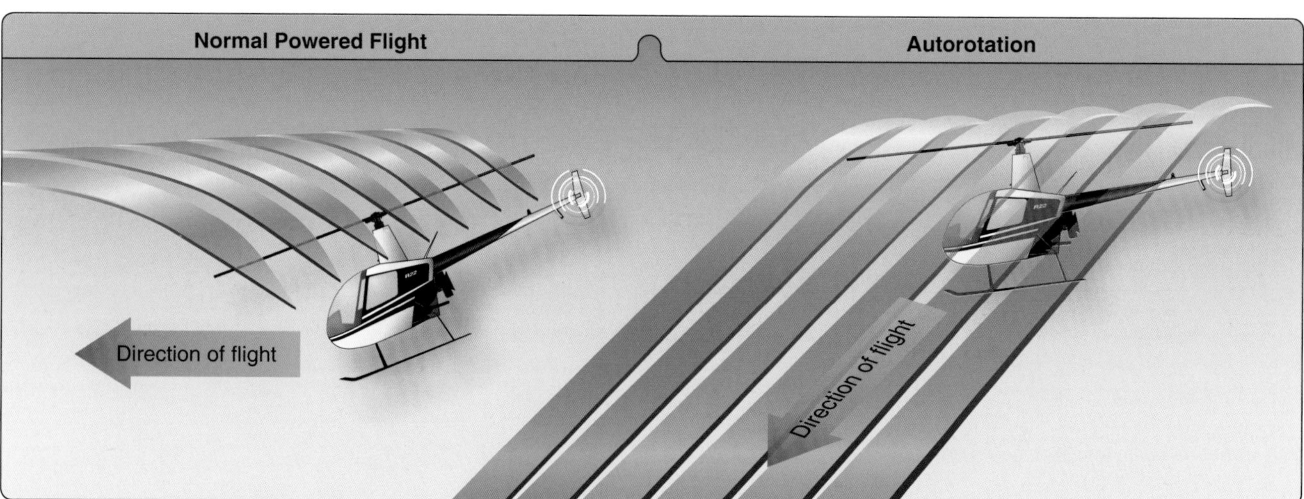

Figure 11-1. *During an autorotation, the upward flow of relative wind permits the main rotor blades to rotate at their normal speed. In effect, the blades are "gliding" in their rotational plane.*

conditions, and high altitude versus sea level flights. For specific autorotation airspeed and rotor rpm combinations for a particular helicopter, refer to the Rotorcraft Flight Manual (RFM). The specific airspeed and rotor rpm for autorotation is established for each type of helicopter based on average weather, calm wind conditions, and normal loading. When the helicopter is operated with heavy loads in high density altitude or gusty wind conditions, best performance is achieved from a slightly increased airspeed in the descent. For autorotation at low density altitude and light loading, best performance is achieved from a slight decrease in normal airspeed. Following this general procedure of fitting airspeed and rotor rpm to existing conditions, a pilot can achieve approximately the same glide angle in any set of circumstances, and thereby estimate the touchdown point accurately.

It is important that pilots experience autorotations from various airspeeds. This provides better understanding of the necessary flight control inputs to achieve the desired airspeed, rotor rpm and autorotation performance, such as the maximum glide or minimum descent airspeed. The decision to use the appropriate airspeed and rotor rpm for the given conditions should be instinctive to reach a suitable landing area. The helicopter glide ratio is much less than that of a fixed-wing aircraft and takes some getting used to. The flare to land at 80 knots indicated airspeed (KIAS) will be significantly greater than that from 55 KIAS. Rotor rpm control is critical at these points to ensure adequate rotor energy for cushioning the landing.

Use collective pitch control to manage rotor rpm. If rotor rpm builds too high during an autorotation, raise the collective sufficiently to decrease rpm back to the normal operating range, then reduce the collective to maintain proper rotor rpm. If the collective increase is held too long, the rotor rpm may decay rapidly. The pilot would have to lower the collective in order to regain rotor rpm. If the rpm begins decreasing, the pilot must again lower the collective. Always keep the rotor rpm within the established recommended range for the helicopter being flown.

RPM Control

Rotor rpm in low inertia rotor systems has been studied in simulator flight evaluations which indicate that the simultaneous application of aft cyclic, down collective, and alignment with the relative wind (trim) at a wide range of airspeeds, including cruise airspeeds, is critical for all operations during the entry of an autorotation. The applicable Rotorcraft Flight Manual (RFM) should be consulted to determine the appropriate procedure(s) for safely entering an autorotation. This is vitally important since the procedure(s) for safely entering an autorotation may vary with specific makes and/or models of helicopters. A basic discussion of the aerodynamics and control inputs for single rotor systems is in order here.

Helicopter pilots must understand the use of the collective for rotor rpm control during power off autorotations in a turn. Upward movement of the collective reduces the rpm and downward movement increases the rpm. Cyclic movement is primarily associated with attitude/airspeed control in powered flight but may not be given the credit appropriate for rotor rpm control during practice and emergency power off autorotations. As long as the line of cyclic movement is parallel with the flight path of the helicopter (trimmed), the aft movement of the cyclic also creates greater air flow up through the bottom of the rotor disk and contributes to an increase in rotor rpm. If the flight path is 10 degrees to the right of the longitudinal axis of the helicopter, theoretically, the cyclic should be moved 10 degrees aft and left of the longitudinal axis to get maximum air up through the rotor system.

As the pilot lowers the collective in reaction to a loss of power during cruise flight there may be a tendency for the nose of the helicopter to pitch down. As a result, the pilot may tend to lean forward slightly, which delays the application of simultaneous aft cyclic to prevent the pitch change and associated loss of rotor rpm. A slight gain in altitude at cruise airspeed during the power off entry into an autorotation should not be of great concern as is the case for the execution of practice or actual quick stops.

Various accident investigations have concluded that, when faced with a real power failure at cruise airspeed, pilots are not simultaneously applying down collective, aft cyclic, and antitorque pedal inputs in a timely manner. Low inertia rotor systems store less kinetic energy during autorotation and, as a result, rotor rpm decays rapidly during deceleration and touchdown. Conversely, less energy is required to regain safe rotor rpm during autorotation entry and autorotative descent. The pilot should immediately apply simultaneous down collective, aft cyclic and trim the helicopter for entry into an autorotation initiated at cruise airspeed. If rotor rpm has been allowed to decrease, or has inadvertently decreased below acceptable limits, an application of aft cyclic may help rebuild rotor rpm. This application of aft cyclic must be made at least at a moderate rate and may be combined with a turn, either left or right, to increase airflow through the rotor system. This will work to increase rotor rpm. Care should be maintained to not over-speed the rotor system as this is attempted.

Risk Management during Autorotation Training

The following sections describe enhanced guidelines for autorotations during rotorcraft/helicopter flight training, as stated in Advisory Circular (AC) 61-140. There are risks inherent in performing autorotations in the training environment, and in particular the 180-degree autorotation. This section describes an acceptable means, but not the only means, of training applicants for a rotorcraft/helicopter airman certificate to meet the qualifications for various rotorcraft/helicopter ratings. You may use alternate methods for training if you establish that those methods meet the requirements of the Helicopter Flying Handbook (HFH), FAA practical test standards (PTS), and the Rotorcraft Flight Manual (RFM).

Straight-In Autorotation

A straight-in autorotation is one made from altitude with no turns. Winds have a great effect on an autorotation. Strong headwinds cause the glide angle to be steeper due to the slower groundspeed. For example, if the helicopter is maintaining 60 KIAS and the wind speed is 15 knots, then the groundspeed is 45 knots. The angle of descent will be much steeper, although the rate of descent remains the same. The speed at touchdown and the resulting ground run depend on the groundspeed and amount of deceleration. The greater the degree of deceleration, or flare, and the longer it is held, the slower the touchdown speed and the shorter the ground run. Caution must be exercised at this point as the tail rotor will be the component of the helicopter closest to the ground. If timing is not correct and a landing attitude not set at the appropriate time, the tail rotor may contact the ground causing a forward pitching moment of the nose and possible damage to the helicopter.

A headwind is a contributing factor in accomplishing a slow touchdown from an autorotative descent and reduces the amount of deceleration required. The lower the speed desired at touchdown, the more accurate the timing and speed of the flare must be, especially in helicopters with low-inertia rotor disks. If too much collective is applied too early during the final stages of the autorotation, the kinetic energy may be depleted, resulting in little or no cushioning effect available. This could result in a hard landing with corresponding damage to the helicopter. It is generally better practice to accept more ground run than a harder landing with minimal groundspeed. As proficiency increases, the amount of ground run may be reduced.

Technique (How to Practice)

Refer to *Figure 11-2* (position 1). From level flight at the appropriate airspeed (cruise or the manufacturer's recommended airspeed), 500–700 feet above ground level (AGL), and heading into the wind, smoothly but firmly lower the collective to the full down position. Use aft cyclic to prevent a nose low attitude while maintaining rotor rpm in the green arc with collective. If the collective is in the full down position, the rotor rpm is then being controlled by the mechanical pitch stops. During maintenance, the rotor stops must be set to allow minimum autorotational rpm with a light loading. This means that collective will still be able to be reduced even under conditions of extreme reduction of vertical loading (e.g., very low helicopter weight, at very low-density altitude). After entering an autorotation, collective pitch must be adjusted to maintain the desired rotor rpm.

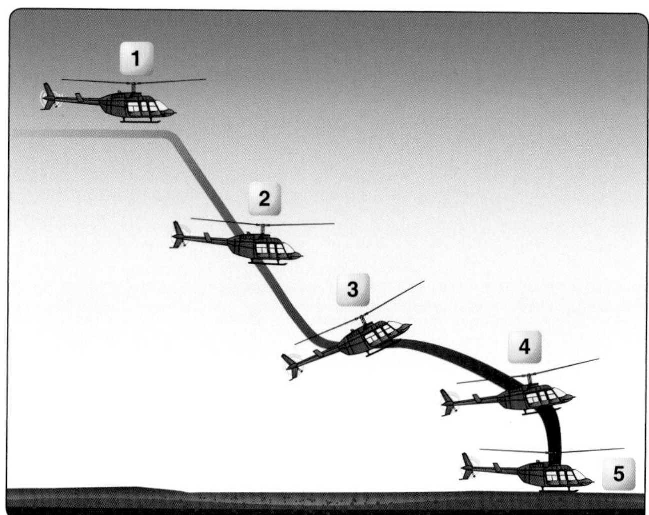

Figure 11-2. *Straight-in autorotation.*

Coordinate the collective movement with proper antitorque pedal for trim, and apply cyclic control to maintain proper airspeed. Once the collective is fully lowered, decrease throttle to ensure a clean split/separation of the needles. This means that the rotor rpm increases to a rate higher than that of the engine—a clear indication that the freewheeling unit has allowed the engine to disconnect. After splitting the needles, readjust the throttle to keep engine rpm above normal idling speed, but not high enough to cause rejoining of the needles. See the RFM for the manufacturer's recommendations for autorotation rate of descent.

At position 2, adjust attitude with cyclic to obtain the manufacturer's recommended autorotation (or best gliding) speed. Adjust collective as necessary to maintain rotor rpm in the lower part of the green arc (see page 11-2). Aft cyclic movements cause an increase in rotor rpm, which is then controlled by a small increase in collective. Avoid a large collective increase, which results in a rapid decay of rotor rpm, and leads to "chasing the rpm." Avoid looking straight down in front of the aircraft. Continually crosscheck attitude, trim, rotor rpm, and airspeed.

At the altitude recommended by the manufacturer (position 3), begin the flare with aft cyclic to reduce forward airspeed and decrease the rate of descent. Maintain heading with the antitorque pedals. During the flare, maintain rotor rpm in the green range. In the execution of the flare, care must be taken that the cyclic be moved rearward neither so abruptly that it causes the helicopter to climb, nor so slowly that it fails to arrest the descent, which may allow the helicopter to settle so rapidly that the tail rotor strikes the ground. In most helicopters, the proper flare attitude is that resulting in a groundspeed of a slow run. When forward motion decreases to the desired groundspeed—usually the lowest possible speed (position 4)—move the cyclic forward to place the helicopter in the proper attitude for landing.

This action gives the student an idea of airframe attitude to avoid, because a pilot should never allow ground contact unless the helicopter is more nose-low than that attitude. Limiting the flare to that attitude may result in slightly faster touchdown speeds but will eliminate the possibility of tail rotor impact on level surfaces.

The landing gear height at this time should be approximately 3–15 feet AGL, depending on the altitude recommended by the manufacturer. As the apparent groundspeed and altitude decrease, the helicopter must be returned to a more level attitude for touchdown by applying forward cyclic. Some helicopters can be landed on the heels in a slightly nose high attitude to help decrease the forward groundspeed, whereas others must land skids or landing gear level, in order to spread the landing loads equally to all of the landing gear. Extreme caution should be used to avoid an excessive nose high and tail low attitude below 10 feet. The helicopter must be close to the landing attitude to keep the tail rotor from contacting the surface.

At this point, if a full touchdown landing is to be performed, allow the helicopter to descend vertically (position 5). This collective application uses some of the kinetic energy in the rotor disk to help slow the descent rate of the helicopter. When the collective is raised, the opposite antitorque pedal used in powered flight will be needed due to the friction within the transmission/drive train. Touch down in a level flight attitude.

Control response with increased pitch angles will be slightly different than normal. With a decrease in main rotor rpm, the antitorque authority is reduced (the pedals react more slowly), requiring larger control inputs to maintain heading at touchdown.

Some helicopters, such as the Schweitzer 300, have a canted tail stabilizer. With a canted stabilizer, it is crucial that the pilot apply the appropriate pedal input at all times during the autorotation. If not the tailboom tends to swing to the right, which allows the canted stabilizer to raise the tail. This can result in a severe nose tuck which is quickly corrected with right pedal application.

A power recovery can be made during training in lieu of a full touchdown landing. Refer to the section on power recovery for the correct technique.

After the helicopter has come to a complete stop after touchdown, lower the collective pitch to the full-down position. Do not try to stop the forward ground run with aft cyclic, as the main rotor blades can strike the tail boom. By lowering the collective slightly during the ground run, an increase in weight is placed on the landing carriage, slowing the helicopter; however, this is dependent on the condition of the landing surface.

One common error is the holding of the helicopter off the surface, versus cushioning it onto the surface during an autorotation. Holding the helicopter in the air by using all of the rotor rpm kinetic energy usually causes the helicopter to have a hard landing, which results in the blades flexing down and contacting the tail boom. The rotor rpm should be used to cushion the helicopter on to the surface for a controlled, smooth landing instead of allowing the helicopter to drop the last few inches.

Common Errors

1. Not understanding the importance of an immediate entry into autorotation upon powerplant or driveline failure.
2. Failing to use sufficient antitorque pedal when power is reduced.
3. Lowering the nose too abruptly when power is reduced, thus placing the helicopter in a dive.
4. Failing to maintain proper rotor rpm during the descent.
5. Applying up-collective pitch at an excessive altitude, resulting in a hard landing, loss of heading control, and possible damage to the tail rotor and main rotor blade stops.
6. Failing to level the helicopter or achieve the manufacturers preferred landing attitude.
7. Failing to minimize or eliminate lateral movement during ground contact. (Similar for items 8 and 9)
8. Failing to maintain ground track in the air and keeping the landing gear aligned with the direction of travel during touchdown and ground contact.
9. Failing (in a practice run) to go around if not within limits and specified criteria for safe autorotation.

Autorotation with Turns

Turns (or a series of turns) can be made during autorotation to facilitate landing into the wind or avoiding obstacles. Turns during autorotation should be made early so that the remainder of the autorotation is flown identically to a straight-in autorotation. The most common turns in an autorotation are 90 degrees and 180 degrees. The following technique describes an autorotation with a 180-degree turn.

The pilot establishes the aircraft on a downwind heading at the recommended airspeed, and parallel to the intended touchdown point. Then, taking the wind into account, the pilot establishes the ground track approximately 200 feet laterally from the desired course line to the touchdown point. In strong crosswind conditions, the pilot should be prepared to adjust the downwind leg closer or farther out, as appropriate. The pilot uses the autorotation entry airspeed recommended by the RFM. When abeam the intended touchdown point, the pilot smoothly reduces collective, then reduces power to the engine to show a split between the rotor rpm and engine rpm and simultaneously applies appropriate anti-torque pedal and cyclic to maintain proper attitude/airspeed. Throughout the autorotation, the pilot should continually crosscheck the helicopter's attitude, rotor rpm, airspeed, and verify that the helicopter is in trim (centered trim ball).

After the descent and autorotation airspeed is established, the pilot initiates the 180-degree turn. For training operations, initially roll into a bank of at least 30 degrees, but no more than 60 degrees. It is important to maintain the proper airspeed, rotor rpm, and trim (centered trim ball) throughout the turn. Changes in the helicopter's attitude and the angle of bank causes a corresponding change in rotor rpm within normal limits. Do not allow the nose to pitch up or down excessively during the maneuver, as it may cause undesirable rotor rpm excursions.

Pitot-static airspeed indications may be unreliable or lag during an autorotational turn. The pilot should exercise caution to avoid using excessive aircraft pitch attitudes and to avoid chasing airspeed indications in an autorotational turn.

Note: Approaching the 90-degree point, check the position of the landing area. The second 90 degrees of the turn should end with a roll-out on a course line to the landing area. If the helicopter is too close, decrease the bank angle (to increase the radius of turn); if too far out, increase the bank angle (to decrease the radius of the turn). A bank angle of no more than 60 degrees should be encountered during this turn. Monitor the trim ball (along with one's kinesthetic sense) and adjust as necessary with cyclic and anti-torque pedal to maintain coordinated flight. Prior to passing through 200 feet above ground level (AGL), if landing or making a surface-level power recovery, the turn should be completed, and the helicopter aligned with the intended touchdown area. Upon reaching the course line, set the appropriate crosswind correction. If the collective pitch was increased to control the rpm, it may need to be lowered on rollout to prevent decay in rotor rpm.

This maneuver should be aborted at any point the following criteria is not met: if the helicopter is not in a stabilized approach to landing profile (i.e., it is not aligned as close as possible into the wind with the touchdown point, after completing the 180-degree turn); if the rotor rpm is not within limits; if the helicopter is not at a proper attitude/airspeed; or if the helicopter is not under proper control at 200 feet AGL. It is essential that the pilot on the controls (or a certificated flight instructor (CFI), when intervening) immediately abort the maneuver and execute a smooth power recovery and go-around. It is important for the CFI who is intervening at this point to remember that the go-around is a far safer option than trying to recover lost rotor rpm and reestablish or recover to the hover or even the preferred hover taxi.

From all entry positions, but particularly true of the 180-degree entry, a primary concern is getting the aircraft into the course line with as much altitude as possible. Once the collective has been lowered and the engine set to flight idle, the helicopter will lose altitude. A delayed turn will result in a lower altitude when arriving on the course line. Additionally, an uncoordinated flight condition (trim-ball not centered) results in an increased sink rate, which may be unrecoverable if not corrected.

During the turn to the course line, the pilot should use a scan pattern to see outside as well as inside the cockpit. Of primary importance outside is maintaining the appropriate descending attitude and a proper turn rate. Essential items to scan inside are rotor rpm and centered trim ball. Rotor rpm will build anytime "G" forces are applied to the rotor system. Usually, this occurs in the turn to the course line and during the deceleration flare.

Throughout the maneuver, rotor rpm should be maintained in the range recommended in the RFM. Rotor rpm outside of the recommended range results in a higher rate of descent and less glide-ratio. When the rotor rpm exceeds the desired value as a result of increased G load in the turn, timely use of up collective will increase the pitch of the blades and slow the rotor to the desired rpm. In an autorotation, rotor rpm is the most critical element, as it provides the lift required to stabilize an acceptable rate of descent and the energy necessary to cushion the landing. Collective should be lowered to the full down position to maintain rotor rpm immediately following a loss of power. However, rapid or abrupt collective movement could lead to mast bumping in some rotorcraft with teetering rotor systems.

Energy is a very important property of all rotating components, and the kinetic energy stored in the rotor system is used to cushion the landing. More lift is produced at the bottom of an autorotation by raising the collective, which increases the angle of attack of the blades. The rotor rpm will also rapidly decay at this point and it is essential to properly time the flare and the final collective pull to fully arrest the descent and cushion the landing. Upon arriving into the course line prior to the flare, the scan should focus almost entirely outside. The scan should include:

- The horizon for attitude, ground track, and nose alignment;
- the altitude to set the flare and for closure (groundspeed); and
- the instrument cross-check of airspeed, rotor rpm, and engine rpm in the descent.

Every autorotational flare will be different depending on the existing wind conditions, airspeed, density altitude (DA), and the aircraft gross weight. A pilot operating a helicopter at a high DA needs to take into account the effects on the control of the helicopter when recovering from an aborted autorotation.

Some effects to consider are:

- Higher rate of descent.
- Reduced rotor rpm builds in autorotation.
- Low initial rotor rpm response in autorotation.
- The requirement for a higher flare height.
- Reduced engine power performance.

Common Errors

The following common errors should be prevented:

1. Entering the maneuver at an improper altitude or airspeed.
2. Entering the maneuver without a level attitude (or not in coordinated flight).
3. Entering the maneuver and not correcting from the initial deceleration to a steady state attitude (which allows excessive airspeed loss in the descent).
4. Improper transition into the descent on entry.
5. Improper use of anti-torque on entry.
6. Failure to establish the appropriate crosswind correction, allowing the aircraft to drift.
7. Failure to maintain coordinated flight through the tum.
8. Failure to maintain rotor rpm within the RFM recommended range.
9. Excessive yaw when increasing collective to slow rate of descent during power recovery autorotations.
10. During power recovery autorotations, a delay in reapplying power.
11. Initial collective pull either too high or too low.
12. Improper flare (too much or not enough).
13. Flaring too low or too high (AGL).
14. Failure to maintain heading when reapplying power.
15. Not landing with a level attitude.
16. Landing with aircraft not aligned with the direction of travel.
17. Insufficient collective cushioning during full autorotations.
18. Abrupt control inputs on touchdown during full autorotations.

Practice Autorotation with a Power Recovery

A power recovery is used to terminate practice autorotations at a point prior to actual touchdown. After the power recovery, a landing can be made or a go-around initiated.

Technique (How to Practice)

At approximately 3–15 feet landing gear height AGL, depending upon the helicopter being used, begin to level the helicopter with forward cyclic control. Avoid excessive nose-high, tail-low attitude below 10 feet. Just prior to achieving level attitude, with the nose still slightly up, coordinate upward collective pitch control with an increase in the throttle to join the needles at operating rpm. The throttle and collective pitch must be coordinated properly.

If the throttle is increased too fast or too much, an engine overspeed can occur; if throttle is increased too slowly or too little in proportion to the increase in collective pitch, a loss of rotor rpm results. Use sufficient collective pitch to stop the descent, but keep in mind that the collective pitch application must be gradual to allow for engine response. Coordinate proper antitorque pedal pressure to maintain heading. When a landing is to be made following the power recovery, bring the helicopter to a hover and then descend to a landing.

In nearly all helicopters, when practicing autorotations with power recovery, the throttle should be at the flight setting at the beginning of the flare. As the rotor disk begins to dissipate its energy, the engine is up to speed as the needles join when the rotor decreases into the normal flight rpm.

Helicopters that do not have the throttle control located on the collective are generally exceptions to basic technique and require some additional prudence. The autorotation should be initiated with the power levers left in the "flight," or normal, position. If a full touchdown is to be practiced, it is common technique to move the power levers to the idle position once the landing area can safely be reached. In most helicopters, the pilot is fully committed at that point to make a power-off landing. However, it may be possible to make a power recovery prior to passing through 100 feet AGL if the powerplant can recover within that time period and the instructor is very proficient. The pilot should comply with the RFM instructions in all cases.

When practicing autorotations to a power recovery, the differences between reciprocating engines and turbines may be profound. The reciprocating powerplant generally responds very quickly to power changes, especially power increases. Some turbines have delay times depending on the type of fuel control or governing system installed. Any reciprocating engine needing turbocharged boost to develop rated horse power may have significant delays to demands for increased power, such as in the power recovery. Power recovery in those helicopters with slower engine response times must have the engines begin to develop enough power to rejoin the needles by approximately 100 feet AGL.

If a go-around is to be made, the cyclic control should be moved forward to resume forward flight. In transition from a practice autorotation to a go-around, exercise caution to avoid an altitude-airspeed combination that would place the helicopter in an unsafe area of its height/velocity diagram.

This is one of the most difficult maneuvers to perform due to the concentration needed when transitioning from powered flight to autorotation and then back again to powered flight. For helicopters equipped with the power control on the collective, engine power must be brought from flight power to idle power and then back to a flight power setting. A delay during any of these transitions can seriously affect rotor rpm placing the helicopter in a situation that cannot be recovered.

The cyclic must be adjusted to maintain the required airspeed without power, and then used for the deceleration flare, followed by the transition to level hovering flight. Additionally, the cyclic must be adjusted to remove the compensation for translating tendency. The tail rotor is no longer needed to produce antitorque thrust until almost maximum power is applied to the rotor disk for hovering flight, when the tail rotor must again compensate for the main rotor torque, which also demands compensation for the tail rotor thrust and translating tendency.

The pedals must be adjusted from a powered flight antitorque trim setting to the opposite trim setting to compensate for transmission drag and any unneeded vertical fin thrust countering the now nonexistent torque and then reset to compensate for the high power required for hovering flight.

All of the above must be accomplished during the 23 seconds of the autorotation, and the quick, precise control inputs must be made in the last 5 seconds of the maneuver.

Common Errors

1. Initiating recovery too late, which requires a rapid application of controls and results in overcontrolling.
2. Failure to obtain and maintain a level attitude near the surface.
3. Failure to coordinate throttle and collective pitch properly, which results in either an engine overspeed or a loss of rotor rpm.
4. Failure to coordinate proper antitorque pedal with the increase in power.
5. Late engine power engagement causing excessive temperature or torque, or rpm drop.
6. Failure to go around if not within limits and specified criteria for safe autorotation.

Practicing Power Failure in a Hover

Power failure in a hover, also called hovering autorotation, is practiced so that a pilot can automatically make the correct response when confronted with engine stoppage or certain other emergencies while hovering. The techniques discussed in this section are for helicopters with a counterclockwise rotor disk and an antitorque rotor.

Technique (How to Practice)

To practice hovering autorotation, establish a normal hovering height (approximately 2–3 feet) for the particular helicopter being used, considering load and atmospheric conditions. Keep the helicopter headed into the wind and hold maximum allowable rpm.

To simulate a power failure, firmly roll the throttle to the engine idle position. This disengages the driving force of the engine from the rotor, thus eliminating torque effect. As the throttle is closed, apply proper antitorque pedal to maintain heading. Usually, a slight amount of right cyclic control is necessary to keep the helicopter from drifting to the left, to compensate for the loss of tail rotor thrust. However, use cyclic control, as required, to ensure a vertical descent and a level attitude. Do not adjust the collective on entry.

Helicopters with low inertia rotor disks settle immediately. Keep a level attitude and ensure a vertical descent with cyclic control while maintaining heading with the pedals. Any lateral movement must be avoided to prevent dynamic rollover. As rotor rpm decays, cyclic response decreases, so compensation for the winds will require more cyclic input. At approximately 1 foot AGL, apply upward collective control, as necessary, to slow the descent and cushion the landing without arresting the rate of descent above the surface. Usually, the full amount of collective is required just as the landing gear touches the surface. As upward collective control is applied, the throttle must be held in the idle detent position to prevent the engine from re-engaging. The idle detention position is a ridged stop position between idle and off in which the idle release button snaps into, prevent accidental throttle off.

Helicopters with high-inertia rotor disks settle more slowly after the throttle is closed. In this case, when the helicopter has settled to approximately 1 foot AGL, apply upward collective control while holding the throttle in the idle detent position to slow the descent and cushion the landing. The timing of collective control application and the rate at which it is applied depend upon the particular helicopter being used, its gross weight, and the existing atmospheric conditions. Cyclic control is used to maintain a level attitude and to ensure a vertical descent. Maintain heading with antitorque pedals.

When the weight of the helicopter is entirely resting on the landing gear, cease application of upward collective. When the helicopter has come to a complete stop, lower the collective pitch to the full-down position.

The timing of the collective movement is a very important consideration. If it is applied too soon, the remaining rpm may not be sufficient to make a soft landing. On the other hand, if it is applied too late, surface contact may be made before sufficient blade pitch is available to cushion the landing. The collective must not be used to hold the helicopter off the surface, causing a blade stall. Low rotor rpm and ensuing blade stall can result in a total loss of rotor lift, allowing the helicopter to fall to the surface and possibly resulting in blade strikes to the tail boom and other airframe damage such as landing gear damage, transmission mount deformation, and fuselage cracking.

Common Errors

1. Failure to use sufficient proper antitorque pedal when power is reduced.
2. Failure to stop all sideward or backward movement prior to touchdown.
3. Failure to apply up-collective pitch properly, resulting in a hard touchdown.
4. Failure to touch down in a level attitude.
5. Failure to roll the throttle completely to idle.
6. Failure to hover at a safe altitude for the helicopter type, atmospheric conditions, and the level of training/ proficiency of the pilot.
7. Failure to go around if not within limits and specified criteria for safe autorotation.

Vortex Ring State

Vortex ring state (formerly referenced as settling-with-power) describes an aerodynamic condition in which a helicopter may be in a vertical descent with 20 percent up to maximum power applied, and little or no climb performance. The previously used term settling-with-power came from the fact that the helicopter keeps settling even though full engine power is applied.

In a normal out-of-ground-effect (OGE) hover, the helicopter is able to remain stationary by propelling a large mass of air down through the main rotor. Some of the air is recirculated near the tips of the blades, curling up from the bottom of the rotor disk and rejoining the air entering the rotor from the top. This phenomenon is common to all airfoils and is known as tip vortices. Tip vortices generate drag and degrade airfoil efficiency. As long as the tip vortices are small, their only effect is a small loss in rotor efficiency. However, when the helicopter begins to descend vertically, it settles into its own downwash, which greatly enlarges the tip vortices. In this vortex ring state, most of the power developed by the engine is wasted in circulating the air in a doughnut pattern around the rotor.

In addition, the helicopter may descend at a rate that exceeds the normal downward induced-flow rate of the inner blade sections. As a result, the airflow of the inner blade sections is upward relative to the disk. This produces a secondary vortex ring in addition to the normal tip vortices. The secondary vortex ring is generated about the point on the blade where the airflow changes from up to down. The result is an unsteady turbulent flow over a large area of the disk. Rotor efficiency is lost even though power is still being supplied from the engine. *[Figure 11-3]*

A fully developed vortex ring state is characterized by an unstable condition in which the helicopter experiences uncommanded pitch and roll oscillations, has little or no collective authority, and achieves a descent rate that may approach 6,000 feet per minute (fpm) if allowed to develop.

A vortex ring state may be entered during any maneuver that places the main rotor in a condition of descending in a column of disturbed air and low forward airspeed. Airspeeds

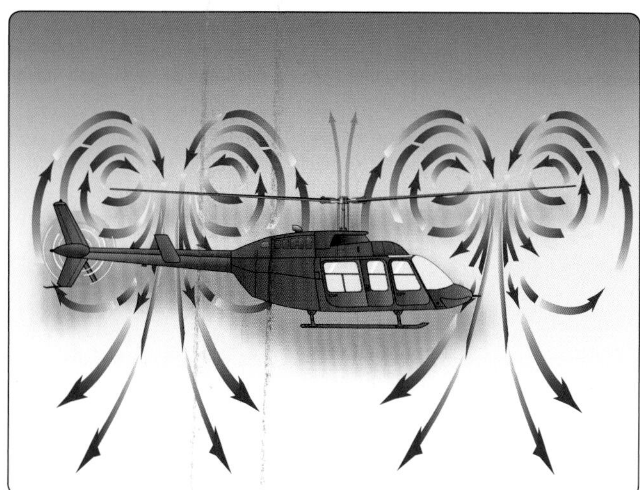

Figure 11-3. *Vortex ring state.*

that are below translational lift airspeeds are within this region of susceptibility to vortex ring state aerodynamics. This condition is sometimes seen during quick-stop type maneuvers or during recovery from autorotation.

The following combination of conditions is likely to cause settling in a vortex ring state in any helicopter:

1. A vertical or nearly vertical descent of at least 300 fpm. (Actual critical rate depends on the gross weight, rpm, density altitude, and other pertinent factors.)
2. The rotor disk must be using some of the available engine power (20–100 percent).
3. The horizontal velocity must be slower than effective translational lift.

Situations that are conducive to a vortex ring state condition are attempting to hover OGE without maintaining precise altitude control, and approaches, especially steep approaches, with a tailwind component.

When recovering from a vortex ring state condition, the pilot tends first to try to stop the descent by increasing collective pitch. However, this only results in increasing the stalled area of the rotor, thereby increasing the rate of descent. Since inboard portions of the blades are stalled, cyclic control may be limited. The traditional recovery is accomplished by increasing airspeed, and/or partially lowering collective to exit the vortex. In most helicopters, lateral cyclic thrust combined with an increase in power and lateral antitorque thrust will produce the quickest exit from the hazard. This technique, known as the Vuichard Recovery (named after the Swiss examiner from the Federal Office of Civil Aviation who developed it) recovers by eliminating the descent rate as opposed to exiting the vortex. If the vortex ring state and the corresponding descent rate is allowed to progress to what is called the windmill brake state, the point where the airflow is completely up through the rotor, the only recovery may be an autorotation.

Tandem rotor helicopters should maneuver laterally to achieve clean air in both rotors at the same time.

For vortex ring state demonstrations and training in recognition and recovery should be performed from a safe altitude to allow recovery no less than 1000 feet AGL or the manufacturer's recommended altitude, whichever is higher.

To enter the maneuver, come to an OGE hover, maintaining little or no airspeed (any direction), decrease collective to begin a vertical descent, and as the turbulence begins, increase collective. Then allow the sink rate to increase to 300 fpm or more as the attitude is adjusted to obtain airspeed of less than 10 knots. When the aircraft begins to shudder, the application of additional up collective increases the vibration and sink rate. As the power is increased, the rate of sink of the aircraft in the column of air will increase.

If altitude is sufficient, some time can be spent in the vortices, to enable the pilot to develop a healthy knowledge of the maneuver. However, helicopter pilots would normally initiate recovery at the first indication of vortex ring state. Recovery should be initiated at the first sign of vortex ring state by applying forward cyclic to increase airspeed and/ or simultaneously reducing collective. The recovery is complete when the aircraft passes through effective translational lift and a normal climb is established.

Common Errors—Traditional Recovery

1. Too much lateral speed for entry into vortex ring state.
2. Excessive decrease of collective.

Common Errors—Vuichard Recovery

1. Excessive lateral cyclic
2. Failure to maintain heading

Retreating Blade Stall

In forward flight, the relative airflow through the main rotor disk is different on the advancing and retreating side. The relative airflow over the advancing side is higher due to the forward speed of the helicopter, while the relative airflow on the retreating side is lower. This dissymmetry of lift increases as forward speed increases.

To generate the same amount of lift across the rotor disk, the advancing blade flaps up while the retreating blade flaps down. This causes the AOA to decrease on the advancing blade, which reduces lift, and increase on the retreating blade, which increases lift. At some point as the forward speed increases, the low blade speed on the retreating blade, and its high AOA cause a stall and loss of lift.

Retreating blade stall is a factor in limiting a helicopter's never-exceed speed (V_{NE}) and its development can be felt by a low frequency vibration, pitching up of the nose, and a roll in the direction of the retreating blade. High weight, low rotor rpm, high density altitude, turbulence and/or steep, abrupt turns are all conducive to retreating blade stall at high forward airspeeds. As altitude is increased, higher blade angles are required to maintain lift at a given airspeed. Thus, retreating blade stall is encountered at a lower forward airspeed at altitude. Most manufacturers publish charts and graphs showing a V_{NE} decrease with altitude.

When recovering from a retreating blade stall condition caused by high airspeed, moving the cyclic aft only worsens the stall as aft cyclic produces a flare effect, thus increasing the AOA. Pushing forward on the cyclic also deepens the stall as the AOA on the retreating blade is increased. While the first step in a proper recovery is usually to reduce collective, RBS should be evaluated in light of the relevant factors discussed in the previous paragraph and addressed accordingly. For example, if a pilot at high weight and high DA is about to conduct a high reconnaissance prior to a confined area operation where rolling into a steep turn causes onset of RBS, the recovery is to roll out of the turn. If the cause is low rotor rpm, then increase the rpm.

Common Errors

1. Failure to recognize the combination of contributing factors leading to retreating blade stall.
2. Failure to compute V_{NE} limits for altitudes to be flown.

Ground Resonance

Helicopters with articulating rotors (usually designs with three or more main rotor blades) are subject to ground resonance, a destructive vibration phenomenon that occurs at certain rotor speeds when the helicopter is on the ground. Ground resonance is a mechanical design issue that results from the helicopter's airframe having a natural frequency that can be intensified by an out-of-balance rotor. The unbalanced rotor disk vibrates at the same frequency (or multiple thereof) of the airframe's resonant frequency, and the harmonic oscillation increases because the engine is adding power to the system, increasing the magnitude (amplitude) of the vibrations until the structure or structures fail. This condition can cause a helicopter to self-destruct in a matter of seconds.

Hard contact with the ground on one corner (and usually with wheel-type landing gear) can send a shockwave to the main rotor head, resulting in the blades of a three-blade rotor disk moving from their normal 120° relationship to each other. This movement occurs along the drag hinge and could result in something like 122°, 122°, and 116° between blades. *[Figure 11-4]* When another part of the landing gear strikes the surface, the unbalanced condition could be further aggravated.

If the rpm is low, the only corrective action to stop ground resonance is to close the throttle immediately and fully lower the collective to place the blades in low pitch. If the rpm is in the normal operating range, fly the helicopter off the ground, and allow the blades to rephase themselves automatically. Then, make a normal touchdown. If a pilot lifts off and allows the helicopter to firmly re-contact the surface before the blades are realigned, a second shock could move the blades again and aggravate the already unbalanced condition. This could lead to a violent, uncontrollable oscillation.

This situation does not occur in rigid or semi-rigid rotor disks because there is no drag hinge. In addition, skid-type landing gear is not as prone to ground resonance as wheel-type landing gear, since the rubber tires' resonant frequency typically can match that of the spinning rotor, unlike the condition of a rigid landing gear.

Dynamic Rollover

A helicopter is susceptible to a lateral rolling tendency, called dynamic rollover, when it is in contact with the surface

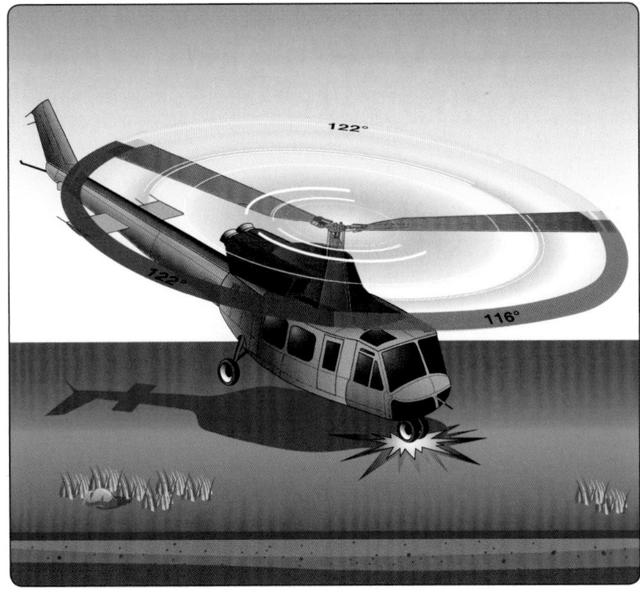

Figure 11-4. *Ground resonance.*

during takeoffs or landings. For dynamic rollover to occur, some factor must first cause the helicopter to roll or pivot around a skid or landing gear wheel, until its critical rollover angle is reached. The angle at which dynamic rollover occurs will vary based on helicopter type. Then, beyond this point, main rotor thrust continues the roll and recovery is impossible. After this angle is achieved, the cyclic does not have sufficient range of control to eliminate the thrust component and convert it to lift. If the critical rollover angle is exceeded, the helicopter rolls on its side regardless of the cyclic corrections made.

Dynamic rollover begins when the helicopter starts to pivot laterally around its skid or wheel. For dynamic rollover to occur the following three factors must be present:

1. A rolling moment
2. A pivot point other than the helicopter's normal CG
3. Thrust greater than weight

This can occur for a variety of reasons, including the failure to remove a tie down or skid-securing device, or if the skid or wheel contacts a fixed object while hovering sideward, or if the gear is stuck in ice, soft asphalt, or mud. Dynamic rollover may also occur if you use an improper landing or takeoff technique or while performing slope operations. Whatever the cause, dynamic rollover is possible if not using the proper corrective technique.

Once started, dynamic rollover cannot be stopped by application of opposite cyclic control alone. For example, the right skid contacts an object and becomes the pivot point while the helicopter starts rolling to the right. Even with full left cyclic applied, the main rotor thrust vector and its moment follows the aircraft as it continues rolling to the right. Quickly reducing collective pitch is the most effective way to stop dynamic rollover from developing. Dynamic rollover can occur with any type of landing gear and all types of rotor disks.

It is important to remember rotor blades have a limited range of movement. If the tilt or roll of the helicopter exceeds that range (5–8°), the controls (cyclic) can no longer command a vertical lift component and the thrust or lift becomes a lateral force that rolls the helicopter over. When limited rotor blade movement is coupled with the fact that most of a helicopter's weight is high in the airframe, another element of risk is added to an already slightly unstable center of gravity. Pilots must remember that in order to remove thrust, the collective must be lowered as this is the only recovery technique available.

Critical Conditions

Certain conditions reduce the critical rollover angle, thus increasing the possibility for dynamic rollover and reducing the chance for recovery. The rate of rolling motion is also a consideration because, as the roll rate increases, there is a reduction of the critical rollover angle at which recovery is still possible. Other critical conditions include operating at high gross weights with thrust (lift) approximately equal to the weight.

Refer to *Figure 11-5.* The following conditions are most critical for helicopters with counterclockwise rotor rotation:

1. Right side skid or landing wheel down, since translating tendency adds to the rollover force.
2. Right lateral center of gravity (CG).
3. Crosswinds from the left.
4. Left yaw inputs.

For helicopters with clockwise rotor rotation, the opposite conditions would be true.

Cyclic Trim

When maneuvering with one skid or wheel on the ground, care must be taken to keep the helicopter cyclic control carefully adjusted. For example, if a slow takeoff is attempted and the cyclic is not positioned and adjusted to account for translating tendency, the critical recovery angle may be exceeded in less than two seconds. Control can be maintained if the pilot maintains proper cyclic position and does not allow the helicopter's roll and pitch rates to become too great. Fly the helicopter into the air smoothly while keeping movements of pitch, roll, and yaw small; do not allow any abrupt cyclic pressures.

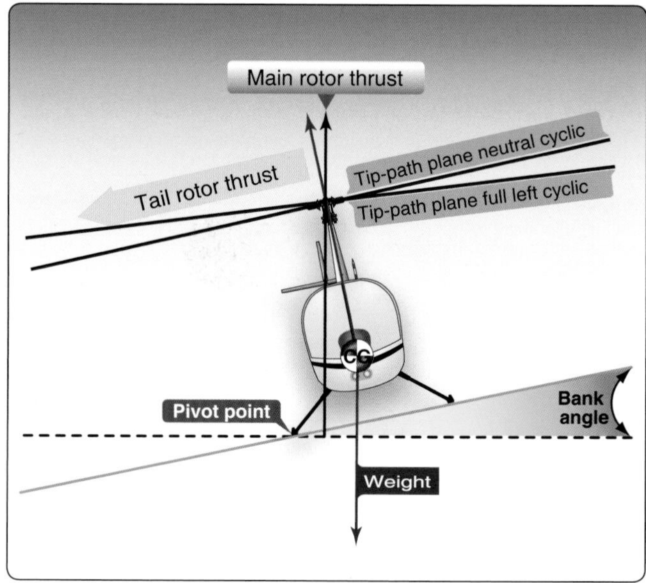

Figure 11-5. *Forces acting on a helicopter with right skid on the ground.*

Normal Takeoffs and Landings

Dynamic rollover is possible even during normal takeoffs and landings on relatively level ground, if one wheel or skid is on the ground and thrust (lift) is approximately equal to the weight of the helicopter. If the takeoff or landing is not performed properly, a roll rate could develop around the wheel or skid that is on the ground. When taking off or landing, perform the maneuver smoothly and carefully adjust the cyclic so that no pitch or roll movement rates build up, especially the roll rate. If the bank angle starts to increase to an angle of approximately 5–8°, and full corrective cyclic does not reduce the angle, the collective should be reduced to diminish the unstable rolling condition. Excessive bank angles can also be caused by landing gear caught in a tie down strap, or a tie down strap still attached to one side of the helicopter. Lateral loading imbalance (usually outside published limits) is another contributing factor.

Slope Takeoffs and Landings

During slope operations, excessive application of cyclic control into the slope, together with excessive collective pitch control, can result in the downslope skid or landing wheel rising sufficiently to exceed lateral cyclic control limits, and an upslope rolling motion can occur. *[Figure 11-6]*

When performing slope takeoff and landing maneuvers, follow the published procedures and keep the roll rates small. Slowly raise the downslope skid or wheel to bring the helicopter level, and then lift off. During landing, first touch down on the upslope skid or wheel, then slowly lower the downslope skid or wheel using combined movements of cyclic and collective. If the helicopter rolls approximately 5–8° to the upslope side, decrease collective to correct the bank angle and return to level attitude, then start the landing procedure again.

Use of Collective

The collective is more effective in controlling the rolling motion than lateral cyclic, because it reduces the main rotor thrust (lift). A smooth, moderate collective reduction, at a rate of less than approximately full up to full down in two seconds, may be adequate to stop the rolling motion. Take care, therefore, not to dump collective at an excessively high rate, as this may cause a main rotor blade to strike the fuselage. Additionally, if the helicopter is on a slope and the roll starts toward the upslope side, reducing collective too fast may create a high roll rate in the opposite direction. When the upslope skid or wheel hits the ground, the dynamics of the motion can cause the helicopter to bounce off the upslope skid or wheel, and the inertia can cause the helicopter to roll about the downslope ground contact point and over on its side. *[Figure 11-7]*

Under normal conditions on a slope, the collective should not be pulled suddenly to get airborne because a large and abrupt rolling moment in the opposite direction could occur. Excessive application of collective can result in the upslope skid or wheel rising sufficiently to exceed lateral cyclic control limits. This movement may be uncontrollable. If the helicopter develops a roll rate with one skid or wheel on the ground, the helicopter can roll over on its side.

Precautions

To help avoid dynamic rollover:

1. Always practice hovering autorotations into the wind, and be wary when the wind is gusty or greater than 10 knots.
2. Use extreme caution when hovering close to fences, sprinklers, bushes, runway/taxi lights, tiedown cables, deck nets, or other obstacles that could catch a skid or wheel. Aircraft parked on hot asphalt overnight might find the landing gear sunk in and stuck as the ramp cooled during the evening.

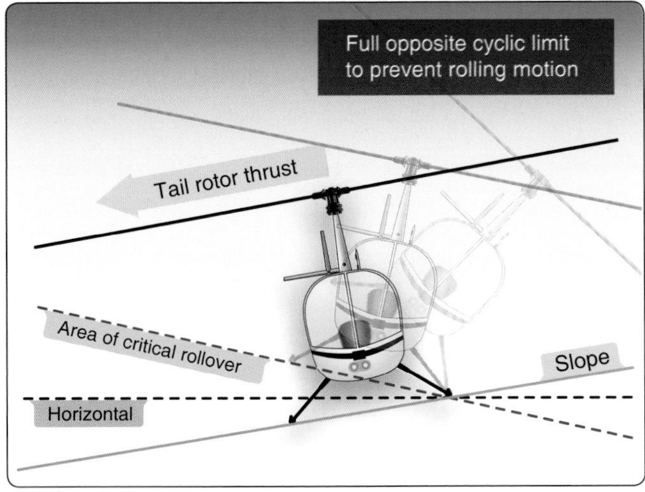

Figure 11-6. *Upslope rolling motion.*

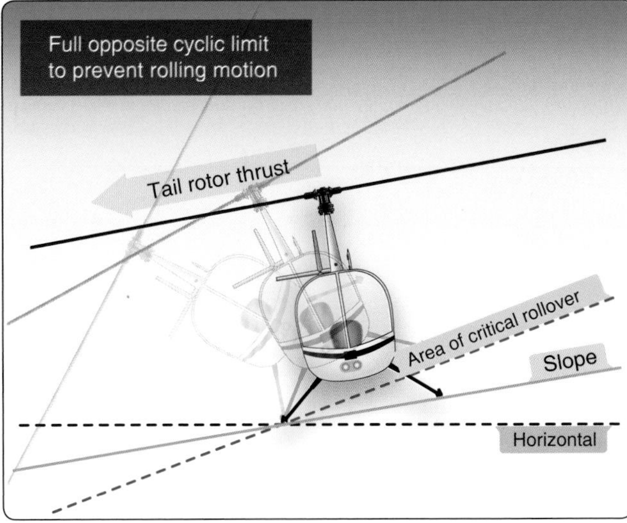

Figure 11-7. *Downslope rolling motion.*

3. Always use a two-step lift-off. Pull in just enough collective pitch control to be light on the skids or landing wheels and feel for equilibrium, then gently lift the helicopter into the air. 4.
Hover high enough to have adequate skid or landing wheel clearance from any obstacles when practicing hovering maneuvers close to the ground, especially when practicing sideways or rearward flight.
5. Remember that when the wind is coming from the upslope direction, less lateral cyclic control is available.
6. Avoid tailwind conditions when conducting slope operations.
7. Remember that less lateral cyclic control is available due to the translating tendency of the tail rotor when the left skid or landing wheel is upslope. (This is true for counterclockwise rotor disks.)
8. Keep in mind that the lateral cyclic requirement changes when passengers or cargo are loaded or unloaded.
9. Be aware that if the helicopter utilizes interconnecting fuel lines that allow fuel to automatically transfer from one side of the helicopter to the other, the gravitational flow of fuel to the downslope tank could change the CG, resulting in a different amount of cyclic control application to obtain the same lateral result.
10. Do not allow the cyclic limits to be reached. If the cyclic control limit is reached, further lowering of the collective may cause mast bumping. If this occurs, return to a hover and select a landing point with a lesser degree of slope.
11. During a takeoff from a slope, begin by leveling the main rotor disk with the horizon or very slightly into the slope to ensure vertical lift and only enough lateral thrust to prevent sliding on the slope. If the upslope skid or wheel starts to leave the ground before the downslope skid or wheel, smoothly and gently lower the collective and check to see if the downslope skid or wheel is caught on something. Under these conditions, vertical ascent is the only acceptable method of lift-off.
12. Be aware that dynamic rollover can be experienced during flight operations on a floating platform if the platform is pitching/rolling while attempting to land or takeoff. Generally, the pilot operating on floating platforms (barges, ships, etc.) observes a cycle of seven during which the waves increase and then decrease to a minimum. It is that time of minimum wave motion that the pilot needs to use for the moment of landing or takeoff on floating platforms. Pilots operating from floating platforms should also exercise great caution concerning cranes, masts, nearby boats (tugs) and nets.

Low-G Conditions and Mast Bumping

"G" is an abbreviation for acceleration due to the earth's gravity. A person standing on the ground or sitting in an aircraft in level flight is experiencing one G. An aircraft in a tight, banked turn with the pilot being pressed into the seat is experiencing more than one G or high-G conditions. A person beginning a downward ride in an elevator or riding down a steep track on a roller coaster is experiencing less than one G or low-G conditions. The best way for a pilot to recognize low G is a weightless feeling similar to the start of a downward elevator ride.

Helicopters rely on positive G to provide much or all of their response to pilot control inputs. The pilot uses the cyclic to tilt the rotor disk, and, at one G, the rotor is producing thrust equal to aircraft weight. The tilting of the thrust vector provides a moment about the center of gravity to pitch or roll the fuselage. In a low-G condition, the thrust and consequently the control authority are greatly reduced.

Although their control ability is reduced, multi-bladed (three or more blades) helicopters can generate some moment about the fuselage independent of thrust due to the rotor hub design with the blade attachment offset from the center of rotation. However, helicopters with two-bladed teetering rotors rely entirely on the tilt of the thrust vector for control. Therefore, low-G conditions can be catastrophic for two-bladed helicopters.

At lower speeds, such as initiation of a takeoff from hover or the traditional recovery from vortex ring state, forward cyclic maneuvers do not cause low G and are safe to perform. However, an abrupt forward cyclic input or pushover in a two-bladed helicopter can be dangerous and must be avoided, particularly at higher speeds. During a pushover from moderate or high airspeed, as the helicopter noses over, it enters a low-G condition. Thrust is reduced, and the pilot has lost control of fuselage attitude but may not immediately realize it. Tail rotor thrust or other aerodynamic factors will often induce a roll. The pilot still has control of the rotor disk, and may instinctively try to correct the roll, but the fuselage does not respond due to the lack of thrust. If the fuselage is rolling right, and the pilot puts in left cyclic to correct, the combination of fuselage angle to the right and rotor disk angle to the left becomes quite large and may exceed the clearances built into the rotor hub. This results in the hub contacting the rotor mast, which is known as mast bumping. *[Figure 11-8]* Low-G mast bumping has been the cause of numerous military and civilian fatal accidents. It was initially encountered during nap-of-the-earth flying, a very low-altitude tactical flight technique used by the military where

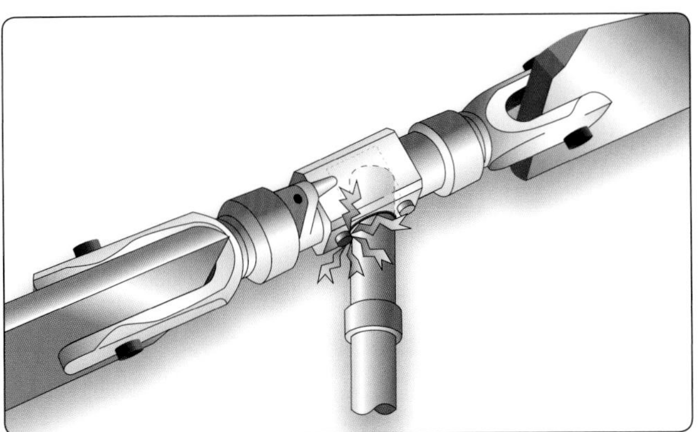

Figure 11-8. *Result of improper corrective action in a low-G condition.*

the aircraft flies following the contours of the geographical terrain. The accident sequence may be extremely rapid, and the energy and inertia in the rotor system can sever the mast or allow rotor blades to strike the tail or other portions of the helicopter.

Turbulence, especially severe downdrafts, can also cause a low-G condition and, when combined with high airspeed, may lead to mast bumping. Typically, helicopters handle turbulence better than a light airplane due to smaller surface area of the rotor blades. During flight in turbulence, momentary excursions in airspeed, altitude, and attitude are to be expected. Pilots should respond with smooth, gentle control inputs and avoid overcontrolling. Most importantly, pilots should slow down, as mast bumping is less likely at lower airspeeds.

Pilots can avoid mast bumping accidents as follows:

- Avoid abrupt forward cyclic inputs in two-bladed helicopters. Airplane pilots may find this a difficult habit to break because pushing the nose down is an accepted collision avoidance maneuver in an airplane. Helicopter pilots would accomplish the same rapid descent by lowering the collective, and airplane pilots should train to make this instinctual.
- Recognize the weightless feeling associated with the onset of low G and quickly take corrective action before the situation becomes critical.
- Recognize that uncommanded right roll for helicopters with main rotors which rotate counter-clockwise when viewed from above indicates that loss of control is imminent, and immediate corrective action must be taken.
- Recover from a low-G situation by first gently applying aft cyclic to restore normal G before attempting to correct any roll.
- If turbulence is expected or encountered, reduce power and use a slower than normal cruise speed. Turbulence (where high rotor flapping angles are already present), and higher airspeeds (where the controls are more sensitive) both increase susceptibility to low-G conditions.
- Use a flight simulator to learn to recognize and experience low G conditions that result in mast bumping, its correct recovery technique, and the consequences of using incorrect recovery actions. Refer to Chapter 14, Simulation.

Multi-bladed rotors may experience a phenomenon similar to mast bumping known as droop stop pounding if flapping clearances are exceeded, but because they retain some control authority at low G, occurrences are less common than for teetering rotors.

Low Rotor RPM and Rotor Stall

Rotor rpm is a critically important parameter for all helicopter operations. Just as airplanes will not fly below a certain airspeed, helicopters will not fly below a certain rotor rpm. Safe rotor rpm ranges are marked on the helicopter's tachometer and specified in the RFM. If the pilot allows the rotor rpm to fall below the safe operating range, the helicopter is in a low rpm situation. If the rotor rpm continues to fall, the rotor will eventually stall.

Rotor stall should not be confused with retreating blade stall, which occurs at high forward speeds and over a small portion of the retreating blade tip. Retreating blade stall causes vibration and control problems, but the rotor is still very capable of providing sufficient lift to support the weight of the helicopter. Rotor stall, however, can occur at any airspeed, and the rotor quickly stops producing enough lift to support the helicopter, causing it to lose lift and descend rapidly.

Rotor stall is very similar to the stall of an airplane wing at low airspeeds. The airplane wing relies on airspeed to produce the required airflow over the wing, whereas the helicopter relies on rotor rpm. As the airspeed of the airplane decreases or the speed of the helicopter rotor slows down, the AOA of the wing/rotor blade must be increased to support the weight of the aircraft. At a critical angle (about 15°), the airflow over the wing or the rotor blade will separate and stall, causing a sudden loss of lift and increase in drag (refer to Chapter 2, Aerodynamics of Flight). An airplane pilot recovers from a stall by lowering the nose to reduce the AOA and adding power to restore normal airflow over the wing. However, the falling helicopter is experiencing upward airflow through the rotor disk, and the resulting AOA is so high that even full down collective will not restore normal airflow. In the helicopter when the rotor stalls, it does not do so symmetrically because any forward airspeed will produce a higher airflow on the advancing side than on the retreating side. This causes the retreating blade to stall first, and its weight makes it descend as it moves aft while the advancing blade is climbing as it goes forward. The resulting low aft blade and high forward blade become a rapid aft tilting of the rotor disc sometimes referred to as rotor "blow back" or "flap back." As the helicopter begins to descend, the upward flow of air acting on the bottom surfaces of the tail boom and any horizontal stabilizers tend to pitch the aircraft nose down. These two effects, combined with any aft cyclic by the pilot attempting to keep the aircraft level, allow the rotor blades to blow back and contact the tail boom, in some cases actually severing the tail boom. Since the tail rotor is geared to the main rotor, in many helicopters the loss of main rotor rpm also causes a significant loss of tail rotor thrust and a corresponding loss of directional control.

Rotor stalls in helicopters are not recoverable. At low altitude, rotor stall will result in an accident with significant damage to the helicopter, and at altitudes above approximately 50 feet the accident will likely be fatal. Consequently, early recognition of the low rotor rpm condition and proper recovery technique is imperative.

Low rotor rpm can occur during power-off and power-on operations. During power-off flight, a low rpm situation can be caused by the failure to quickly lower the collective after an engine failure or by raising the collective at too great a height above ground at the bottom of an autorotation. However, more common are power-on rotor stall accidents. These occur when the engine is operating normally but the pilot demands more power than is available by pulling up too much on the collective. Known as "overpitching," this can easily occur at higher density altitudes where the engine is already producing its maximum horsepower and the pilot raises the collective. The corresponding increased AOA of the blades requires more engine horsepower to maintain the speed of the blades; however, the engine cannot produce any additional horsepower, so the speed of the blades decreases. A similar situation can occur with a heavily loaded helicopter taking off from a confined area. Other causes of a power-on low rotor rpm condition include the pilot rolling the throttle the wrong way in helicopters not equipped with a governor or a governor failure in helicopters so equipped.

As the rpm decreases, the amount of horsepower the engine can produce also decreases. Engine horsepower is directly proportional to its rpm, so a 10 percent loss in rpm due to overpitching, or one of the other scenarios above, will result in a 10 percent loss in the engine's ability to produce horsepower, making recovery even slower and more difficult than it would otherwise be. With less power from the engine and less lift from the decaying rotor rpm, the helicopter will start to settle. If the pilot raises the collective to stop the settling, the situation will feed upon itself rapidly leading to rotor stall.

There are a number of ways the pilot can recognize the low rotor rpm situation. Visually, the pilot can not only see the rotor rpm indicator decrease but also the change in torque will produce a yaw; there will also be a noticeable decrease in engine noise, and at higher airspeeds or in turns, an increase in vibration. Many helicopters have a low rpm warning system that alerts the pilot to the low rotor rpm condition.

To recover from the low rotor rpm condition the pilot must simultaneously lower the collective, increase throttle if available and apply aft cyclic to maintain a level attitude. At higher airspeeds, additional aft cyclic may be used to help recover lost rpm. Recovery should be accomplished immediately before investigating the problem and must be practiced to become a conditioned reflex.

System Malfunctions

By following the manufacturer's recommendations regarding operating limits and procedures and periodic maintenance and inspections, many system and equipment failures can be eliminated. Certain malfunctions or failures can be traced to some error on the part of the pilot; therefore, appropriate flying techniques and use of threat and error management may help to prevent an emergency

Antitorque System Failure

Antitorque failure usually falls into one of two categories. One is failure of the power drive portion of the tail rotor disk resulting in a complete loss of antitorque. The other category covers mechanical control failures prohibiting the pilot from changing or controlling tail rotor thrust even though the tail rotor may still be providing antitorque thrust.

Tail rotor drive system failures include driveshaft failures, tail rotor gearbox failures, or a complete loss of the tail rotor itself. In any of these cases, the loss of antitorque normally results in an immediate spinning of the helicopter's nose. The helicopter spins to the right in a counterclockwise rotor disk and to the left in a clockwise system. This discussion is for a helicopter with a counterclockwise rotor disk. The severity of the spin is proportionate to the amount of power being used and the airspeed. An antitorque failure with a high-power setting at a low airspeed results in a severe spinning to the right. At low power settings and high airspeeds, the spin is less severe. High airspeeds tend to streamline the helicopter and keep it from spinning.

If a tail rotor failure occurs, power must be reduced in order to reduce main rotor torque. The techniques differ depending on whether the helicopter is in flight or in a hover, but ultimately require an autorotation. If a complete tail rotor failure occurs while hovering, enter a hovering autorotation by rolling off the throttle. If the failure occurs in forward flight, enter a normal autorotation by lowering the collective and rolling off the throttle. If the helicopter has enough forward airspeed (close to cruising speed) when the failure occurs, and depending on the helicopter design, the vertical stabilizer may provide enough directional control to allow the pilot to maneuver the helicopter to a more desirable landing sight. Applying slight cyclic control opposite the direction of yaw compensates for some of the yaw. This helps in directional control, but also increases drag. Care must be taken not to lose too much forward airspeed because the streamlining effect diminishes as airspeed is reduced. Also, more altitude is required to accelerate to the correct airspeed if an autorotation is entered at a low airspeed.

The throttle or power lever on some helicopters is not located on the collective and readily available. Faced with the loss of antitorque, the pilot of these models may need to achieve forward flight and let the vertical fin stop the yawing rotation. With speed and altitude, the pilot will have the time to set up for an autorotative approach and set the power control to idle or off as the situation dictates. At low altitudes, the pilot may not be able to reduce the power setting and enter the autorotation before impact.

A mechanical control failure limits or prevents control of tail rotor thrust and is usually caused by a stuck or broken control rod or cable. While the tail rotor is still producing antitorque thrust, it cannot be controlled by the pilot. The amount of antitorque depends on the position at which the controls jam or fail. Once again, the techniques differ depending on the amount of tail rotor thrust, but an autorotation is generally not required.

The specific manufacturer's procedures should always be followed. The following is a generalized description of procedures when more specific procedures are not provided.

Landing—Stuck Left Pedal

A stuck left pedal (high power setting), which might be experienced during takeoff or climb conditions, results in the left yaw of the helicopter nose when power is reduced. Rolling off the throttle and entering an autorotation only makes matters worse. The landing profile for a stuck left pedal is best described as a normal-to-steep approach angle to arrive approximately 2–3 feet landing gear height above the intended landing area as translational lift is lost. The steeper angle allows for a lower power setting during the approach and ensures that the nose remains to the right.

Upon reaching the intended touchdown area and at the appropriate landing gear height, increase the collective smoothly to align the nose with the landing direction and cushion the landing. A small amount of forward cyclic is helpful to stop the nose from continuing to the right and directs the aircraft forward and down to the surface. In certain wind conditions, the nose of the helicopter may remain to the left with zero to near zero groundspeed above the intended touchdown point. If the helicopter is not turning, simply lower the helicopter to the surface. If the nose of the helicopter is turning to the right and continues beyond the landing heading, roll the throttle toward flight idle, which is the amount necessary to stop the turn while landing. Flight idle is an engine rpm in flight at a given altitude with the throttle set to the minimum, or idle, position. The flight idling rpm typically increase with an increase in altitude. If the helicopter is beginning to turn left, the pilot should be able to make the landing prior to the turn rate becoming excessive. However, if the turn rate begins to increase prior to the landing, simply add power to make a go-around and return for another landing.

Landing—Stuck Neutral or Right Pedal

The landing profile for a stuck neutral or a stuck right pedal is a low-power approach terminating with a running or roll-on landing. The approach profile can best be described as a shallow to normal approach angle to arrive approximately 2–3 feet landing gear height above the intended landing area with a minimum airspeed for directional control. The minimum airspeed is one that keeps the nose from continuing to yaw to the right.

Upon reaching the intended touchdown area and at the appropriate landing gear height, reduce the throttle as necessary to overcome the yaw effect if the nose of the helicopter remains to the right of the landing heading. The amount of throttle reduction will vary based on power applied and winds. The higher the power setting used to cushion the landing, the more the throttle reduction will be. A coordinated throttle reduction and increased collective will result in a very smooth touchdown with some forward groundspeed. If the nose of the helicopter is to the left of the landing heading, a slight increase in collective or aft cyclic may be used to align the nose for touchdown. The decision to land or go around has to be made prior to any throttle reduction. Using airspeeds slightly above translational lift may be helpful to ensure that the nose does not continue yawing to the right. If a go-around is required, increasing the collective too much or too rapidly with airspeeds below translational lift may cause a rapid spinning to the right.

Once the helicopter has landed and is sliding/rolling to a stop, the heading can be controlled with a combination of collective, cyclic and throttle. To turn the nose to the right, raise the collective or apply aft cyclic. The throttle may be increased as well if it is not in the full open position. To turn the nose to the left, lower the collective or apply forward cyclic. The throttle may be decreased as well if it is not already at flight idle.

Loss of Tail Rotor Effectiveness (LTE)

Loss of tail rotor effectiveness (LTE) or an unanticipated yaw is defined as an uncommanded, rapid yaw towards the advancing blade which does not subside of its own accord. It can result in the loss of the aircraft if left unchecked. It is very important for pilots to understand that LTE is caused by an aerodynamic interaction between the main rotor and tail rotor and not caused from a mechanical failure. Some helicopter types are more likely to encounter LTE due to the normal certification thrust produced by having a tail rotor that, although meeting certification standards, is not always able to produce the additional thrust demanded by the pilot.

A helicopter is a collection of compromises. Compare the size of an airplane propeller to that of a tail rotor. Then, consider the horsepower required to run the propeller. For example, a Cessna 172P is equipped with a 160-horsepower (HP) engine. A Robinson R-44 with a comparably sized tail rotor is rated for a maximum of 245 HP. If you assume the tail rotor consumes 50 HP, only 195 HP remains to drive the main rotor. If the pilot were to apply enough collective to require 215 HP from the engine, and enough left pedal to require 50 HP for the tail rotor, the resulting engine overload would lead to one of two outcomes: slow down (reduction in rpm) or premature failure. In either outcome, antitorque would be insufficient and total lift might be less than needed to remain airborne.

Every helicopter design requires some type of antitorque system to counteract main rotor torque and prevent spinning once the helicopter lifts off the ground. A helicopter is heavy, and the powerplant places a high demand on fuel. Weight penalizes performance, but all helicopters must have an antitorque system, which adds weight. Therefore, the tail rotor is certified for normal flight conditions. Environmental forces can overwhelm any aircraft, rendering the inherently unstable helicopter especially vulnerable.

As with any aerodynamic condition, it is very important for pilots to not only to understand the definition of LTE, but more importantly, how and why it happens, how to avoid it, and lastly, how to correct it once it is encountered. We must first understand the capabilities of the aircraft or even better what it is not capable of doing. For example, if you were flying a helicopter with a maximum gross weight of 5,200 lb, would you knowingly try to take on fuel, baggage and passengers causing the weight to be 5,500 lb? A wise professional pilot should not ever exceed the certificated maximum gross weight or performance flight weight for any aircraft. The manuals are written for safety and reliability. The limitations and emergency procedures are stressed because lapses in procedures or exceeding limitations can result in aircraft damage or human fatalities. At the very least, exceeding limitations will increase the costs of maintenance and ownership of any aircraft and especially helicopters.

Overloaded parts may fail before their designed lifetime. There are no extra parts in helicopters. The respect and discipline pilots exercise in following flight manuals should also be applied to understanding aerodynamic conditions. If flight envelopes are exceeded, the end results can be catastrophic.

LTE is an aerodynamic condition and is the result of a control margin deficiency in the tail rotor. It can affect all single-rotor helicopters that utilize a tail rotor. The design of main and tail rotor blades and the tail boom assembly can affect the characteristics and susceptibility of LTE but will not nullify the phenomenon entirely. Translational lift is obtained by any amount of clean air through the main rotor disk. Chapter 2, Aerodynamics of Flight, discusses translational lift with respect to the main rotor blade, explaining that the more clean air there is going through the rotor disk, the more efficient it becomes. The same holds true for the tail rotor. As the tail rotor works in less turbulent air, it reaches a point of translational thrust. At this point, the tail rotor becomes aerodynamically efficient and the improved efficiency produces more antitorque thrust. The pilot can determine when the tail rotor has reached translational thrust. As more antitorque thrust is produced, the nose of the helicopter yaws to the left (opposite direction of the tail rotor thrust), forcing the pilot to correct with right pedal application (actually decreasing the left pedal). This, in turn, decreases the AOA in the tail rotor blades. Pilots should be aware of the characteristics of the helicopter they fly and be particularly aware of the amount of tail rotor pedal typically required for different flight conditions.

LTE is a condition that occurs when the flow of air through a tail rotor is altered in some way, by altering the angle or speed at which the air passes through the rotating blades of the tail rotor disk. As discussed in the previous paragraph, an effective tail rotor relies on a stable and relatively undisturbed airflow in order to provide a steady and constant antitorque reaction. The pitch and AOA of the individual blades will determine the thrust. A change to either of these alters the amount of thrust generated. A pilot's yaw pedal input causes a thrust reaction from the tail rotor. Altering the amount of thrust delivered for the same yaw input creates an imbalance. Taking this imbalance to the extreme will result in the loss of effective control in the yawing plane, and LTE will occur.

This alteration of tail rotor thrust can be affected by numerous external factors. The main factors contributing to LTE are:

1. Airflow and downdraft generated by the main rotor blades interfering with the airflow entering the tail rotor assembly.
2. Main blade vortices developed at the main blade tips entering the tail rotor disk.
3. Turbulence and other natural phenomena affecting the airflow surrounding the tail rotor.
4. A high-power setting, hence large main rotor pitch angle, induces considerable main rotor blade downwash and hence more turbulence than when the helicopter is in a low power condition.
5. A slow forward airspeed, typically at speeds where translational lift and translational thrust are in the process of change and airflow around the tail rotor will vary in direction and speed.
6. The airflow relative to the helicopter;
   a. Worst case—relative wind within ±15° of the 10 o'clock position, generating vortices that can blow directly into the tail rotor. This is dictated by the characteristics of the helicopters aerodynamics of tailboom position, tail rotor size and position relative to the main rotor and vertical stabilizer, size and shape. *[Figure 11-9]*
   b. Weathercock stability—tailwinds from 120° to 240° *[Figure 11-10]*, such as left crosswinds, causing high pilot workload.
   c. Tail rotor vortex ring state (210° to 330°). *[Figure 11-11]* Winds within this region will result in the development of the vortex ring state of the tail rotor.
7. Combinations (a, b, c) of these factors in a particular situation can easily require more antitorque than the helicopter can generate and in a particular environment LTE can be the result.

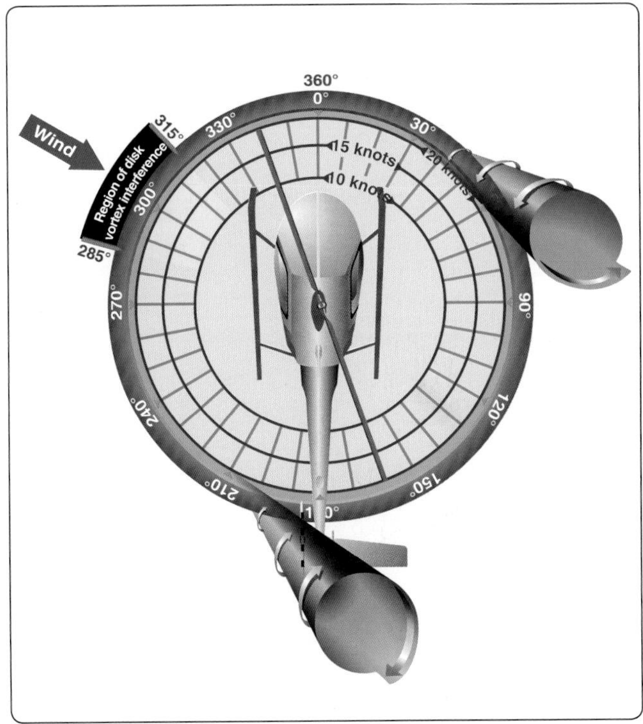

Figure 11-9. *Main rotor disk vortex interference.*

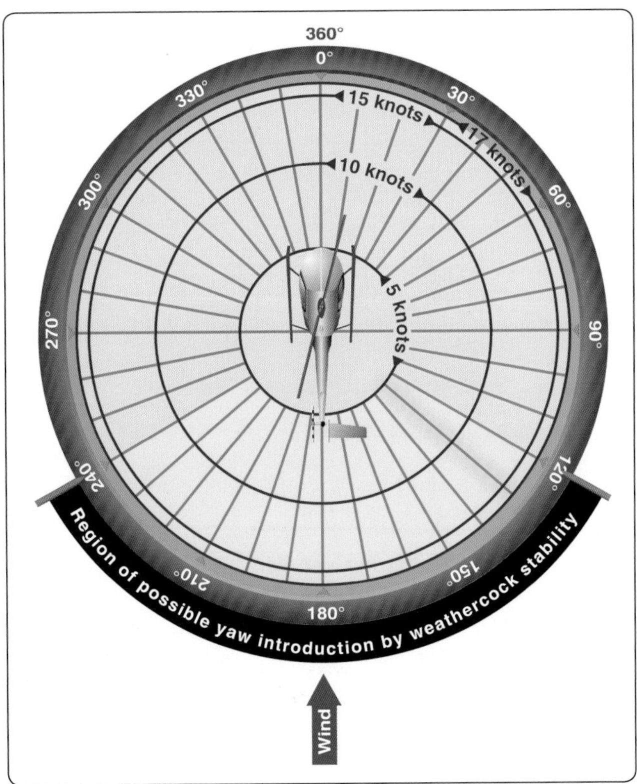

Figure 11-10. *Weathercock stability.*

Certain flight activities lend themselves to being at higher risk of LTE than others. For example, power line and pipeline patrol sectors, low speed aerial filming/photography as well as in the Police and Helicopter Emergency Medical Services (EMS) environments can find themselves in low-and-slow situations over geographical areas where the exact wind speed and direction are hard to determine.

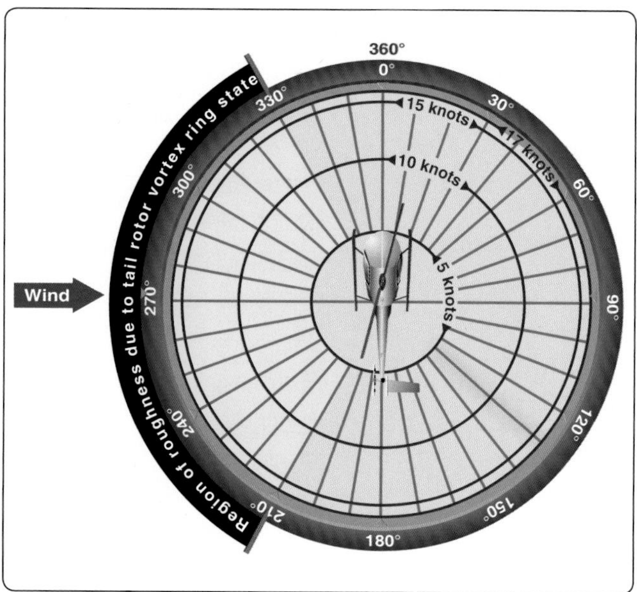

Figure 11-11. *Tail rotor vortex ring state.*

Unfortunately, the aerodynamic conditions that a helicopter is susceptible to are not explainable in black and white terms. LTE is no exception. There are a number of contributing factors, but what is more important in preventing LTE is to note them, and then to associate them with situations that should be avoided. Whenever possible, pilots should learn to avoid the following combinations:

1. Low and slow flight outside of ground effect.
2. Winds from ±15° of the 10 o'clock position and probably on around to 5 o'clock position *[Figure 11-9]*
3. Tailwinds that may alter the onset of translational lift and translational thrust, and hence induce high power demands and demand more anti-torque (left pedal) than the tail rotor can produce.
4. Low speed downwind turns.
5. Large changes of power at low airspeeds.
6. Low speed flight in the proximity of physical obstructions that may alter a smooth airflow to both the main rotor and tail rotor.

Pilots who put themselves in situations where the combinations above occur should know that they are likely to encounter LTE. The key is not to put the helicopter in a compromising condition, while at the same time being educated enough to recognize the onset of LTE and being prepared to react quickly to it before the helicopter cannot be controlled.

Early detection of LTE, followed by the immediate flight control application of corrective action, applying forward cyclic to regain airspeed, applying right pedal not left as necessary to maintain rotor rpm, and reducing the collective (thus reducing the high-power demand on the tail rotor), is the key to a safe recovery. Pilots should always set themselves up when conducting any maneuver to have enough height and space available to recover in the event they encounter an aerodynamic situation such as LTE.

Understanding the aerodynamic phenomenon of LTE is by far the most important factor in preventing an LTE-related accident, and maintaining the ability and option either to go around if making an approach or pull out of a maneuver safely and re-plan, is always the safest option. Having the ability to fly away from a situation and re-think the possible options should always be part of a pilot's planning process in all phases of flight. Unfortunately, there have been many pilots who have idled a good engine and fully functioning tail rotor disk and autorotated a perfectly airworthy helicopter to the crash site because they misunderstood or misperceived both the limitations of the helicopter and the aerodynamic situation.

Main Rotor Disk Interference (285–315°)

Refer to *Figure 11-9.* Winds at velocities of 10–30 knots from the left front cause the main rotor vortex to be blown into the tail rotor by the relative wind. This main rotor disk vortex causes the tail rotor to operate in an extremely turbulent environment. During a right turn, the tail rotor experiences a reduction of thrust as it comes into the area of the main rotor disk vortex. The reduction in tail rotor thrust comes from the airflow changes experienced at the tail rotor as the main rotor disk vortex moves across the tail rotor disk.

The effect of the main rotor disk vortex initially increases the AOA of the tail rotor blades, thus increasing tail rotor thrust. The increase in the AOA requires that right pedal pressure be added to reduce tail rotor thrust in order to maintain the same rate of turn. As the main rotor vortex passes the tail rotor, the tail rotor AOA is reduced. The reduction in the AOA causes a reduction in thrust and right yaw acceleration begins. This acceleration can be surprising, since previously adding right pedal to maintain the right turn rate. This thrust reduction occurs suddenly, and if uncorrected, develops into an uncontrollable rapid rotation about the mast. When operating within this region, be aware that the reduction in tail rotor thrust can happen quite suddenly, and be prepared to react quickly to counter this reduction with additional left pedal input.

Weathercock Stability (120–240°)

In this region, the helicopter attempts to weathervane, or weathercock, its nose into the relative wind. *[Figure 11-10]* Unless a resisting pedal input is made, the helicopter starts a slow, uncommanded turn either to the right or left, depending upon the wind direction. If the pilot allows a right yaw rate to develop and the tail of the helicopter moves into this region, the yaw rate can accelerate rapidly. In order to avoid the onset of LTE in this downwind condition, it is imperative to maintain positive control of the yaw rate and devote full attention to flying the helicopter.

Tail Rotor Vortex Ring State (210–330°)

Winds within this region cause a tail rotor vortex ring state to develop. *[Figure 11-11]* The result is a nonuniform, unsteady flow into the tail rotor. The vortex ring state causes tail rotor thrust variations, which result in yaw deviations. The net effect of the unsteady flow is an oscillation of tail rotor thrust. Rapid and continuous pedal movements are necessary to compensate for the rapid changes in tail rotor thrust when hovering in a left crosswind. Maintaining a precise heading in this region is difficult, but this characteristic presents no significant problem unless corrective action is delayed. However, high pedal workload, lack of concentration, and overcontrolling can lead to LTE.

When the tail rotor thrust being generated is less than the thrust required, the helicopter yaws to the right. When hovering in left crosswinds, concentrate on smooth pedal coordination and do not allow an uncommanded right yaw to develop. If a right yaw rate is allowed to build, the helicopter can rotate into the wind azimuth region where weathercock stability then accelerates the right turn rate. Pilot workload during a tail rotor vortex ring state is high. Do not allow a right yaw rate to increase.

LTE at Altitude

At higher altitudes where the air is thinner, tail rotor thrust and efficiency are reduced. Because of the high-density altitude, powerplants may be much slower to respond to power changes. When operating at high altitudes and high gross weights, especially while hovering, the tail rotor thrust may not be sufficient to maintain directional control, and LTE can occur. In this case, the hovering ceiling is limited by tail rotor thrust and not necessarily power available. In these conditions, gross weights need to be reduced and/or operations need to be limited to lower density altitudes. This may not be noted as criteria on the performance charts.

Reducing the Onset of LTE

To help reduce the onset of LTE, follow these steps:

1. Maintain maximum power-on rotor rpm. If the main rotor rpm is allowed to decrease, the antitorque thrust available is decreased proportionally.
2. Avoid tailwinds below airspeeds of 30 knots. If loss of translational lift occurs, it results in an increased power demand and additional antitorque pressures.
3. Avoid OGE operations and high-power demand situations below airspeeds of 30 knots at low altitudes.
4. Be especially aware of wind direction and velocity when hovering in winds of about 8–12 knots. A loss of translational lift results in an unexpected high power demand and an increased antitorque requirement.
5. Be aware that if a considerable amount of left pedal is being maintained, a sufficient amount of left pedal may not be available to counteract an unanticipated right yaw.
6. Be alert to changing wind conditions, which may be experienced when flying along ridge lines and around buildings.
7. Execute right turns slowly. This limits the effects of rotating inertia, and decreases loading on the tailrotor to control yawing.

Recovery Technique (Uncontrolled Right Yaw)

If a sudden unanticipated right yaw occurs, the following recovery technique should be performed. Apply full left pedal. Simultaneously, apply forward cyclic control to increase speed. If altitude permits, reduce power. As recovery is affected, adjust controls for normal forward flight. A recovery path must always be planned, especially when terminating to an OGE hover and executed immediately if an uncommanded yaw is evident.

Collective pitch reduction aids in arresting the yaw rate but may cause an excessive rate of descent. Any large, rapid increase in collective to prevent ground or obstacle contact may further increase the yaw rate and decrease rotor rpm. The decision to reduce collective must be based on the pilot's assessment of the altitude available for recovery.

If the rotation cannot be stopped and ground contact is imminent, an autorotation may be the best course of action. Maintain full left pedal until the rotation stops, then adjust to maintain heading. For more information on LTE, see Advisory Circular (AC) 90-95, Unanticipated Right Yaw in Helicopters.

Main Drive Shaft or Clutch Failure

The main drive shaft, located between the engine and the main rotor transmission, provides engine power to the main rotor transmission. In some helicopters, particularly those with piston engines, a drive belt is used instead of a drive shaft. A failure of the drive shaft clutch or belt has the same effect as an engine failure because power is no longer provided to the main rotor and an autorotation must be initiated. There are a few differences, however, that need to be taken into consideration. If the drive shaft or belt breaks, the lack of any load on the engine results in an overspeed. In this case, the throttle must be closed in order to prevent any further damage. In some helicopters, the tail rotor drive system continues to be powered by the engine even if the main drive shaft breaks. In this case, when the engine unloads, a tail rotor overspeed can result. If this happens, close the throttle immediately and enter an autorotation. The pilot must be knowledgeable of the specific helicopter's system and failure modes.

Pilots should keep in mind that when there is any suspected mechanical malfunction, first and foremost they should always attempt to maintain rotor rpm. If the rotor rpm is at the normal indication with normal power settings, an instrument failure might be occurring, and it would be best to fly the helicopter to a safe landing area. If the rotor rpm is in fact decreasing or low, then there is a drive line failure.

Hydraulic Failure

Many helicopters incorporate the use of hydraulic actuators to overcome high control forces. A hydraulic system consists of actuators, also called servos, on each flight control; a pump, which is usually driven by the main rotor transmission; and a reservoir to store the hydraulic fluid. A switch in the cockpit can turn the system off, although it is left on during normal conditions. A pressure indicator in the cockpit may be installed to monitor the system.

An impending hydraulic failure can be recognized by a grinding or howling noise from the pump or actuators, increased control forces and feedback, and limited control movement. The required corrective action is stated in detail in the RFM. In most cases, airspeed needs to be reduced in order to reduce control forces. The hydraulic switch and circuit breaker should be checked and recycled. If hydraulic power is not restored, make a shallow approach to a running or roll-on landing. This technique is used because it requires less control force and pilot workload. Additionally, the hydraulic system should be disabled by placing the switch in the off position. The reason for this is to prevent an inadvertent restoration of hydraulic power, which may lead to overcontrolling near the ground.

In those helicopters in which the control forces are so high that they cannot be moved without hydraulic assistance, two or more independent hydraulic systems are installed. Some helicopters use hydraulic accumulators to store pressure that can be used for a short time while in an emergency if the hydraulic pump fails. This gives enough time to land the helicopter with normal control.

Governor or Fuel Control Failure

Governors and fuel control units automatically adjust engine power to maintain rotor rpm when the collective pitch is changed. If the governor or fuel control unit fails, any change in collective pitch requires manual adjustment of the throttle to maintain correct rpm. In the event of a high side failure, the engine and rotor rpm tend to increase above the normal range due to the engine being commanded to put out too much power. If the rpm cannot be reduced and controlled with the throttle, close the throttle and enter an autorotation. If the failure is on the low side, the engine output is allowed to go below the collective and normal rpm may not be attainable, even if the throttle is manually controlled. In this case, the collective has to be lowered to maintain rotor rpm. A running or roll-on landing may be performed if the engine can maintain sufficient rotor rpm. If there is insufficient power, enter an autorotation. As stated previously in this chapter, before responding to any type of mechanical failure, pilots should confirm that rotor rpm is not responding to flight control inputs. If the rotor rpm can be maintained in the green operating range, the failure is in the instrument, and not mechanical.

Abnormal Vibration

With the many rotating parts found in helicopters, some vibration is inherent. A pilot needs to understand the cause and effect of helicopter vibrations because abnormal vibrations cause premature component wear and may even result in structural failure. With experience, a pilot learns what vibrations are normal and those that are abnormal and can then decide whether continued flight is safe or not. Helicopter vibrations are categorized into low, medium, or high frequency.

Low-Frequency Vibrations

Low-frequency vibrations (100–500 cycles per minute) usually originate from the main rotor disk. The main rotor operational range, depending on the helicopter, is usually between 320 and 500 rpm. A rotor blade that is out of track or balance will cause a cycle to occur with every rotation. The vibration may be felt through the controls, the airframe, or a combination of both. The vibration may also have a definite direction of push or thrust. It may be vertical, lateral, horizontal, or even a combination of these. Normally, the direction of the vibration can be determined by concentrating on the feel of the vibration, which may push a pilot up and down, backwards and forwards, or in the case of a blade being out of phase, from side to side. The direction of the vibration and whether it is felt in the controls or the airframe is important information for the mechanic when he or she troubleshoots the source. Out-of-track or out-of-balance main rotor blades, damaged blades, worn bearings, dampers out of adjustment, or worn parts are possible causes of low frequency vibrations.

Medium- and High-Frequency Vibrations

Medium-frequency vibrations (1,000–2,000 cycles per minute) range between the low frequencies of the main rotor (100–500 cycles per minute) and the high frequencies (2,100 cycles per minute or higher) of the engine and tail rotor. Depending on the helicopter, medium-frequency vibration sources may be engine and transmission cooling fans, and accessories such as air conditioner compressors, or driveline components. Medium-frequency vibrations are felt through the entire airframe, and prolonged exposure to the vibrations will result in greater pilot fatigue.

Most tail rotor vibrations fall into the high-frequency range (2,100 cycles per minute or higher) and can be felt through the tail rotor pedals as long as there are no hydraulic actuators to dampen out the vibration. This vibration is felt by the pilot through his or her feet, which are usually "put to sleep" by the vibration. The tail rotor operates at approximately a 6:1 ratio with the main rotor, meaning for every one rotation of the main rotor the tail rotor rotates 6 times. A main rotor operating rpm of 350 means the tail rotor rpm would be 2,100 rpm. Any imbalance in the tail rotor disk is very harmful as it can cause cracks to develop and rivets to work loose. Piston engines usually produce a normal amount of high-frequency vibration, which is aggravated by engine malfunctions, such as spark plug fouling, incorrect magneto timing, carburetor icing and/or incorrect fuel/air mixture. Vibrations in turbine engines are often difficult to detect as these engines operate at a very high rpm. Turbine engine vibration can be at 30,000 rpm internally, but common transmission speeds are in the 1,000 to 3,000 rpm range for the output shaft. The vibrations in turbine engines may be short lived as the engine disintegrates rapidly when damaged due to high rpm and the forces present.

Tracking and Balance

Modern equipment used for tracking and balancing the main and tail rotor blades can also be used to detect other vibrations in the helicopter. These systems use accelerometers mounted around the helicopter to detect the direction, frequency, and intensity of the vibration. The built-in software can then analyze the information, pinpoint the origin of the vibration, and suggest the corrective action.

The use of a system such as a health and usage monitoring system (HUMS) provides the operator the ability to record engine and transmission performance and provide rotor track and balance. This system has been around for over 30 years and is now becoming more affordable, more capable, and more commonplace in the rotorcraft industry.

Multiengine Emergency Operations

Single-Engine Failure

When one engine has failed, the helicopter can often maintain altitude and airspeed until a suitable landing site can be selected. Whether or not this is possible becomes a function of such combined variables as aircraft weight, density altitude, height above ground, airspeed, phase of flight, and single-engine capability. Environmental response time and control technique may be additional factors. Caution must be exercised to correctly identify the malfunctioning engine since there is no telltale yawing as occurs in most multiengine airplanes. Shutting down the wrong engine could be disastrous!

Even when flying multiengine powered helicopters, rotor rpm must be maintained at all costs, because fuel contamination has been documented as the cause for both engines failing in flight.

Dual-Engine Failure

The flight characteristics and the required crew member control responses after a dual-engine failure are similar to those during a normal power-on descent. Full control of the helicopter can be maintained during autorotational descent. In autorotation, as airspeed increases above 70–80 KIAS, the rate of descent and glide distance increase significantly. As airspeed decreases below approximately 60 KIAS, the rate of descent increases and glide distance decreases.

Lost Procedures

Pilots become lost while flying for a variety of reasons, such as disorientation, flying over unfamiliar territory, or visibility that is low enough to render familiar terrain unfamiliar. When a pilot becomes lost, the first order of business is to fly the aircraft; the second is to implement lost procedures. Keep in mind that the pilot workload will be high, and increased concentration will be necessary. If lost, always remember to look for the practically invisible hazards, such as wires, by searching for their support structures, such as poles or towers, which are almost always near roads.

If lost, follow common sense procedures.

- Try to locate any large landmarks, such as lakes, rivers, towers, railroad tracks, or Interstate highways. If a landmark is recognized, use it to find the helicopter's location on the sectional chart. If flying near a town or city, a pilot may be able to read the name of the town on a water tower or even land to ask for directions.
- If no town or city is nearby, the first thing a pilot should do is climb. An increase in altitude increases radio and navigation reception range as well as radar coverage.

- Navigation aids, dead reckoning, and pilotage are skills that can be used as well.
- Do not forget air traffic control (ATC)—controllers assist pilots in many ways, including finding a lost helicopter. Once communication with ATC has been established, follow their instructions.

These common-sense procedures can be easily remembered by using the four Cs: Climb, Communicate, Confess, and Comply.

- Climb for a better view, improved communication and navigation reception, and terrain avoidance.
- Communicate by calling the nearest flight service station (FSS)/automated flight service station (AFSS) on 122.2 MHz. If the FSS/AFSS does not respond, call the nearest control tower, center, or approach control. For frequencies, check the chart in the vicinity of the last known position. If that fails, switch to the emergency radio frequency (121.5 MHz) and transponder code (7700).
- Report the lost situation to ATC and request help.
- Comply with controller instructions.

Pilots should understand the services provided by ATC and the resources and options available. These services enable pilots to focus on aircraft control and help them make better decisions in a time of stress.

When contacting ATC, pilots should provide as much information as possible because ATC uses the information to determine what kind of assistance it can provide with available assets and capabilities. Information requirements vary depending on the existing situation, but at a minimum a pilot should provide the following information:

- Aircraft identification and type
- Nature of the emergency
- Aviator's desires

To reduce the chances of getting lost in the first place, use flight following through active contact with an aircraft during flight either by radio or through automated flight following systems when it is available, monitor checkpoints no more than 25 miles apart, keep navigation aids such as Very High-Frequency Omni-Directional Range (VOR) tuned in, and maintain good situational awareness. Flight following provides ongoing surveillance information to assist pilots in avoiding collisions with other aircraft.

Getting lost is a potentially dangerous situation for any aircraft, especially when low on fuel. Due to the helicopter's unique ability to land almost anywhere, pilots have more flexibility than other aircraft as to landing site. An inherent risk associated with being lost is waiting too long to land in a safe area. Helicopter pilots should land before fuel exhaustion occurs because maneuvering with low fuel levels could cause the engine to stop due to fuel starvation as fuel sloshes or flows away from the pickup port in the tank.

If lost and low on fuel, it is advisable to make a precautionary landing. Preferably, land near a road or in an area that would allow space for another helicopter to safely land and provide assistance. Having fuel delivered is a minor inconvenience when compared to having an accident. Once on the ground, pilots may seek assistance.

VFR Flight into Instrument Meteorological Conditions

Helicopters, unlike airplanes, generally operate under Visual Flight Rules (VFR) and require pilots to maintain aircraft control by visual cues. However, when unforecast weather leads to degraded visibility, the pilot may be at increased risk of Inadvertent flight into Instrument Meteorological Conditions (IIMC). During an IIMC encounter, the pilot may be unprepared for the loss of visual reference, resulting in a reduced ability to continue safe flight. IIMC is a life-threatening emergency for any pilot. To capture these IIMC events, the Commercial Aviation Safety Team (CAST) and International Civil Aviation Organization (ICAO) Common Taxonomy Team (CICTT) categorizes this occurrence as Unintended flight in Instrument Meteorological Conditions (UIMC). This term is also recognized by the National Transportation Safety Board (NTSB) and Federal Aviation Administration (FAA). It is used to classify occurrences (accidents and incidents) at a high level to improve the capacity to focus on common safety issues and complete analysis of the data in support of safety initiatives.

The onset of IIMC may occur gradually or suddenly, has no simple procedural exit, and is unlike flight training by reference to while in Visual Meteorological Conditions (VMC). Most training helicopters are not equipped or certified to fly under Instrument Flight Rules (IFR). Therefore, General Aviation (GA) helicopter pilots may not have the benefit of flight in actual Instrument Meteorological Conditions (IMC) during their flight training. Helicopter pilots that encounter IIMC may experience physiological illusions which can lead to spatial disorientation and loss of aircraft control. Even with some instrument training, many available and accessible helicopters are not equipped with the proper augmented safety systems or autopilots, which would significantly aid in helicopter control during an IIMC emergency. The need to use outside visual references is natural for helicopter pilots because much of their flight training is based upon visual cues, not on flight instruments. This primacy can only be overcome through significant instrument training. Additionally, instrument flight may be intimidating to some and too costly for others. As a result, many helicopter pilots choose not to seek an instrument rating.

While commercial helicopter operators often prefer their pilots to be instrument rated, fatal accidents still occur as a result of IIMC. Many accidents can be traced back to the pilot's inability to recover the helicopter after IIMC is encountered, even with adequate equipment installed. Therefore, whether instrument rated or not, all pilots should understand that avoiding IIMC is critical.

A good practice for any flight is to set and use personal minimums, which should be more conservative than those required by regulations for VFR flight. In addition, a thorough preflight and understanding of weather conditions that may contribute to the risk of IMC developing along a planned route of flight is essential for safety. Pilots should recognize deteriorating weather conditions so the route of flight can be changed or a decision made to terminate the flight and safely land at a suitable area, well before IIMC occurs. If weather conditions deteriorate below the pilot's personal minimums during flight, a pilot who understands the risks of IIMC knows that he or she is at an en route decision point, where it is necessary to either turn back to the departure point or immediately land somewhere safe to wait until the weather has cleared. Pilots should recognize that descent below a predetermined minimum altitude above ground level (AGL) (for example, 500 feet AGL) to avoid clouds or, slowing the helicopter to a predetermined minimum airspeed (for example, slowing to 50 KIAS) to reduce the rate of closure from the deteriorating weather conditions, indicates the decision point had been reached. Ceilings that are lower than reported and/or deteriorating visibility along the route of flight should trigger the decision to discontinue and amend the current route to avoid IIMC.

If the helicopter pilot is instrument rated, it is advisable to maintain instrument currency and proficiency as this may aid the pilot in a safe recovery from IIMC. A consideration for instrument rated pilots when planning a VFR flight should include a review of published instrument charts for safe operating altitudes, e.g. minimum safe altitude (MSA), minimum obstruction clearance altitude (MOCA), minimum in VMC throughout a flight: off-route altitude (MORA), etc. If IIMC occurs, the pilot may consider a climb to a safe altitude. Once the helicopter is stabilized, the pilot should declare an emergency with air traffic control (ATC). It is imperative that the pilot commit to controlling the helicopter and remember to aviate, navigate, and finally communicate. Often communication is attempted first, as it is natural to look for help in stressful situations. This may distract the pilot from maintaining control of the helicopter.

If the pilot is not instrument rated, instrument current nor proficient, or is flying a non-IFR equipped helicopter, remaining in VMC is paramount. Pilots who are not trained or proficient in flight solely by reference to instruments have a tendency to attempt to maintain flight by visual ground reference, which tends to result in flying at lower altitudes, just above the trees or by following roads. The thought process is that, "as long as I can see what is below me, I can continue to my intended destination." Experience and statistical data indicate that attempting to continue VFR flight into IMC can often lead to a fatal outcome as pilots often fixate on what they see below them and are unable to see the hazards ahead of them (e.g., power lines, towers, rising terrain, etc.). By the time the pilot sees the hazard, it is either too late to avoid a collision, or while successfully maneuvering to avoid an obstacle, the pilot becomes disoriented.

Flying at night involves even more conservative personal minimums to ensure safety and avoidance of IIMC than daytime flying. At night, deteriorating weather conditions may be difficult to detect. Therefore, pilots should ensure that they not only receive a thorough weather briefing, but that they remain vigilant for unforecasted weather during their flight. The planned route should include preselected landing sites that will provide options to the pilot in the event a precautionary landing is required to avoid adverse weather conditions. As a pilot gains night flight experience their ability to assess weather during a flight will improve.

Below are some basic guidelines to assist a pilot to remain in VMC throughout a flight:

1. Slowly turn around if threatened by deteriorating visual cues and proceed back to VMC or to the first safe landing area if the weather ahead becomes questionable. Remember that prevention is paramount.
2. Do not proceed further on a course when the terrain ahead is not clearly discernible.
3. Delay or consider cancelling the flight if weather conditions are already questionable, could deteriorate significantly based on forecasts, or if you are uncertain whether the flight can be conducted safely. Often, a gut feeling can provide a warning that unreasonable risks are present.
4. Always have a safe landing area (such as large open areas or airports) in mind for every route of flight.

There are five basic steps that every pilot should be familiar with, and which should be executed immediately at the onset of IIMC, if applicable. However, remember that if you are not trained to execute the following maneuvers solely by reference to instruments, or your aircraft is not equipped with such instruments, this guidance may be less beneficial to you and loss of helicopter control may occur:

1. Level the "wings" – level the bank angle using the attitude indicator.
2. Attitude – set a climb attitude that achieves a safe climb speed appropriate to your type of helicopter. This is often no more than 10° of pitch up on the attitude indicator.
3. Airspeed – verify that the attitude selected has achieved the desired airspeed. It is critical to recognize that slower airspeeds, closer to effective translational lift, may require large control inputs and will decrease stability, making recover impossible while in UIMC.
4. Power – adjust to a climb power setting relative to the desired airspeed. This should be executed concurrent with steps 2 and 3.
5. Heading and Trim – pick a heading known to be free of obstacles and maintain it. This will likely be the heading you were already on, which was planned and briefed. Set the heading bug, if installed, to avoid over-controlling your bank. Maintain coordinated flight so that an unusual attitude will not develop.

Try to avoid immediately turning 180°. Turning around is not always the safest route and executing a turn immediately after UIMC may lead to spatial disorientation. If a 180° turn is the safest option, first note the heading you are on then begin the turn to the reciprocal heading, but only after stable flight is achieved (items 1 through 5 above) and maintain a constant rate of turn appropriate to the selected airspeed.

Each encounter with UIMC is unique, and no single procedure can ensure a safe outcome. Considerations in determining the best course of action upon encountering UIMC should include, at a minimum, terrain, obstructions, freezing levels, aircraft performance and limitations, and availability of ATC services.

There are new technologies being developed regarding aircraft design, enhanced and lower-cost technologies, and aircraft certification. Because of this promising future, much of the discussion and guidance in this chapter may one day become irrelevant. As helicopters integrate more into the National Airspace System, the IFR infrastructure and instrument training will become more prevalent. In the future, UIMC may no longer be the emergency that ends with a fatality but rather associated with proper prevention, skilled recovery techniques along with the aid of emerging new life saving avionics technology. A helicopter instrument rating may be a life-saving addition to a pilot's level of certification. Please refer to the Instrument Flying Handbook (FAA-H-8083-15, as revised); Advanced Avionics Handbook (FAA-H-8083-6, as revised); and the Pilot's Handbook of Aeronautical Knowledge (FAA-H-8083-25, as revised) for further exploration of IFR operations and how to obtain an instrument rating.

When faced with deteriorating weather, planning and prevention, not recovery, are the best strategies to eliminate UIMC-related accidents and fatalities.

EMERGENCY EQUIPMENT AND SURVIVAL GEAR
Food cannot be subject to deterioration due to heat or cold. There should be at least 10,000 calories for each person on board, and it should be stored in a sealed waterproof container. It should have been inspected within the previous 6 months, verifying the amount and satisfactory condition of the contents.
A supply of water
Cooking utensils
Matches in a waterproof container
A portable compass
An ax weighing at least 2.5 pounds with a handle not less than 28 inches in length
A flexible saw blade or equivalent cutting tool
30 feet of snare wire and instructions for use
Fishing equipment, including still-fishing bait and gill net with not more than a two-inch mesh
Mosquito nets or netting and insect repellent sufficient to meet the needs of all persons aboard, when operating in areas where insects are likely to be hazardous
A signaling mirror
At least three pyrotechnic distress signals
A sharp, quality jackknife or hunting knife
A suitable survival instruction manual
Flashlight with spare bulbs and batteries
Portable emergency locator transmitter (ELT) with spare batteries
Stove with fuel or a self-contained means of providing heat for cooking
Tent(s) to accommodate everyone on board
Additional items for winter operations: • Winter sleeping bags for all persons when the temperature is expected to be below 7 °C • Two pairs of snow shoes • Spare ax handle • Ice chisel • Snow knife or saw knife

Figure 11-12. *Emergency equipment and survival gear.*

Emergency Equipment and Survival Gear

Both Canada and Alaska require pilots to carry survival gear. Always carry survival gear when flying over rugged and desolate terrain. The items suggested in *Figure 11-12* are both weather and terrain dependent. The pilot also needs to consider how much storage space the helicopter has and how the equipment being carried affects the overall weight and balance of the helicopter.

Chapter Summary

Emergencies should always be anticipated. Knowledge of the helicopter, possible malfunctions and failures, and methods of recovery can help the pilot avoid accidents and be a safer pilot. Helicopter pilots should always expect the worse hazards and possible aerodynamic effects and plan for a safe exit path or procedure to compensate for the hazard.

Chapter 12

Night Operations

Introduction

Pilots rely more on vision than on any other sense to orient themselves in flight. The following visual factors contribute to flying performance: good depth perception for safe landings, good visual acuity to identify terrain features and obstacles in the flightpath, and good color vision. Although vision is the most accurate and reliable sense, visual cues can be misleading, contributing to incidents occurring within the flight environment. Pilots should be aware of and know how to compensate effectively for the following:

- Physical deficiency or self-imposed stress, such as smoking, which limits night-vision capability
- Visual cue deficiencies
- Limitations in visual acuity, dark adaptation, and color and depth perception

For example, at night, the unaided eye has degraded visual acuity. For more information on night operations, reference Chapter 17, Aeromedical Factors, of the Pilot's Handbook of Aeronautical Knowledge (FAA-H-8083-25, as revised).

Visual Deficiencies

Night Myopia

At night, blue wavelengths of light prevail in the visible portion of the spectrum. Therefore, slightly nearsighted (myopic) individuals viewing blue-green light at night may experience blurred vision. Even pilots with perfect vision find that image sharpness decreases as pupil diameter increases. For individuals with mild refractive errors, these factors combine to make vision unacceptably blurred unless they wear corrective glasses. Another factor to consider is "dark focus." When light levels decrease, the focusing mechanism of the eye may move toward a resting position and make the eye more myopic. These factors become important when pilots rely on terrain features during unaided night flights. Practicing good light discipline is very important and helps pilots to retain their night adaptation. Keeping the cockpit lighting on dim allows the pilot to better identify outside details, unmarked hazards such as towers less than 200' AGL, and unimproved landing sites with no hazard lighting.

A simple exercise that shows the effect of high versus low light contrast would be to go out to a very dark road and turn the dash board lights down very low or off and let your eyes adjust to the ambient light level. Then, turn the dash board lights up and note how the outside features disappear. The same concept applies to cockpit lighting and being able to see the surrounding terrain and obstacles. *[Figure 12-1]* Special corrective lenses can be prescribed to pilots who experience night myopia.

The eye automatically adjusts for the light level experienced. During night flight, the cockpit and instrument lights should be as dim as possible. The eye can then adjust for the outside lighting conditions (ambient lighting) to see outside. The dimmer the inside lighting is, the better you can see outside.

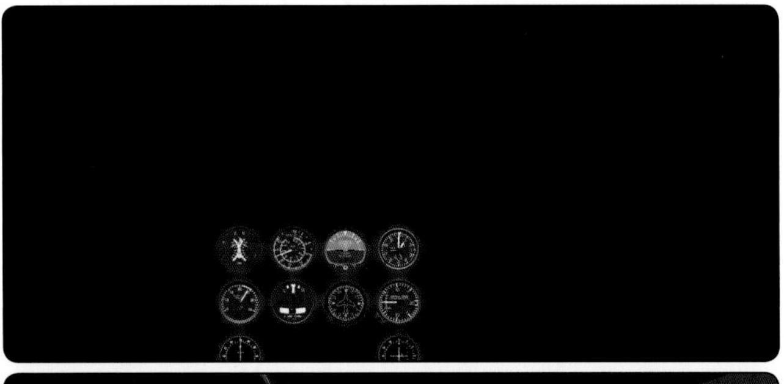

Figure 12-1. *Effects of dimming cockpit lighting during night flight to better see surrounding terrain.*

Hyperopia

Hyperopia is also caused by an error in refraction. In a hyperopic state, when a pilot views a near image, the actual focal point of the eye is behind the retinal plane (wall), causing blurred vision. Objects that are nearby are not seen clearly; only more distant objects are in focus. This problem, is referred to as farsightedness.

Astigmatism

An unequal curvature of the cornea or lens of the eye causes this condition. A ray of light is spread over a diffused area in one meridian. In normal vision, a ray of light is sharply focused on the retina. Astigmatism is the inability to focus different meridians simultaneously. If, for example, astigmatic individuals focus on power poles (vertical), the wires (horizontal) are out of focus for most of them. *[Figure 12-2]*

Presbyopia

This condition is part of the normal aging process, which causes the lens to harden. Beginning in the early teens, the human eye gradually loses the ability to accommodate for and focus on nearby objects. When people are about 40 years old, their eyes are unable to focus at normal reading distances without reading glasses. Reduced illumination interferes with focus depth and accommodation ability. Hardening of the lens may also result in clouding of the lens (cataract formation). Aviators with early cataracts may see a standard eye chart clearly under normal daylight but have difficulty seeing under bright light conditions. This problem is due to light scattering as it enters the eye. This glare sensitivity is disabling under certain circumstances. Glare disability, related to contrast sensitivity, is the ability to detect objects against varying shades of backgrounds. Other visual functions decline with age and affect the aircrew member's performance:

- Dynamic acuity
- Recovery from glare
- Function under low illumination
- Information processing

Vision in Flight

The visual sense is especially important in collision avoidance and depth perception. Due to the structure of the human eye, illusions and blindspots occur. The more pilots understand the eye and how it functions, the easier it is to compensate for these illusions and blindspots. *Figure 12-3* shows the basic anatomy of the human eye and how it is like

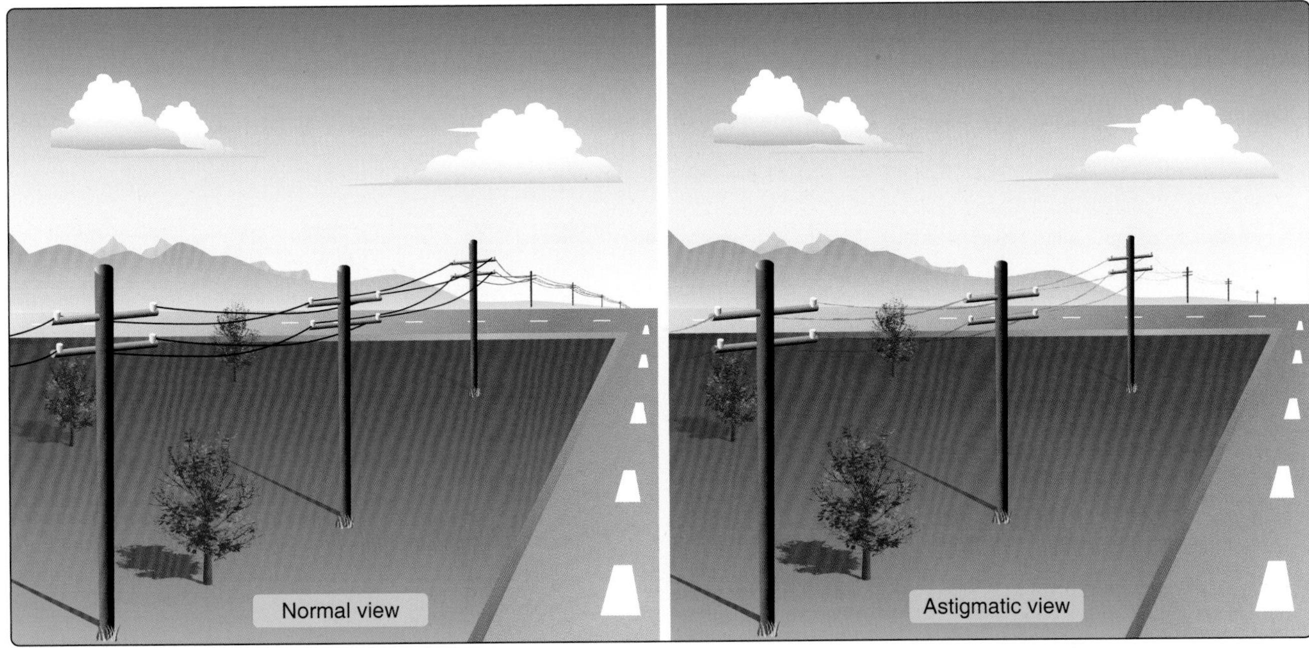

Figure 12-2. *Example of a view that might be experienced by someone with astigmatism.*

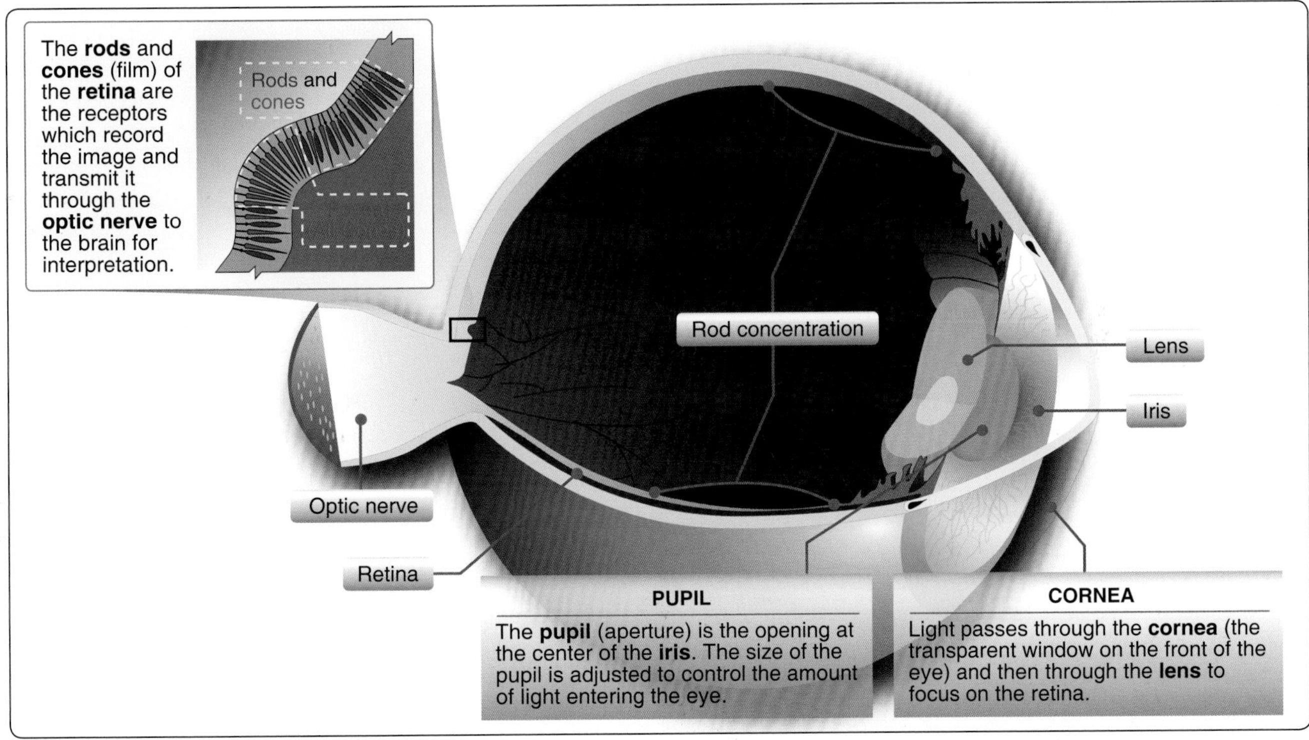

Figure 12-3. *The human eye.*

a camera. A camera is able to focus on near and far objects by changing the distance between the lens and the film. With the eye on the other hand, objects can be seen clearly at various distances because the shape of its lens is changed automatically by small muscles.

Visual Acuity

Normal visual acuity, or sharpness, is 20/20. A value of 20/80 indicates that an individual reads at 20 feet the letters that an individual with normal acuity (20/20) reads at 80 feet away. The human eye functions like a camera. It has an instantaneous field of view, which is oval and typically measures 120° vertically by 150° horizontally. When both eyes are used for viewing, the overall field of vision measures about 120° vertically by 200° horizontally.

The Eye

Vision is primarily the result of light striking a photosensitive layer, called the retina, at the back of the eye. The retina is composed of light-sensitive cones and rods. The cones in the eye perceive an image best when the light is bright, while the rods work best in low light. The pattern of light that strikes the cones and rods is transmitted as electrical impulses by the optic nerve to the brain where these signals are interpreted as an image.

Cones

Cones are concentrated around the center of the retina. They gradually diminish in number as the distance from the center increases. Cones allow color perception by sensing red, blue, and green light. Directly behind the lens, on the retina, is a small, notched area called the fovea. This area contains only a high concentration of cone receptors. The best vision in daylight is obtained by looking directly at the object. This focuses the image on the fovea, where detail is best seen. The cones, however, do not function well in darkness, which explains why color is not seen as vividly at night as it is during the day.

Rods

Concentrated outside the fovea area, the rods are the dim light and night receptors. The number of rods increases as the distance from the fovea increases. Rods sense images only in black and white. Because the rods are not located directly behind the pupil, they are responsible for most peripheral vision. Images that move are perceived more easily by the rod areas than by the cones in the fovea. If you have ever seen something move out of the corner of your eye, it was most likely detected by rod receptors.

In low light, the cones lose much of their function, while rods become more receptive. The eye sacrifices sharpness for sensitivity. The ability to see an object directly in front of you is reduced, and much depth perception is lost, as well as judgment of size. The concentration of cones in the fovea can make a night blindspot at the center of vision. How well a person sees at night is determined by the rods in the eyes, as well as by the amount of light allowed into the eyes. At night, the wider the pupil is open at night, the better night vision becomes.

Night Vision

Diet and general physical health have an impact on how well a person can see in the dark. Deficiencies in vitamins A and C have been shown to reduce night acuity. Other factors, such as carbon monoxide poisoning, smoking, alcohol, and certain drugs can greatly decrease night vision. Lack of oxygen can also decrease night vision as the eye requires more oxygen per unit weight than any other part of the body.

Night Scanning

Good night visual acuity is needed for collision avoidance. Night scanning, like day scanning, uses a series of short, regularly spaced eye movements in 10° sectors. Unlike day scanning, however, off-center viewing is used to focus objects on the rods rather than the fovea blindspot. *[Figure 12-4]* When looking at an object, avoid staring at it too long. If staring at an object without moving the eyes, the retina becomes accustomed to the light intensity and the image begins to fade. To keep it clearly visible, new areas in the retina must be exposed to the image. Small, circular eye movements help eliminate the fading. Also, move the eyes more slowly from sector to sector than during the day to prevent blurring.

During daylight, objects can be perceived at a great distance with good detail. At night, range is limited, and detail is poor. Objects along the flight path can be more readily identified at night, by using the proper techniques to scan the terrain. To

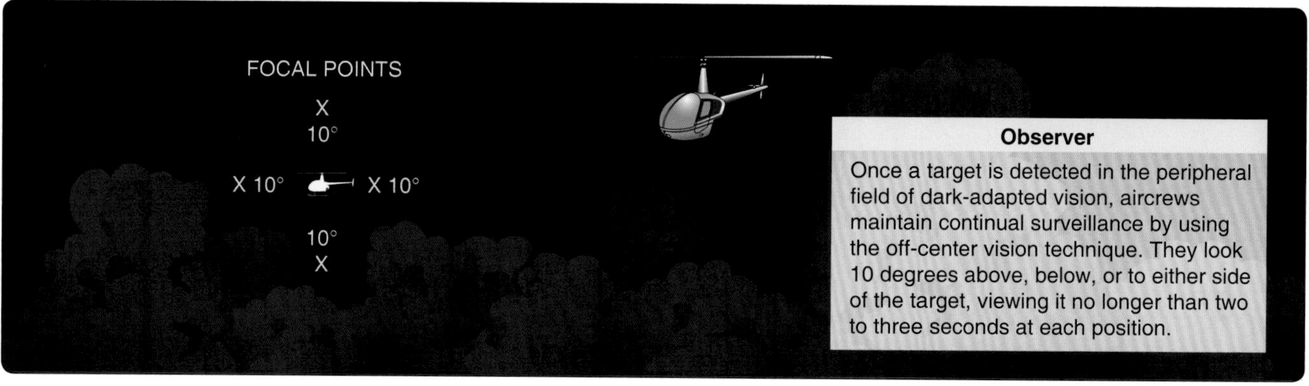

Figure 12-4. *Off-center vision technique.*

scan effectively, pilots look from side to side. They should begin scanning at the greatest distance at which an object can be perceived high on the horizon, thence moving inward toward the position of the aircraft. *Figure 12-5* shows this scanning pattern. Because the light-sensitive elements of the retina are unable to perceive images that are in motion, a stop-turn-stop-turn motion should be used. For each stop, an area about 30 degrees wide should be scanned. This viewing angle includes an area about 250 meters wide at a distance of 500 meters. The duration of each stop is based on the degree of detail that is required, but no stop should last more than two or three seconds. When moving from one viewing point to the next, pilots should overlap the previous field of view by 10 degrees. This scanning technique allows greater clarity in observing the periphery. Other scanning techniques, as illustrated in *Figure 12-6,* may be developed to fit the situation.

Obstruction Detection

Obstructions having poor reflective surfaces, such as wires and small tree limbs, are difficult to detect. The best way to

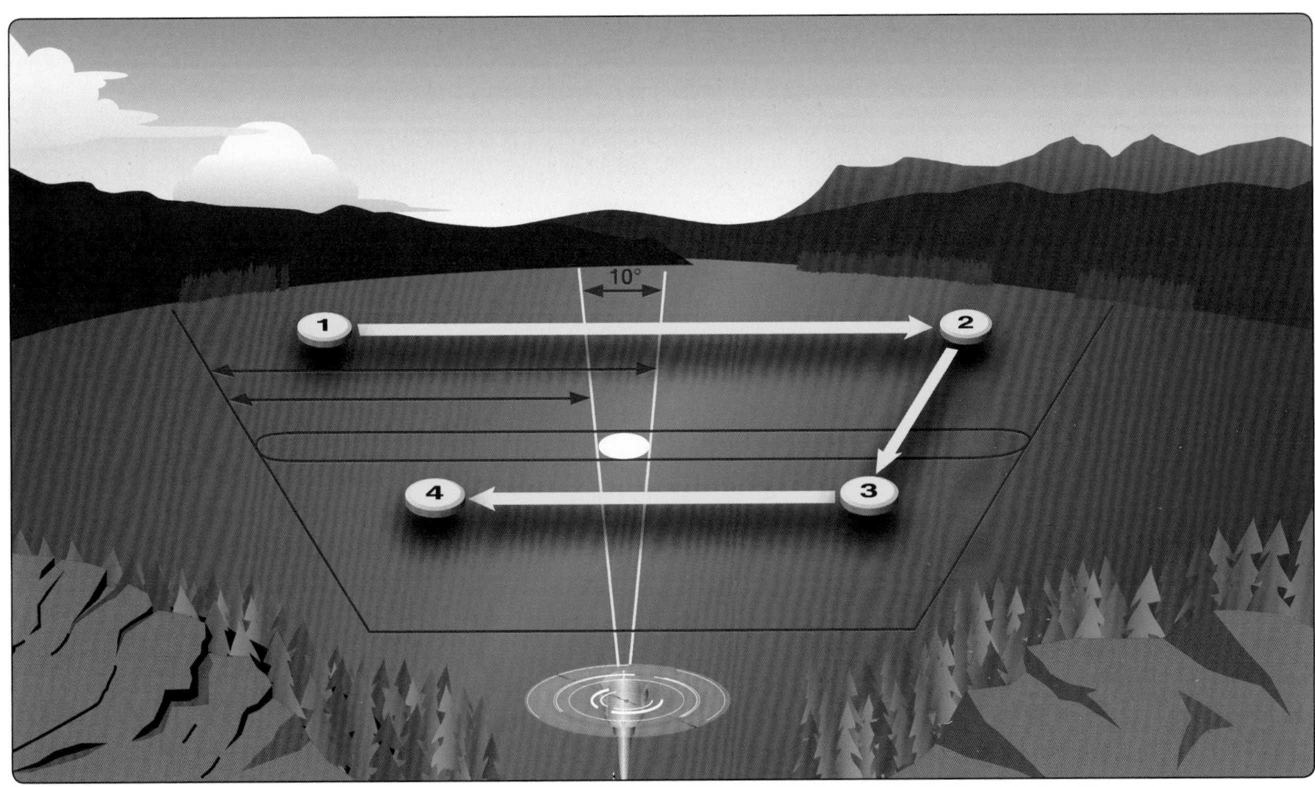

Figure 12-5. *Scanning pattern.*

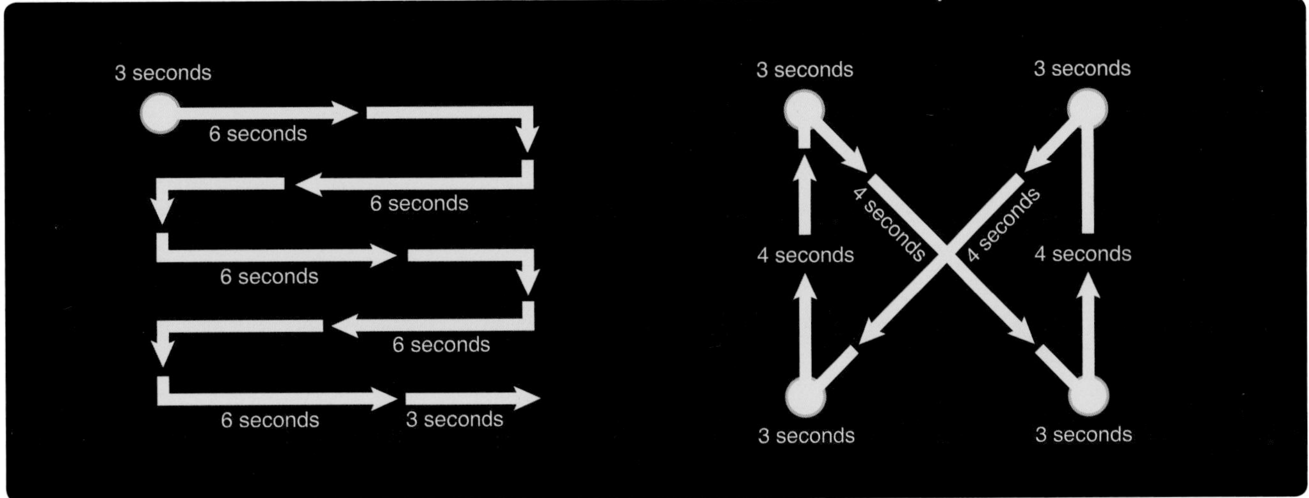

Figure 12-6. *Night vision.*

locate wires is by looking for the support structures. However, pilots should review the most current hazard maps with known wire locations before night flights.

Aircraft Lighting

In order to see other aircraft more clearly, regulations require that all aircraft operating during the night hours have special lights and equipment. The requirements for operating at night are found in Title 14 of the Code of Federal Regulations (14 CFR) part 91. In addition to aircraft lighting, the regulations also provide a definition of night flight in accordance with 14 CFR part 91, currency requirements, fuel reserves, and necessary electrical systems.

Position lights enable a pilot to locate another aircraft, as well as help determine its direction of flight. The approved aircraft lights for night operations are a green light on the right cabin side or wingtip, a red light on the left cabin side or wingtip, and a white position light on the tail. In addition, flashing aviation red or white anticollision lights are required for all flights, if equipped on the aircraft and in an operable condition (in accordance with 14 CFR Section 91.209(b), which aids in the identification during night conditions). These flashing lights can be in a number of locations but are most commonly found on the top and bottom of the cabin.

Figure 12-7 shows examples of aircraft lighting. By interpreting the position lights on other aircraft, the pilot in aircraft 3 can determine whether the aircraft is flying in the opposite direction or is on a collision course. If a red position light is seen to the right of a green light, such as shown by aircraft 1, it is flying toward aircraft 3. A pilot should watch this aircraft closely and be ready to change course. Aircraft 2, on the other hand, is flying away from aircraft 3, as indicated by the white position light.

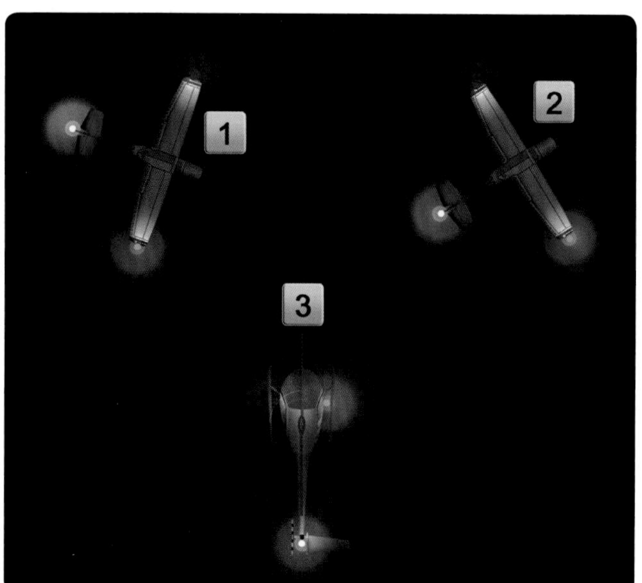

Figure 12-7. *Aircraft position lights.*

Visual Illusions

Illusions give false impressions or misconceptions of actual conditions; therefore, pilots must understand the type of illusions that can occur and the resulting disorientation. Although the eye is the most reliable of the senses, some illusions can result from misinterpreting what is seen; what is perceived is not always accurate. Even with the references outside the cockpit and the display of instruments inside, pilots must be on guard to interpret information correctly.

Relative-Motion Illusion

Relative motion is the falsely perceived self-motion in relation to the motion of another object. The most common example is as follows. An individual in a car is stopped at a traffic light and another car pulls alongside. The individual who was stopped at the light perceives the forward motion of the second car as his or her own motion rearward. This results in the individual applying more pressure to the brakes unnecessarily. This illusion can be encountered during flight in situations such as formation flight, hover taxi, or hovering over water or tall grass.

Confusion with Ground Lights

Confusion with ground lights occurs when a pilot mistakes ground lights for stars. The pilot can place the helicopter in an extremely dangerous flight attitude if he or she aligns it with the wrong lights. In *Figure 12-8A*, the helicopter is aligned with a road and not with the horizon. Isolated ground lights can appear as stars and could lead to the illusion that the helicopter is in a nose-high attitude.

When no stars are visible because of overcast conditions, unlighted areas of terrain can blend with the dark overcast to create the illusion that the unlighted terrain is part of the sky in *Figure 12-8B.* In this illusion, the shoreline is mistaken for the horizon. In an attempt to correct for the apparent nose-high attitude, a pilot may lower the collective and attempt to fly "beneath the shore." This illusion can be avoided by referencing the flight instruments and establishing a true horizon and attitude.

Reversible Perspective Illusion

At night, an aircraft or helicopter may appear to be moving away when it is actually approaching. If the pilot of each aircraft has the same assumption, and the rate of closure is significant, by the time each pilot realizes his or her own error in assumption, it may be too late to avoid a mishap. This illusion is called reversible perspective and is often experienced when a pilot observes another aircraft

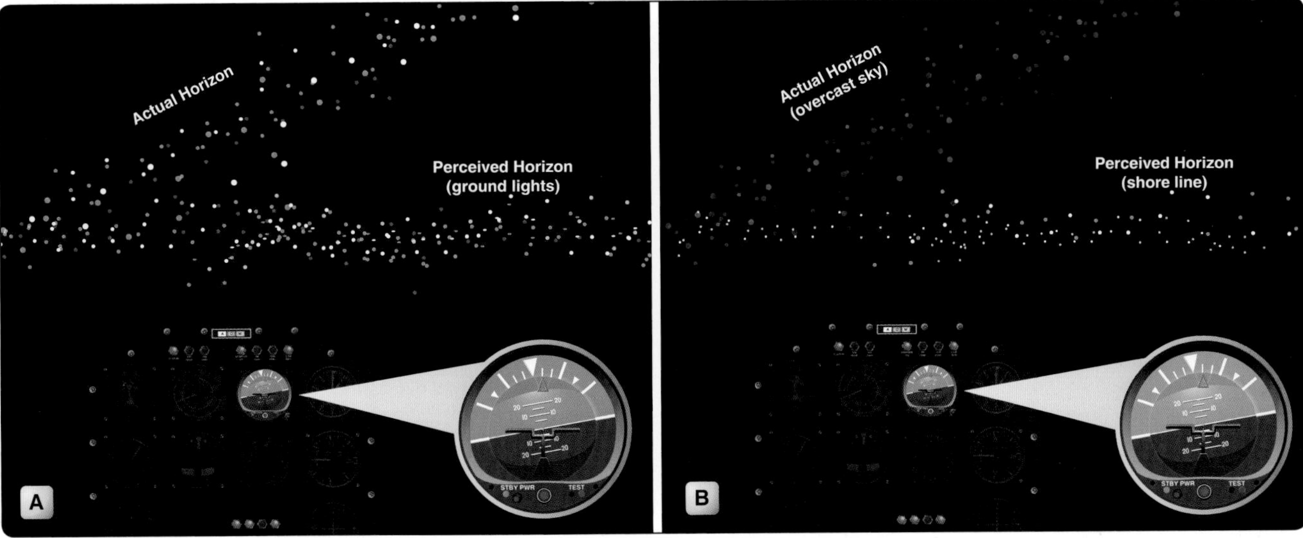

Figure 12-8. *At night, the horizon may be hard to discern due to dark terrain and misleading light patterns on the ground.*

or helicopter flying an approaching, parallel course. To determine the direction of flight, the pilot should observe the other aircraft's position lights. Remember the following: red on right returning; that is, if an aircraft is seen with the red position light on the right and the green position light on the left, the observed aircraft is traveling in the opposite direction.

Flicker Vertigo

Flicker vertigo is technically not an illusion; however, as most people are aware from personal experience, viewing a flickering light can be both distracting and annoying. Flicker vertigo may be created by helicopter rotor blades or airplane propellers interrupting direct sunlight at a rate of 4 to 20 cycles per second. Flashing anticollision strobe lights, especially while the aircraft is in the clouds, can also produce this effect. One should also be aware that photic stimuli at certain frequencies could produce seizures in those rare individuals who are susceptible to flicker-induced epilepsy.

Night Flight

The night flying environment and the techniques used when flying at night depend on outside conditions. Flying on a bright, clear, moonlit evening when the visibility is good, and the wind is calm is not much different from flying during the day. However, if flying on an overcast night over a sparsely populated area, with few or no outside lights on the ground, the situation is quite different. Visibility is restricted, so be more alert in steering clear of obstructions and low clouds. Options are also limited in the event of an emergency, as it is more difficult to find a place to land and determine wind direction and speed. At night, rely more heavily on the aircraft systems, such as lights, flight instruments, and navigation equipment. As a precaution, if visibility is limited or outside references are inadequate, strongly consider delaying the flight until conditions improve, unless proper instrument flight training has been received and the helicopter has the appropriate instrumentation and equipment.

Preflight

Aircraft preflight inspection is a critical aspect of flight safety. It must comply with the appropriate rotorcraft flight manual (RFM). Preflight should be scheduled as early as possible in the flight planning sequence, preferably during daylight hours, allowing time for maintenance assistance and correction. If a night preflight is necessary, a flashlight with an unfiltered lens (white light) should be used to supplement lighting. Oil and hydraulic fluid levels and leaks are difficult to detect with a blue-green or red lens. Windscreens should be checked to ensure they are clean and relatively free of scratches. Slight scratches are acceptable for day flight but may not be for night flight. The search light or landing light should be positioned for the best possible illumination during an emergency descent.

Careful attention must be paid to the aircraft electrical system. In helicopters equipped with fuses, a spare set is required by regulation, and by common sense, so make sure they are on board. If the helicopter is equipped with circuit breakers, check to see that they are not tripped. A tripped circuit breaker may be an indication of an equipment malfunction and should be left for maintenance to troubleshoot before flying.

All aircraft operating between sunset and sunrise are required to have operable navigation (position) lights. Turn these lights on during the preflight to inspect them visually for proper operation. Between sunset and sunrise, these lights must be on any time the helicopter is operating.

All recently manufactured aircraft certificated for night flight must have an anticollision light that makes the aircraft more visible to other pilots. This light is either a red or white flashing light and may be in the form of a rotating beacon or a strobe. While anticollision lights are required for night visual flight rules (VFR) flights, they may be turned off any time they create a distraction for the pilot.

One of the first steps in preparation for night flight is to become thoroughly familiar with the helicopter's cockpit, instrumentation, and control layout. It is recommended that a pilot practice locating each instrument, control, and switch, both with and without cabin lights. Since the markings on some switches and circuit breaker panels may be difficult to read at night, be able to locate and use these devices, and read the markings in poor light conditions. Before starting the engine, make sure all necessary equipment and supplies needed for the flight, such as charts, notepads, and flashlights, are accessible and ready for use.

Cockpit Lights

Check all interior lights with special attention to the instrument and panel lights. The panel lighting can usually be controlled with a rheostat or dimmer switch, allowing the pilot to adjust the intensity. If a particular light is too bright or causes reflection or glare off the windshield, it should be adjusted or turned off. As ambient light level decreases from twilight to darkness, intensity of the cockpit lights is reduced to a low, usable intensity level that reduces any glare or reflection off the windshield. The light level should be adjusted to as close to the ambient light level as possible. A flashlight, with red or blue-green lens filter, or map light can supplement the available light in the cockpit. Always carry a flashlight with fresh batteries to provide an alternate source of light if the interior lights malfunction. If an existing map/utility light is used, it should be hand-held or remounted to a convenient location. In order to retain night adaptation, use low level light when using your checklist. Brief your passengers on the importance of light discipline during night flight so the pilot is not blinded, causing loss of dark adaptation.

Engine Starting and Rotor Engagement

Use extra caution when starting the engine and engaging the rotors, especially in dark areas with little or no outside lights. In addition to the usual call of "clear," turn on the position and anticollision lights. If conditions permit, also turn the landing light on momentarily to help warn others that the engine is about to start and engage the rotors.

Taxi Technique

Landing lights usually cast a beam that is narrow and concentrated ahead of the helicopter, so illumination to the side is minimal. Therefore, slow the taxi at night, especially in congested ramp and parking areas. Some helicopters have a hover light in addition to a landing light, which illuminates a larger area under the helicopter.

When operating at an unfamiliar airport at night, ask for instructions or advice concerning local conditions, so as to avoid taxiing into areas of construction, or unlighted, unmarked obstructions. Ground controllers or UNICOM operators are usually cooperative in furnishing this type of information.

Night Traffic Patterns

Traffic patterns are covered in Chapter 9, Basic Flight Maneuvers, but the following additional considerations should be taken into account when flying a helicopter in a night traffic pattern:

1. The minimum recommended pattern height at night is 1,000 feet when able.
2. If possible, consider taking the right hand night pattern with fixed wing in the left hand pattern for extra separation, but if needed, conform and integrate with the fixed wing using the same pattern height.
3. Be extra vigilant on abiding with noise abatement procedures at night.
4. Always plan to use the lit runway at night for unaided (no night vision equipment) approaches and departures.
5. Avoid downwind and crosswind approaches at night when able.

Takeoff

Before takeoff, make sure that there is a clear, unobstructed takeoff path. At airports, this is accomplished by taking off over a runway or taxi way, however, if operating off-airport, pay more attention to the surroundings. Obstructions may also be difficult to see if taking off from an unlighted area. Once a suitable takeoff path is chosen, select a point down the takeoff path to use for directional reference. The landing light should be positioned in order to illuminate the tallest obstacles in the takeoff path. During a night takeoff, notice a lack of reliable outside visual references after becoming airborne. This is particularly true at small airports and off-airport landing sites located in sparsely populated areas. To compensate for the lack of outside references, use the available flight instruments as an aid. Check the altimeter and the airspeed indicator to verify the proper climb attitude. An attitude indicator, if installed, can enhance attitude reference.

The first 500 feet of altitude after takeoff is considered to be the most critical period in transitioning from the comparatively well-lit airport or heliport into what sometimes appears to be total darkness. A takeoff at night is usually an "altitude over airspeed" maneuver, meaning a pilot most likely performs a nearly maximum performance takeoff. This improves the chances for obstacle clearance and enhances safety.

En Route Procedures

In order to provide a higher margin of safety, it is recommended that a cruising altitude somewhat higher than normal be selected. There are three reasons for this. First, a higher altitude gives more clearance between obstacles, especially those that are difficult to see at night, such as high-tension wires and unlighted towers. Second, in the event of an engine failure, there is more time to set up for a landing and the greater gliding distance gives more options for a safe landing. Third, radio reception is improved, particularly if using radio aids for navigation.

During preflight planning, when possible, it is recommended that a route of flight be selected that is within reach of an airport, or any safe landing site. It is also recommended that pilots fly as close as possible to a populated or lighted area, such as a highway or town. Not only does this offer more options in the event of an emergency, but also makes navigation a lot easier. A course comprised of a series of slight zigzags to stay close to suitable landing sites and well-lit areas, only adds a little more time and distance to an otherwise straight course.

In the event of a forced landing at night, use the same procedure recommended for day time emergency landings. If available, turn on the landing light during the final descent to help in avoiding obstacles along the approach path.

Collision Avoidance at Night

Because the quantity and quality of outside visual references are greatly reduced, a pilot tends to focus on a single point or instrument, making him or her less aware of the other traffic around. Make a special effort to devote enough time to scan for traffic. As discussed previously in this chapter, effective scanning is accomplished with a series of short, regularly spaced eye movements that bring successive areas of the sky into the central visual field. Contrary to the 30-degree scan used to view the ground in the case of scanning for other aircraft, each movement in this case should not exceed 10 degrees, and each area should be observed for at least 1 second to enable detection. If the pilot detects a dimly lit object in a certain direction, the pilot should not look directly at the object, but scan the area adjacent to it, called off-center viewing. This will decrease the chances of fixating on the light and allow focusing more on the objects (e.g., tower, aircraft, ground lights). Short stops of a few seconds in duration in each scan will help to detect the light and its movement. A pilot can determine another aircraft's direction of flight by interpreting the position and anticollision lights, as previously described. When scanning, pilots should also remember to move their heads, not just their eyes. Ground obstructions can cover a considerable amount of sky, and the area can easily be uncovered by a small head movement.

Approach and Landing

Night approaches and landings do have some advantages over daytime approaches, as the air is generally smoother, and the disruptive effects of turbulence and excessive crosswinds are often absent. However, there are a few special considerations and techniques that apply to approaches at night. For example, when landing at night, especially at an unfamiliar airport, make the approach to a lighted runway and then use the taxiways to avoid unlighted obstructions or equipment.

Carefully controlled studies have revealed that pilots have a tendency to make lower approaches at night than during the day. This is potentially dangerous as there is a greater chance of hitting an obstacle, such as an overhead wire or fence, that is difficult to see. It is good practice to make steeper approaches at night, increasing the probability of clearing obstacles. Monitor altitude and rate of descent using the altimeter.

Another pilot tendency during night flight is to focus too much on the landing area and not pay enough attention to airspeed. If too much airspeed is lost, a vortex ring state condition may result. Maintain the proper attitude during the approach, and ensure that you keep some forward airspeed and movement until close to the ground. Outside visual references for airspeed and rate of closure may not be available, especially when landing in an unlit area, so pay special attention to the airspeed indicator.

Although the landing light is a helpful aid when making night approaches, there is an inherent disadvantage. The portion of the landing area illuminated by the landing light seems higher than the dark area surrounding it. This effect can cause a pilot to terminate the approach at an altitude that is too high, which may result in a vortex ring state condition and a hard landing.

Illusions Leading to Landing Errors

Various surface features and atmospheric conditions encountered in night landing can create illusions of incorrect height above and distance from the runway threshold. Landing errors from these illusions can be prevented by anticipating them during approaches, conducting an aerial visual inspection of unfamiliar airports before landing, using electronic glideslope or VASI systems when available, and maintaining optimum proficiency in landing procedures.

Featureless Terrain Illusion

An absence of ground features, as when landing over water, darkened areas, and terrain made featureless by snow, can create the illusion that the aircraft is at a higher altitude than it actually is. The pilot who does not recognize this illusion will fly a lower approach.

Atmospheric Illusions

Rain on the windscreen can create the illusion of greater height, and atmospheric haze can create the illusion of being at a greater distance from the runway. The pilot who does not recognize these illusions flies a higher approach. Penetration of fog can create the illusion of pitching up. The pilot who does not recognize this illusion steepens the approach, often quite abruptly.

Ground Lighting Illusions

Lights along a straight path can be mistaken for runway and approach lights. This might include street lights along a roadside or even the internal lights of a moving train. Another illusion may occur with very intense runway and approach lighting. Due to the relative brightness of these lights, the pilot may perceive them to be closer than they really are. Assuming that the lights are as close as they appear, the pilot may attempt an approach that is actually lower than glideslope. Conversely, the pilot flying over terrain with few lights may make a lower than normal approach.

Helicopter Night VFR Operations

While ceiling and visibility significantly affect safety in night VFR operations, lighting conditions also have a profound effect on safety. Even in conditions in which visibility and ceiling are determined to be visual meteorological conditions, the ability to discern unlit or low contrast objects and terrain at night may be compromised. The ability to discern these objects and terrain is referred to as the "seeing condition," and is related to the amount of natural and man-made lighting available, and the contrast, reflectivity, and texture of surface terrain and obstruction features. In order to conduct operations safely, seeing conditions must be accounted for in the planning and execution of night VFR operations.

Night VFR seeing conditions can be described by identifying high lighting conditions and low lighting conditions.

High lighting conditions exist when one of two sets of conditions are present:

1. The sky cover is less than broken (less than ⅝ cloud cover), the time is between the local moon rise and moon set, and the lunar disk is at least 50 percent illuminated; or
2. The aircraft is operated over surface lighting that, at least, provides lighting of prominent obstacles, the identification of terrain features (shorelines, valleys, hills, mountains, slopes) and a horizontal reference by which the pilot may control the helicopter. For example, this surface lighting may be the result of:
    a. Extensive cultural lighting (manmade, such as a built-up area of a city),
    b. Significant reflected cultural lighting (such as the illumination caused by the reflection of a major metropolitan area's lighting reflecting off a cloud ceiling), or
    c. Limited cultural lighting combined with a high level of natural reflectivity of celestial illumination, such as that provided by a surface covered by snow or a desert surface.

Low lighting conditions are those that do not meet the high lighting conditions requirements.

Some areas may be considered a high lighting environment only in specific circumstances. For example, some surfaces, such as a forest with limited cultural lighting, normally have little reflectivity, requiring dependence on significant moonlight to achieve a high lighting condition. However, when that same forest is covered with snow, its reflectivity may support a high lighting condition based only on starlight. Similarly, a desolate area, with little cultural lighting, such as a desert, may have such inherent natural reflectivity that it may be considered a high lighting conditions area regardless of season, provided the cloud cover does not prevent starlight from being reflected from the surface. Other surfaces, such as areas of open water, may never have enough reflectivity or cultural lighting to ever be characterized as a high lighting area.

Through the accumulation of night flying experience in a particular area, the pilot develops the ability to determine, prior to departure, which areas can be considered supporting high or low lighting conditions. Without that pilot experience, low lighting considerations should be applied by pilots for both preflight planning and operations until high lighting conditions are observed or determined to be regularly available. Even if the aircraft is certified for day and night VFR conditions, night flight should only be conducted if adequate celestial illumination is assured during the entirety of the flight.

Chapter Summary

Knowledge of the basic anatomy and physiology of the eye is helpful in the study of helicopter night operations. Adding to that knowledge a study of visual illusions gives the pilot ways to overcome those illusions. Techniques for preflight, engine start-up, collision avoidance, and night approach and landings help teach the pilot safer ways to conduct flight at night. More detailed information on the subjects discussed in this chapter is available in the Aeronautical Information Manual (AIM) and online at www.faa.gov.

Chapter 13

Effective Aeronautical Decision-Making

Introduction

The accident rate for helicopters has traditionally been higher than the accident rate of fixed-wing aircraft, probably due to the helicopter's unique capabilities to fly and land in more diverse situations than fixed-wing aircraft and pilot attempts to fly the helicopter beyond the limits of his or her abilities or beyond the capabilities of the helicopter. With no significant improvement in helicopter accident rates for the last 20 years, the Federal Aviation Administration (FAA) has joined with various members of the helicopter community to improve the safety of helicopter operations.

According to National Transportation Safety Board (NTSB) statistics, approximately 80 percent of all aviation accidents are caused by pilot error, the human factor. Many of these accidents are the result of the failure of instructors to incorporate single-pilot resource management (SRM) and risk management into flight training instruction of aeronautical decision-making (ADM).

SRM is defined as the art of managing all the resources (both on board the aircraft and from outside sources) available to a pilot prior to and during flight to ensure a successful flight. When properly applied, SRM is a key component of ADM. Additional discussion includes integral topics such as, the concepts of risk management, workload or task management, situational awareness, controlled flight into terrain (CFIT) awareness, and automation management.

ADM is all about learning how to gather information, analyze it, and make decisions. It helps the pilot accurately assess and manage risk and make accurate and timely decisions. Although the flight is coordinated by a single person, the use of available resources, such as air traffic control (ATC) and flight service stations (FSS)/automated flight service stations (AFSS), replicates the principles of crew resource management (CRM) (see page 14-7).

References on SRM and ADM include:

- FAA-H-8083-2, Risk Management Handbook.
- Aeronautical Information Manual (AIM).
- Advisory Circular (AC) 60-22, Aeronautical Decision Making, which provides background information about ADM training in the general aviation (GA) environment.
- FAA-H-8083-25, Pilot's Handbook of Aeronautical Knowledge.

Aeronautical Decision-Making (ADM)

Making good choices sounds easy enough. However, there are a multitude of factors that come into play when these choices, and subsequent decisions, are made in the aeronautical world. Many tools are available for pilots to become more self-aware and assess the options available, along with the impact of their decision. Yet, with all the available resources, accident rates are not being reduced. Poor decisions continue to be made, frequently resulting in lives being lost and/or aircraft damaged or destroyed. The Risk Management Handbook discusses ADM and SRM in detail and should be thoroughly read and understood.

While progress is continually being made in the advancement of pilot training methods, aircraft equipment and systems, and services for pilots, accidents still occur. Historically, the term "pilot error" has been used to describe the causes of these accidents. Pilot error means an action or decision made by the pilot was the cause of, or a contributing factor that led to, the accident. This definition also includes the pilot's failure to make a decision or take action. From a broader perspective, the phrase "human factors related" more aptly describes these accidents since it is usually not a single decision that leads to an accident, but a chain of events triggered by a number of factors. *[Figure 13-1]*

The poor judgment chain, sometimes referred to as the "error chain," is a term used to describe this concept of contributing factors in a human factors related accident. Breaking one link in the chain is often the only event necessary to change the outcome of the sequence of events. The following is an example of the type of scenario illustrating the poor judgment chain.

Scenario

A Helicopter Air Ambulance (HAA) pilot is nearing the end of his shift when he receives a request for a patient pickup at a roadside vehicle accident. The pilot has started to feel the onset of a cold; his thoughts are on getting home and getting a good night's sleep. After receiving the request, the pilot checks the accident location and required flightpath to determine if he has time to complete the flight to the scene, then on to the hospital before his shift expires. The pilot checks the weather and determines that, although thunderstorms are approaching, the flight can be completed prior to their arrival.

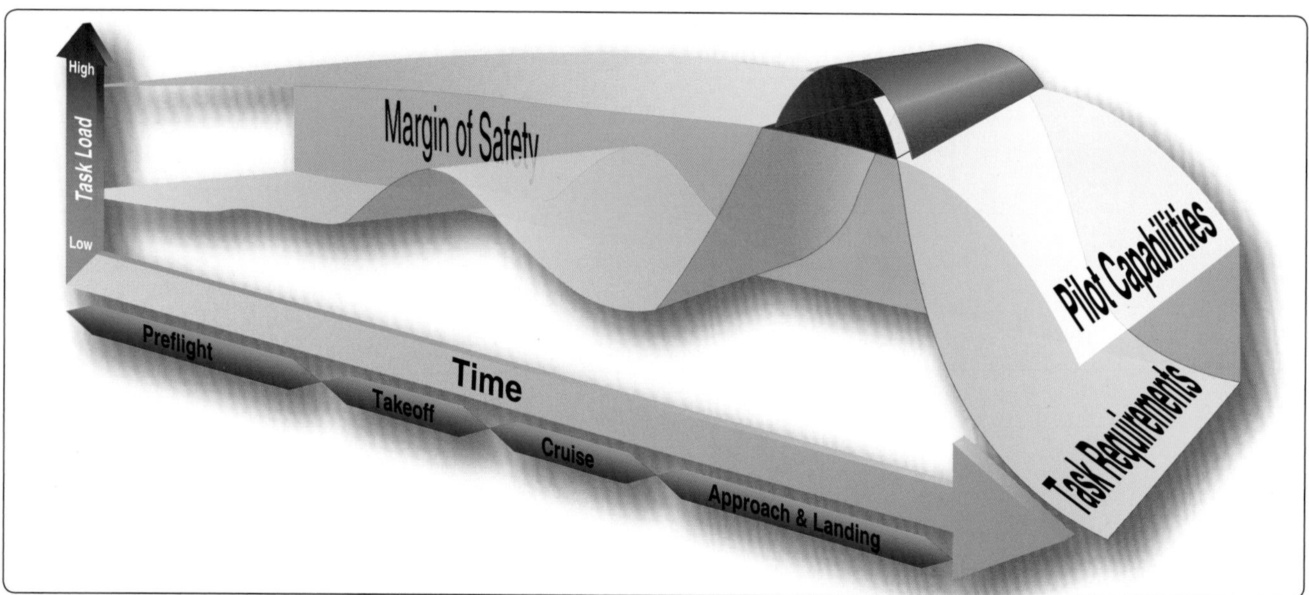

Figure 13-1. *The pilot has a limited capacity of doing work and handling tasks, meaning there is a point at which the tasking exceeds the pilot's capability. When this happens, either tasks are not done properly or some are not done at all.*

The pilot and on-board medical crews depart the home location and arrive overhead, at the scene of the vehicular accident. The pilot is not comfortable with the selected landing area due to tall trees in all quadrants of the confined area. The pilot searches for a secondary landing area. Unable to find one nearby, the pilot then returns to the initial landing area and decides he can make it work.

After successfully landing the aircraft, he is told that there will be a delay before the patient is loaded because more time is needed to extricate the patient from the wreckage. Knowing his shift is nearly over, the pilot begins to feel pressured to "hurry up" or he will require an extension for his duty day.

After 30 minutes, the patient is loaded, and the pilot ensures everyone is secure. He notes that the storm is now nearby and that winds have picked up considerably. The pilot thinks, "No turning back now, the patient is on board and I'm running out of time." The pilot knows he must take off almost vertically to clear the obstacles and chooses his departure path based on the observed wind during landing. Moments later, prior to clearing the obstacles, the aircraft begins an uncontrollable spin and augers back to the ground, seriously injuring all on board and destroying the aircraft.

What could the pilot have done differently to break this error chain? More important—what would you have done differently? By discussing the events that led to this accident, you should develop an understanding of how a series of judgmental errors contributed to the final outcome of this flight.

For example, the pilot's decision to fly the aircraft knowing that the effects of an illness were present was the initial contributing factor. The pilot was aware of his illness, but, was he aware of the impact of the symptoms—fatigue, general uneasy feeling due to a slight fever, perhaps?

Next, knowing the shift was about to end, the pilot based his time required to complete the flight on ideal conditions, and did not take into consideration the possibility of delays. This led to a feeling of being time limited.

Even after determining the landing area was unsuitable, the pilot forced the landing due to time constraints. At any time during this sequence, the pilot could have aborted the flight rather than risk crew lives. Instead, the pilot became blinded by a determination to continue.

After landing, and waiting 30 minutes longer than planned, the pilot observed the outer effects of the thunderstorm, yet still attempted to depart. The pilot dispelled any available options by thinking the only option was to go forward; however, it would have been safer to discontinue the flight. Using the same departure path selected under different wind conditions, the pilot took off and encountered winds that led to loss of aircraft control. Once again faced with a self-imposed time constraint, the pilot improperly chose to depart the confined area. The end result: instead of one patient to transport by ground (had the pilot aborted the flight at any point), there were four patients to be transported.

On numerous occasions leading to and during the flight, the pilot could have made effective decisions that could have prevented this accident. However, as the chain of events unfolded, each poor decision left him with fewer options. Making sound decisions is the key to preventing accidents. Traditional pilot training emphasizes flying skills, knowledge of the aircraft, and familiarity with regulations. SRM and ADM training focus on the decision-making process and on the factors that affect a pilot's ability to make effective choices.

Trescott Tips

Max Trescott, Master Certificated Flight Instructor (CFI) and Master Ground Instructor and winner of the 2008 CFI of the year, has published numerous safety tips that every pilot should heed. He believes that the word "probably" should be purged from our flying vocabulary. Mr. Trescott contends that "probably" means we've done an informal assessment of the likelihood of an event occurring and have assigned a probability to it. He believes the term implies that we believe things are likely to work out, but there's some reasonable doubt in our mind. He further explains that if you ever think that your course of action will "probably work out," you need to choose a new option that you know will work out.

Another safety tip details the importance of accumulating flight hours in one specific airframe type. He explains that "statistics have shown that accidents are correlated more with the number of hours of experience a pilot has in a particular aircraft model and not with his or her total number of flight hours. Accidents tend to decrease after a pilot accumulates at least 100 hours of experience in the aircraft he or she is flying. Thus, when learning to fly, or when transitioning into a new model, your goal should be to concentrate your flying hours in that model." He suggests waiting until you reach 100 hours of experience in one particular model before attempting a dual rating with another model. In addition, if you only fly a few hours per year, maximize your safety by concentrating those hours in just one aircraft model.

The third safety tip that is well worth mentioning is what Mr. Trescott calls "building experience from the armchair." Armchair flying is simply closing your eyes and mentally practicing exactly what you do in the aircraft. This is an excellent way to practice making radio calls, departures, approaches and even visualizing the parts and pieces of the aircraft. This type of flying does not cost a dime and will make you a better prepared and more proficient pilot.

All three of Max Trescott's safety tips incorporate the ADM process and emphasize the importance of how safety and good decision-making is essential to aviation.

The Decision-Making Process

An understanding of the decision-making process provides a pilot with a foundation for developing ADM skills. Some situations, such as engine failures, require a pilot to respond immediately using established procedures with little time for detailed analysis. Called automatic decision-making, it is based upon training, experience, and recognition. Traditionally, pilots have been well trained to react to emergencies, but are not as well prepared to make decisions that require a more reflective response when greater analysis is necessary. They often overlook the phase of decision-making that is accomplished on the ground: the preflight, flight planning, performance planning, weather briefing, and weight/center of gravity configurations. Thorough and proper completion of these tasks provides increased awareness and a base of knowledge available to the pilot prior to departure and once airborne. Typically during a flight, a pilot has time to examine any changes that occur, gather information, and assess risk before reaching a decision. The steps leading to this conclusion constitute the decision-making process.

Defining the Problem

Defining the problem is the first step in the decision-making process and begins with recognizing that a change has occurred or that an expected change did not occur. A problem is perceived first by the senses, then is distinguished through insight (self-awareness) and experience. Insight, experience, and objective analysis of all available information are used to determine the exact nature and severity of the problem. One critical error that can be made during the decision-making process is incorrectly defining the problem.

While going through the following example, keep in mind what errors lead up to the event. What planning could have been completed prior to departing that may have led to avoiding this situation? What instruction could the pilot have had during training that may have better prepared the pilot for this scenario? Could the pilot have assessed potential problems based on what the aircraft "felt like" at a hover? All these factors go into recognizing a change and the timely response.

While doing a hover check after picking up firefighters at the bottom of a canyon, a pilot realized that she was only 20 pounds under maximum gross weight. What she failed to realize was that the firefighters had stowed some of their heaviest gear in the baggage compartment, which shifted the center of gravity (CG) slightly behind the aft limits. Since weight and balance had never created any problems for her in the past, she did not bother to calculate CG and power required. She did try to estimate it by remembering the figures from earlier in the morning at the base camp. At a 5,000-foot density altitude (DA) and maximum gross weight, the performance charts indicated the helicopter had plenty of excess power. Unfortunately, the temperature was 93 °F and the pressure altitude at the pickup point was 6,200 feet (DA = 9,600 feet). Since there was enough power for the hover check, the pilot decided there was sufficient power to takeoff.

Even though the helicopter accelerated slowly during the takeoff, the distance between the helicopter and the ground continued to increase. However, when the pilot attempted to establish the best rate of climb speed, the nose tended to pitch up to a higher-than-normal attitude, and the pilot noticed that the helicopter was not gaining enough altitude in relation to the canyon wall approximately 200 yards ahead.

Choosing a Course of Action

After the problem has been identified, a pilot must evaluate the need to react to it and determine the actions to take to resolve the situation in the time available. The expected outcome of each possible action should be considered and the risks assessed before a pilot decides on a response to the situation.

The pilot's first thought was to pull up on the collective and pull back on the cyclic. After weighing the consequences of possibly losing rotor revolutions per minute (rpm) and not being able to maintain the climb rate sufficiently to clear the canyon wall, which was then only a hundred yards away, she realized the only course was to try to turn back to the landing zone on the canyon floor.

Implementing the Decision and Evaluating the Outcome

Although a decision may be reached and a course of action implemented, the decision-making process is not complete. It is important to think ahead and determine how the decision could affect other phases of the flight. As the flight progresses, a pilot must continue to evaluate the outcome of the decision to ensure that it is producing the desired result.

As the pilot made the turn to the downwind, the airspeed dropped nearly to zero, and the helicopter became very difficult to control. (At this point, the pilot must increase airspeed in order to maintain translational lift.) Since the CG was aft of limits, she needed to apply more forward cyclic than usual. As she approached the landing zone with a high rate of descent, she realized that she would be in a potential vortex ring state situation if she tried to trade airspeed for altitude and lost effective translational lift (ETL). Therefore, it did not appear that she would be able to terminate the approach in a hover. The pilot decided to make the shallowest approach possible and perform a run-on landing.

Pilots sometimes have trouble not because of deficient basic skills or system knowledge, but because of faulty decision-making skills. Although aeronautical decisions may appear to be simple or routine, each individual decision in aviation often defines the options available for the next decision the pilot must make, and the options (good or bad) it provides.

Therefore, a poor decision early in a flight can compromise the safety of the flight at a later time. It is important to make appropriate and decisive choices because good decision-making early in an emergency provide greater latitude for later options.

Decision-Making Models

The decision-making process normally consists of several steps before a pilot chooses a course of action. A variety of structured frameworks for decision-making provide assistance in organizing the decision process. These models include but are not limited to the 5P (Plan, Plane, Pilot, Passengers, Programming), the OODA Loop (Observation, Orientation, Decision, Action), and the DECIDE (Detect, Estimate, Choose, Identify, Do, and Evaluate) models. *[Figure 13-2]* All these models and their variations are discussed in detail in the Pilot's Handbook of Aeronautical Knowledge section covering aeronautical decision-making.

Whichever model is used, the pilot learns how to define the problem, choose a course of action, implement the decision, and evaluate the outcome. Remember, there is no one right answer in this process: a pilot analyzes the situation in light of experience level, personal minimums, and current physical and mental readiness levels, and then makes a decision.

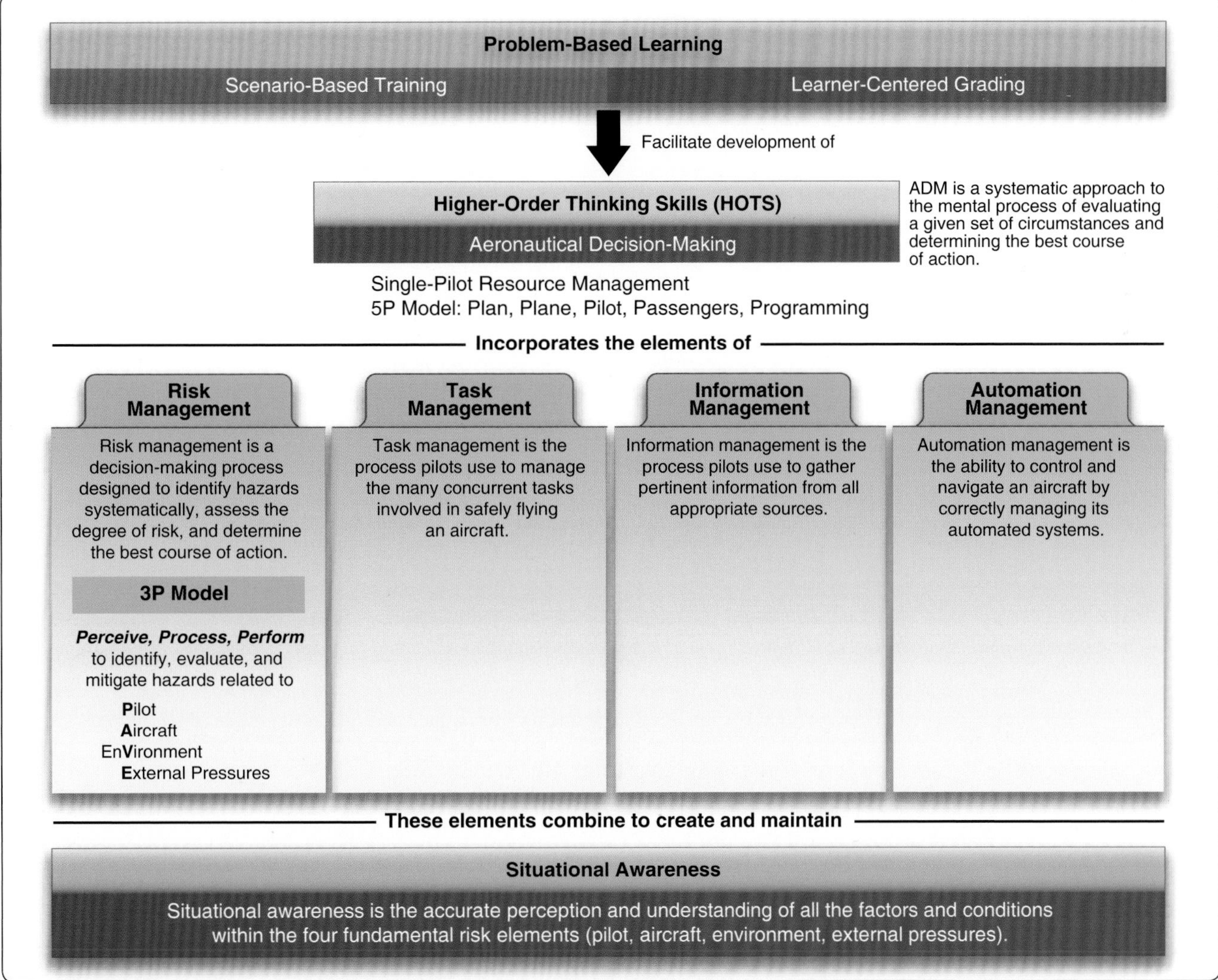

Figure 13-2. *Various models of decision-making are used in problem solving.*

Pilot Self-Assessment

The pilot in command (PIC) of an aircraft is directly responsible for and is the final authority for the operation of that aircraft. The list of PIC responsibilities is long, and nothing should be overlooked. To exercise those responsibilities effectively and make effective decisions regarding the outcome of a flight, a pilot must have an understanding of personal limitations. Pilot performance from planning the flight to execution of the flight is affected by many factors, such as health, experience, knowledge, skill level, and attitude.

Exercising good judgment begins prior to taking the controls of an aircraft. Often, pilots thoroughly check their aircraft to determine airworthiness, yet do not evaluate their own fitness for flight. Just as a checklist is used when preflighting an aircraft, a personal checklist based on such factors as experience, currency, and comfort level can help determine if a pilot is prepared for a particular flight. Specifying when refresher training should be accomplished and designating weather minimums, which may be higher than those listed in Title 14 of the Code of Federal Regulations (14 CFR) part 91, are elements that may be included on a personal checklist. Over confidence can kill just as fast as inexperience. In addition to a review of personal limitations, a pilot should use the I'M SAFE checklist to further evaluate fitness for flight. *[Figure 13-3]*

Curiosity: Healthy or Harmful?

The roots of aviation are firmly based on curiosity. Where would we be today had it not been for the dreams of Leonardo da Vinci, the Wright Brothers, and Igor Sikorsky? They all were infatuated with flight, a curiosity that led to the origins of aviation. The tale of aviation is full of firsts: first flight, first helicopter, first trans-Atlantic flight, and so on. But, along the way there were many setbacks, fatalities, and lessons learned.

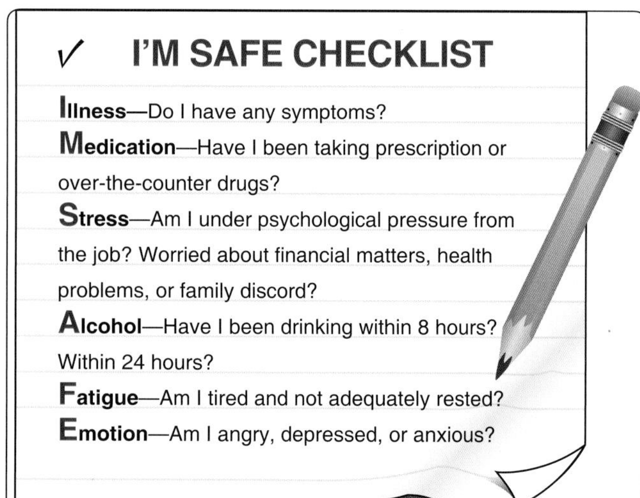

Figure 13-3. *I'M SAFE checklist.*

Today, we continue to learn and investigate the limits of aviation. We've been to the moon, and soon beyond. Our curiosity will continue to drive us to search for the next challenge.

However, curiosity can also have catastrophic consequences. Despite over 100 years of aviation practice, we still see accidents that are caused by impaired judgment formed from curious behavior. Pilots commonly seek to determine the limits of their ability as well as the limits of the aircraft. Unfortunately, too often this leads to mishaps with deadly results. Inquisitive behavior must be harnessed and displayed within personal and material limits.

Deadly curiosity may not seem as obvious to some as it is to others. Simple thoughts such as, "Is visibility really as bad as what the ATIS is reporting?" or "Will the 20-minute fuel light really indicate only 20 minutes worth of fuel?" can lead to poor decisions and disastrous outcomes.

Some aviators blatantly violate rules and aircraft limitations without thinking through the consequences. "What indications and change in flight characteristics will I see if I fly this helicopter above its maximum gross weight?" or "I've heard this helicopter can do aerobatic flight. Why is it prohibited?" are examples of extremely harmful curiosity. Even more astounding is their ignoring to the fact that the damage potentially done to the aircraft will probably manifest later in the aircraft's life, affecting other crews. Spontaneous excursions in aviation can be deadly.

Curiosity is natural and promotes learning. Airmen should abide by established procedures until proper and complete hazard assessment and risk management can be completed.

The PAVE Checklist

As found in the Pilot's Handbook of Aeronautical Knowledge, the FAA has designed a personal minimums checklist. To help pilots with self-assessment, which in turn helps mitigate risk, the acronym PAVE divides the risks of flight into four categories. For each category, think of the applicability specific to helicopter operations:

- Pilot (pilot in command)
  - Physical, emotional readiness.
  - Flight experience, recency, currency, total time in type.
- Aircraft
  - Is the helicopter capable of performing the task?
  - Can it carry the necessary fuel?

- Does it provide adequate power margins for the task to be accomplished?
- Can it carry the weight and remain within CG?
- Will there be external loads?

- Environment
  - Helicopters are susceptible to the impact of changing weather conditions.
  - How will the change in moderating temperatures and DA affect performance?
  - Will controllability be jeopardized by winds, terrain, and turbulence?
- External pressures
  - Do not let the notion to accomplish "the mission" override good judgment and safety.
  - Many jobs include time lines. How often do we hear "time is money" or "time is wasting"? Don't sacrifice safety for an implied or actual need to meet the deadline!
  - Do not allow yourself to feel pressured by coworkers, family events, or friends.

Incorporated into preflight planning, the PAVE checklist provides the pilot with a simple way to remember each category to examine for risk prior to each flight. Once the pilot identifies the risks of a flight, he or she needs to decide whether the risk or combination of risks can be managed safely and successfully. Remember, the PIC is responsible for deciding about canceling the flight. If the pilot decides to continue with the flight, he or she should develop strategies to mitigate the risks.

One way to control risk is by setting personal minimums for items in each risk category. Remember, these are limits unique to an individual pilot's current level of experience and proficiency. They should be reevaluated periodically based upon experience and proficiency.

Single-Pilot Resource Management

Many of the concepts utilized in CRM have been successfully applied to single-pilot operations which led to the development of SRM. Defined as the art and science of managing all the resources (both on board the aircraft and from outside resources) available to a single pilot (prior to and during flight), SRM helps to ensure the successful outcome of the flight. As mentioned earlier, this includes risk management, situational awareness (SA), and CFIT awareness.

SRM training helps the pilot maintain SA by managing automation, associated control, and navigation tasks. This enables the pilot to accurately assess hazards, manage resulting risk potential, and make good decisions.

To make informed decisions during flight operations, a pilot must be aware of the resources found both inside and outside the cockpit. Since useful tools and sources of information may not always be readily apparent, learning to recognize these resources is an essential part of SRM training. The pilot must not only identify the available resources, but he or she must also assess whether sufficient time is available to use a particular one, and the impact its use will have upon the safety of the flight.

If a pilot is flying alone into a confined area with no wind sock or access to a current wind report, should that pilot pick an approach path based on the direction of wind information received from an earlier weather brief? Making an approach into a confined area with a tailwind is a bad decision and can be avoided. Prior to landing, the pilot should use outside resources such a smoke, trees, and water on a pond to help him or her accurately determine which direction the winds are coming from. Pilots should never leave flying up to chance and hope for the best. Many accidents could and should be avoided by simply using the resources, internal and external that are available.

Internal resources are found in the cockpit during flight. Since some of the most valuable internal resources are ingenuity, knowledge, and skill, a pilot can expand cockpit resources immensely by improving these capabilities. This can be accomplished by frequently reviewing flight information publications, such as 14 CFR and the AIM, as well as by pursuing additional training.

No other internal resource is more important than the pilot's own ability to control the situation, thereby controlling the aircraft. Helicopter pilots quickly learn that it is not possible to hover, single pilot, and pick up the checklist, a chart, or publication without endangering themselves, the aircraft, or those nearby.

Checklists are essential cockpit resources used to verify the aircraft instruments and systems are checked, set, and operating properly. They also ensure proper procedures are performed if there is a system malfunction or inflight emergency. Pilots at all levels of experience refer to checklists. The more advanced the aircraft is, the more crucial checklists are.

Therefore, have a plan on how to use the checklist (and other necessary publications) before you begin the flight. Always control the helicopter first. When hovering in an airport environment, the pilot can always land the aircraft to access the checklist or a publication, or have a passenger assist with holding items. There is nothing more unsettling than being in flight and not having a well thought-out plan for managing the necessary documents and data. This lack of planning often leads to confusion, distractions and aircraft mishaps.

Another way to avoid a potentially complex and confusing situation is to remove yourself from the situation. The following is an example of how proper resource management and removal from a situation are vital to safe flight.

A single pilot is conducting a helicopter cross-country flight. He frequently goes to and is familiar with the final destination airport. Weather is briefed to be well above the minimum weather needed, but with isolated thunderstorms possible. For the pilot, this is a routine run-of-the-mill flight. He has done this many times before and has memorized the route, checkpoints, frequencies, fuel required and knows exactly what to expect.

However, once within 30 miles of the destination airport the pilot observes that weather is deteriorating, and a thunderstorm is nearby. The pilot assesses the situation and determines the best course of action is to reroute to another airport. The closest airport is an airport within Class C airspace. At this point, the pilot realizes the publications with the required alternate airport information are in the back of the helicopter out of reach. Now what?

The pilot continues toward the alternate airport while using the onboard equipment to access the information. He struggles to obtain the information because he or she is not thoroughly familiar with its operation. Finally, the information is acquired and the pilot dials in the appropriate alternate airfield information. Upon initial contact ARTCC (Air Route Traffic Control Center) notifies the pilot that he has entered the airspace without the required clearance; in effect the pilot has violated airspace regulations.

Things have gone from bad to worse for him. When did the trouble begin for this pilot and what options were available? Without a doubt, problems began during the planning phase, as the necessary resources were placed in the back of the aircraft, unavailable to the pilot during flight. Additional training with the available automated systems installed on the helicopter would have expedited access to the necessary information. What if they hadn't been installed or were inoperative?

Next, a poor decision to continue towards the Class C airspace was made. The pilot could have turned away from the Class C airspace, removing himself from the situation until the frequencies were entered and contact established. Remember, when possible, choose an option that gives more time to determine a course of action. Proper resource management could have negated this airspace violation.

The example also demonstrates the need to have a thorough understanding of all the equipment and systems in the aircraft. As is often the case, the technology available today is seldom used to its maximum capability. It is necessary to become as familiar as possible with this equipment to utilize all resources fully. For example, advanced navigation and autopilot systems are valuable resources. However, if pilots do not fully understand how to use this equipment, or they rely on it so much they become complacent, the equipment can become a detriment to safe flight.

Another internal resource is the Rotorcraft Flight Manual (RFM). *[Figure 13-4]* The RFM:

- Must be on board the aircraft.
- Is indispensable for accurate flight planning.
- Plays a vital role in the resolution of inflight equipment malfunctions.

Other valuable flight deck resources include current aeronautical charts and publications, such as the Airport/Facility Directory (A/FD).

As stated previously, passengers can also be a valuable resource. Passengers can help watch for traffic and may be able to provide information in an irregular situation, especially if they are familiar with flying. Crew briefs to passengers should always include some basic helicopter terminology. For example, explain that in the event you ask them if you are clear to hover to the right, their response should be either "yes, you are clear to hover to the right" or "no you are not clear." A simple yes or no answer can be ambiguous. A strange smell or sound may alert a passenger to a potential problem. As PIC, a pilot should brief passengers before the flight to make sure that they are comfortable voicing any concerns.

Figure 13-4. *Rotorcraft Flying Manual (RFM).*

Instruction that integrates Single-Pilot Resource Management into flight training teaches aspiring pilots how to be more aware of potential risks in flying, how to identify those risks clearly, and how to manage them successfully. The importance of integrating available resources and learning effective SRM skills cannot be overemphasized. Ignoring safety issues can have fatal results.

Risk Management

Risk management is a formalized way of dealing with hazards. It is the logical process of weighing the potential cost of risks from hazards against the possible benefits of allowing those risks from hazards to stand unmitigated. It is a decision-making process designed to identify hazards systematically, assess the degree of risk, and determine the best course of action. Once risks are identified, they must be assessed. The risk assessment determines the degree of risk (negligible, low, medium, or high) and whether the degree of risk is worth the outcome of the planned activity. If the degree of risk is "acceptable," the planned activity may then be undertaken. Once the planned activity is started, consideration must then be given whether to continue. Pilots must have preplanned, viable alternatives available in the event the original flight cannot be accomplished as planned.

Two defining elements of risk management are hazard and risk.

- A hazard is a present condition, event, object, or circumstance that could lead to or contribute to an unplanned or undesired event, such as an accident. It is a source of danger. For example, binding in the antitorque pedals represents a hazard.
- Risk is the future impact of a hazard that is not controlled or eliminated. It is the possibility of loss or injury. The level of risk is measured by the number of people or resources affected (exposure), the extent of possible loss (severity), and the likelihood of loss (probability).

A hazard can be a real or perceived condition, event, or circumstance that a pilot encounters. Learning how to identify hazards, assess the degree of risk they pose, and determine the best course of action is an important element of a safe flight.

Four Risk Elements

During each flight, decisions must be made regarding events that involve interactions between the four risk elements—the PIC, the aircraft, the environment, and the operation. The decision-making process involves an evaluation of each of these risk elements to achieve an accurate perception of the flight situation. *[Figure 13-5]*

One of the most important decisions that a PIC must make is the go/no-go decision. Evaluating each of these risk elements can help a pilot decide whether a flight should be conducted or continued. In the following situations, the four risk elements and how they affect decision-making are evaluated.

Pilot—A pilot must continually make decisions about personal competency, condition of health, mental and emotional state, level of fatigue, and many other variables. A situation to consider: a pilot is called early in the morning to make a long flight. With only a few hours of sleep and congestion that indicates the possible onset of a cold, is that pilot safe to fly?

Aircraft—A pilot frequently bases decisions to fly on

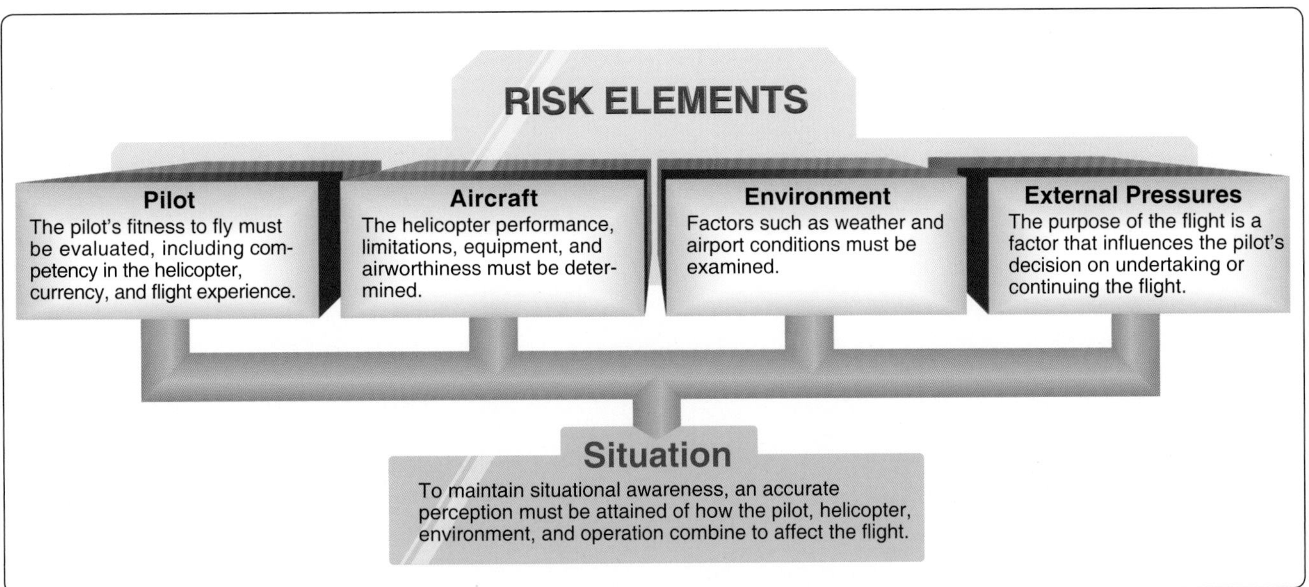

Figure 13-5. *Risk elements to evaluate in decision-making.*

personal evaluations of the aircraft, such as its powerplant, performance, equipment, fuel state, or airworthiness. A situation to consider: en route to an oil rig an hour's flight from shore, having just passed the shoreline, the pilot notices the oil temperature at the high end of the caution range. Should the pilot continue out to sea or return to the nearest suitable heliport/airport?

Environment—This encompasses many elements unrelated to the pilot or aircraft. It can include such factors as weather, ATC, navigational aids (NAVAID), terrain, takeoff and landing areas, and surrounding obstacles. Weather is one element that can change drastically over time and distance. A situation to consider: a pilot is ferrying a helicopter cross-country and encounters unexpected low clouds and rain in an area of rising terrain. Does the pilot try to stay under them and scud run, or turn around, stay in the clear, and obtain current weather information?

External Pressures—The interaction between the pilot, the aircraft, and the environment is greatly influenced by the purpose of each flight operation. A pilot must evaluate the three previous areas to decide on the desirability of undertaking or continuing the flight as planned. It is worth asking why the flight is being made, how critical it is to maintain the schedule, and if the trip is worth the risks. A situation to consider: a pilot is tasked to take some technicians into rugged mountains for a routine survey in marginal weather. Would it be preferable to wait for better conditions to ensure a safe flight? How would the priorities change if a pilot were tasked to search for cross-country skiers who had become lost in deep snow and radioed for help?

Assessing Risk

It is important for a pilot to learn how to assess risk. Before a pilot can begin to assess risk, he or she must first perceive the hazard and attendant risk(s). In aviation, experience, training, and education help a pilot learn how to spot hazards quickly and accurately. During flight training, the instructor should point out the hazards and attendant risks to help the student pilot learn to recognize them.

Once a hazard is identified, determining the probability and severity of an accident (level of risk associated with it) becomes the next step. For example, the hazard of binding in the antitorque pedals poses a risk only if the helicopter is flown. If the binding leads to a loss of directional control, the risk is high that it could cause catastrophic damage to the helicopter and the passengers. The pilot learns to identify hazards and how to deal with them when they are incorporated into the training program.

Every flight has hazards and some level of risk associated with it. It is critical that pilots be able to:

- Differentiate, in advance, between a low-risk flight and a high-risk flight.
- Establish a review process and develop risk mitigation strategies to address flights throughout that range.

Examining NTSB reports and other accident research can help a pilot to assess risk more effectively. For example, the accident rate decreases by nearly 50 percent once a pilot obtains 100 hours and continues to decrease until the 1,000-hour level. The data suggest that for the first 500 hours, pilots flying visual flight rules (VFR) at night should establish higher personal limitations than are required by the regulations and, if applicable, apply instrument flying skills in this environment.

Individuals training to be helicopter pilots should remember that the helicopter accident rate is 30 percent higher than the accident rate for fixed-wing aircraft. While many factors contribute to this, students must recognize the small margin of error that exists for helicopter pilots in making critical decisions. In helicopters, certain emergency actions require immediate action by the pilot. In the event of an engine malfunction, failure to immediately lower the collective results in rotor decay and failed autorotation. Fixed wing pilots may have slightly more time to react and establish a controllable descent. According to the General Aviation (GA) Joint Steering Committee, the leading causes of accidents in GA are CFIT (see p.14-15), weather, runway incursions, pilot decision-making, and loss of control. These causes are referred to as pilot-error, or human factors related, accidents. CFIT, runway incursions, and loss of control type accidents typically occur when the pilot makes a series of bad judgments, which leads to these events. For example, when the pilot has not adequately planned the flight and the pilot subsequently fails to maintain adequate situational awareness to avoid the terrain, a CFIT accident occurs.

While the reasons for individual helicopter incidents vary, it can be argued that it is the helicopter's flight mode and operational complexity that directly contributes to each incident. By nature of its purpose, a helicopter usually flies closer to terrain than does a fixed-wing aircraft. Subsequently, minimal time exists to avoid CFIT, weather related, or loss of control type incidents that require quick and accurate assessments. Fixed-wing aircraft normally fly at higher altitudes and are flown from prepared surface to prepared surface. Helicopters are often operated in smaller, confined area-type environments and require continuous pilot control. Helicopter pilots must be aware of what rotor wash can do when landing to a dusty area or prior to starting where loose debris may come in contact with the rotor blades.

Often, the loss of control occurs when the pilot exceeds design or established operating standards, and the resulting situation exceeds pilot capability to handle it successfully. The FAA generally characterizes these occurrences as resulting from poor judgment. Likewise, most weather-related accidents are not a result of the weather per se, but of a failure of the pilot to avoid a weather phenomenon for which the aircraft is not equipped, or the pilot is not trained to handle. That is, the pilot decides to fly or to continues into conditions beyond pilot capability, an action commonly considered to be demonstrating bad judgment.

It cannot be emphasized enough that the helicopter's unique capabilities come with increased risk. Since most helicopter operations are conducted by a single pilot, the workload is increased greatly. Low-level maneuvering flight (a catch-all category for different types of flying close to terrain or obstacles, such as power line patrol, wildlife control, crop dusting, air taxiing, and maneuvering for landing after an instrument approach), is one of the largest single categories of fatal accidents.

Fatal accidents that occur during approach often happen at night or in instrument flight rules (IFR) conditions. Takeoff/initial climb accidents are frequently due to the pilot's lack of awareness of the effects of density altitude on aircraft performance or other improper takeoff planning that results in loss of control during or shortly after takeoff. One of the most lethal types of GA flying is attempting VFR flight into instrument meteorological conditions (IMC). Accidents involving poor weather decision-making account for about 4 percent of the total accidents but 14 percent of the fatal mishaps. While weather forecast information has been gradually improving, weather should remain a high priority for every pilot assessing risk.

Using the 3P Model to Form Good Safety Habits

As discussed in the Pilot's Handbook of Aeronautical Knowledge, the Perceive, Process, Perform (3P) model helps a pilot assess and manage risk effectively in the real world. *[Figure 13-6]*

To use this model, the pilot will:

- Perceive hazards
- Process level of risk
- Perform risk management

Let's put this to use through a common scenario, involving a common task, such as a confined area approach. As is often the case, the continuous loop consists of several elements; each element must be addressed through the 3P process.

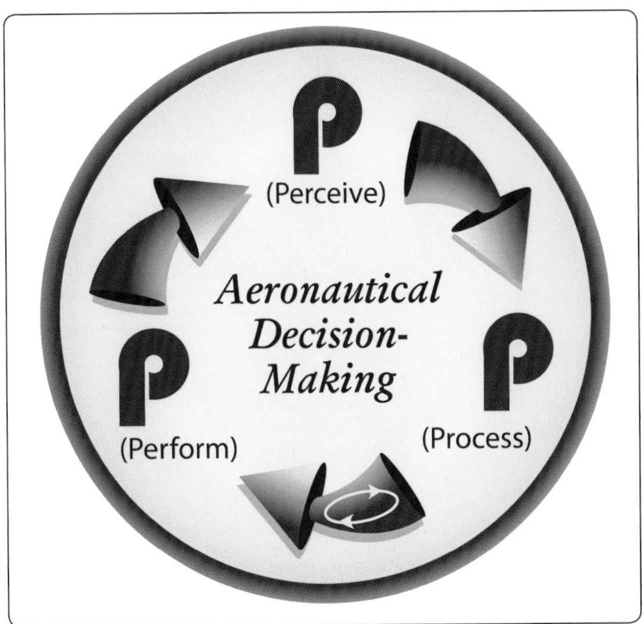

Figure 13-6. *3P Model.*

A utility helicopter pilot receives the task of flying four passengers into a remote area for a hunting expedition. The passengers have picked the location where they would like to be dropped off based on the likelihood of wildlife being in the area. The area has steep, rugged terrain in a series of valleys and canyons leading up to large mountains.

Upon arrival at the location, the pilot locates a somewhat large confined area near the base of one of the mountains. The pilot begins the 3P process by quickly noting (or perceiving) the hazards that affect the approach, landing, and takeoff. Through thorough assessment the pilot takes into consideration:

- *Current aircraft weight/power available,*
- *Required approach angle to clear the trees for landing in the confined area,*
- *Wind direction and velocity,*
- *Limited approach and departure paths (due to constricting terrain),*
- *Escape routes should the approach need to be terminated prior to landing,*
- *Possible hazards, such as wires or structures either around the landing site or inside of the confined area, and*
- *The condition of the terrain at the landing site. Mud, dust, and snow can be extreme hazards if the pilot is not properly trained to land in those particular conditions.*

The pilot reviews the 3P process for each hazard. The pilot has perceived the risk associated for each of the bullets listed above. Now, the pilot assesses the risk level of each and what to do to manage or mitigate the risk.

The aircraft weight/power risk is assessed as low. While performing power checks, the pilot verified adequate out of ground effect (OGE) power exists. The pilot is also aware that, in this scenario, the departure DA (6,500 feet) is greater than the arrival location DA (6,000 feet) and that several hundred pounds of fuel have been burned off en route. Furthermore, once the passengers have disembarked, more power will be available for departure.

The pilot estimates that the highest obstacles along the approach path are 70–80 feet in height. With the size of the confined area, a normal approach angle can be maintained to clear these obstacles, giving this a low risk level. To further mitigate this risk the pilot has selected mental checkpoints along the approach path that will serve as go/no-go points should the pilot feel any assessed parameter is being exceeded.

Wind direction and velocity are assessed as a medium risk because (for this scenario) the direction of the wind is slightly offset from the chosen approach path, creating a 15–20° crosswind with a steady 10-knot wind. The pilot also takes into consideration that, due to the terrain, the wind direction and velocity may change during the approach. The pilot's experience and awareness of the complexity of mountain flow wind provide a management tool for risk reduction.

From an approach and departure standpoint, the risk is assessed to be medium. There is only one viable approach and departure path. Given the size of the confined area and the wind direction, the approach and departure path is deemed acceptable.

The pilot assigns a medium risk level to the selection of an escape route. The pilot is aware of the constricting terrain on either side. Although adequate area exists for maneuvering, the pilot realizes there are physical boundaries and that they can affect the options available should the pilot need to conduct a go-around or abort the approach. Again, the pilot uses mental checkpoints to ensure an early decision is made to conduct a go-around, if needed. The selected go-around or escape route will be in line with the selected approach/departure path and generally into the wind.

As you may have noticed, one identified hazard and its correlating risk management action may have subsequent impact on other factors. This demonstrates the need for continuous assessment and evaluation of the impact of chosen courses of action.

The 3P model offers three good reasons for its use. First, it is fairly simple to remember. Second, it offers a structured, efficient, and systematic way to identify hazards, assess risk, and implement effective risk controls. Third, practicing risk management needs to be as automatic as basic aircraft control. As is true for other flying skills, risk management thinking habits are best developed through repetition and consistent adherence to specific procedures.

Once the pilot completes the 3P decision process and selects a course of action, the process begins anew as the set of circumstances brought about by the selected course of action requires new analysis. Thus, the decision-making process is a continuous loop of perceiving, processing, and performing.

Workload or Task Management

One component of SRM is workload or task management. Research shows that humans have a limited capacity for information. Once information flow exceeds the person's ability to mentally process the information, any additional information becomes unattended or displaces other tasks and information already being processed. Once this situation occurs, only two alternatives exist: shed the unimportant tasks or perform all tasks at a less than optimal level. Like an overloaded electrical circuit, either the consumption must be reduced or a circuit failure is experienced.

Effective workload management ensures essential operations are accomplished by planning and then placing them in a sequence that avoids work overload. As a pilot gains experience, he or she learns to recognize future workload requirements and can prepare for high workload periods during times of low workload.

Reviewing the appropriate chart and setting radio frequencies well in advance of need help reduce workload as a flight nears the airport. In addition, a pilot should listen to Automatic Terminal Information Service (ATIS), Automated Surface Observing System (ASOS), or Automated Weather Observing System (AWOS), if available, and then monitor the tower frequency or Common Traffic Advisory Frequency (CTAF) to get a good idea of what traffic conditions to expect. Checklists should be performed well in advance so there is time to focus on traffic and ATC instructions. These procedures are especially important prior to entering a high-density traffic area, such as Class B airspace.

To manage workload, items should be prioritized. For example, during any situation, and especially in an emergency, a pilot should remember the phrase "aviate, navigate, and communicate." This means that the first thing a pilot should do is make sure the helicopter is under control, then begin flying to an acceptable landing area. Only after the first two items are assured should a pilot try to communicate with anyone.

Another important part of managing workload is recognizing a work overload situation. The first effect of high workload is that a pilot begins to work faster. As workload increases, attention cannot be devoted to several tasks at one time, and a pilot may begin to focus on one item. When a pilot becomes task saturated, there is no awareness of additional inputs from various sources, so decisions may be made on incomplete information, and the possibility of error increases.

A very good example of this is inadvertent IMC. Once entering into bad weather, work overload can occur immediately. Mentally, the pilot must transition from flying outside of the aircraft to flying inside the aircraft. Losing all visual references can cause sensory overload and the ability to think rationally can be lost. Instead of trusting the aircraft's instruments, pilots may try to hang onto the few visual references that they have, and forget all about all other factors surrounding them. Instead of slowing the helicopter down they increase airspeed. This can be caused by an oculogravic illusion. This type of illusion occurs when an aircraft accelerates and decelerates. Inertia from linear accelerations and decelerations cause the otolith organ to sense a nose-high or nose-low attitude. Pilots falsely perceive that the aircraft is in a nose-high attitude. Therefore, pilots increase airspeed. Pilots can also be looking down for visual references and forget about the hazards in front of them. Finally, since the pilots are not looking at the flight instruments, the aircraft is not level. All of this can be avoided by proper training and proper planning. If going inadvertent IMC is your only course of action, pilots must commit to it and fly the helicopter using only the flight instruments and not trying to follow the few visual references they have.

When a work overload situation exists, a pilot needs to:

- Stop,
- Think,
- Slow down, and then
- Prioritize.

It is important for a pilot to understand how to decrease workload by:

- Placing a situation in the proper perspective,
- Remaining calm, and
- Thinking rationally.

These key elements reduce stress and increase the pilot's ability to fly safely. They depend upon the experience, discipline, and training that each safe flight earns. It is important to understand options available to decrease workload. For example, setting a radio frequency may be delegated to another pilot or to a passenger, freeing the pilot to perform higher-priority tasks.

Situational Awareness

In addition to learning to make good aeronautical decisions, and learning to manage risk and flight workload, SA is an important element of ADM. SA is the accurate perception and understanding of all the factors and conditions within the four fundamental risk elements (PAVE) that affect safety before, during, and after the flight. SA involves being aware of what is happening around you, in order to understand how information, events, and your own actions will impact your goals and objectives, both now and in the near future. Lacking SA or having inadequate SA has been identified as one of the primary factors in accidents attributed to human error.

SA in a helicopter can be quickly lost. Understanding the significance and impact of each risk factor independently and cumulatively aid in safe flight operations. It is possible, and all too likely, that we forget flying while at work. Our occupation, or work, may be conducting long line operations, maneuvering around city obstacles to allow a film crew access to news events, spraying crops, ferrying passengers or picking up a patient to be flown to a hospital. In each case we are flying a helicopter. The moment we fail to account for the aircraft systems, the environment, other aircraft, hazards, and ourselves, we lose SA.

To maintain SA, all of the skills involved in SRM are used. For example, an accurate perception of pilot fitness can be achieved through self-assessment and recognition of hazardous attitudes. A clear assessment of the status of navigation equipment can be obtained through workload management, while establishing a productive relationship with ATC can be accomplished by effective resource use.

Obstacles to Maintaining Situational Awareness

What distractions interfere with our focus or train of thought? There are many. A few examples pertinent to aviation, and helicopters specifically, follow.

Fatigue, frequently associated with pilot error, is a threat to aviation safety because it impairs alertness and performance. *[Figure 13-7]* The term is used to describe a range of experiences from sleepy or tired to exhausted. Two major physiological phenomena create fatigue: circadian rhythm disruption and sleep loss.

Many helicopter jobs require scheduling flexibility, frequently affecting the body's circadian rhythm. You

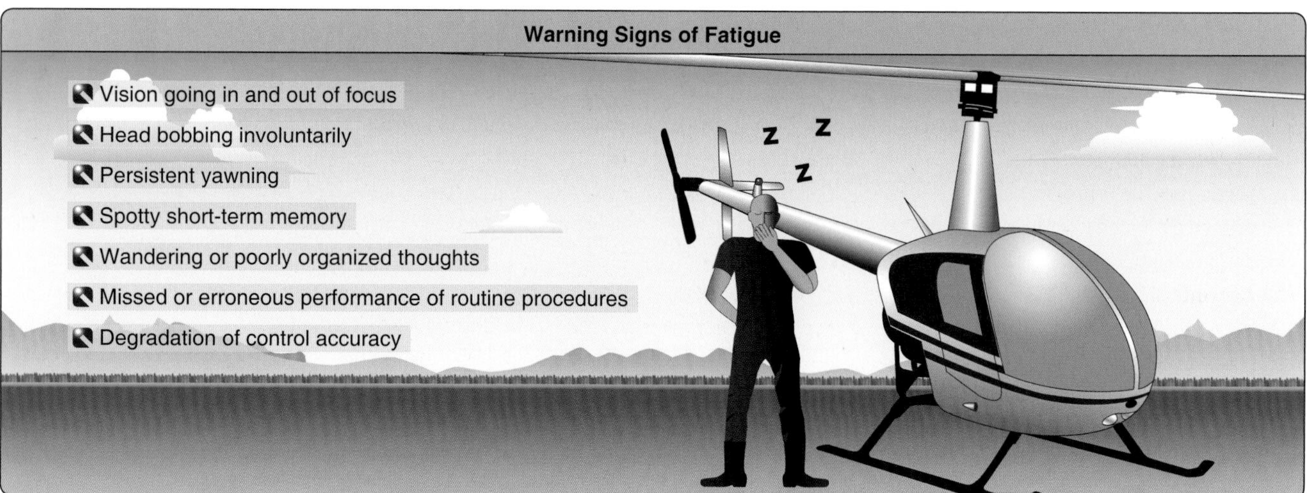

Figure 13-7. Warning signs of *fatigue* according to the FAA Civil Aerospace Medical Institute (CAMI).

may be flying a day flight Monday and then at night on Tuesday. Your awareness of how your body and mind react to this variation in schedule is vital to safety. This disruptive pattern may result in degradation of attention and concentration, impaired coordination, and decreased ability to communicate.

Physical fatigue results from sleep loss, exercise, or physical work. Factors such as stress and prolonged performance of cognitive work result in mental fatigue. Consecutive days of flying the maximum allowable flight time can fatigue a pilot, mentally and physically. It is important to take breaks within the workday, as well as days off when possible. When you find yourself in this situation, take an objective, honest assessment of your state of mind. If necessary, use rest periods to allow rejuvenation of the mind and body. *[Figure 13-8]*

Fatigue also occurs under circumstances in which there is anticipation of flight followed by inactivity. For instance, a pilot is given a task requiring a specific takeoff time. In anticipation of the flight, the pilot's adrenaline kicks in and SA is elevated. After a delay (weather, maintenance, or any other unforeseen delay), the pilot feels a letdown, in effect, becoming fatigued. Then, upon resuming the flight, the pilot does not have that same level of attention.

Complacency presents another obstacle to maintaining SA. Defined as overconfidence from repeated experience with a specific activity, complacency has been implicated as a contributing factor in numerous aviation accidents and incidents. When activities become routine, a pilot may have a tendency to relax and not put as much effort into performance. Like fatigue, complacency reduces a pilot's effectiveness on the flight deck. However, complacency is more difficult to recognize than fatigue, since everything seems to be progressing smoothly.

Since complacency seems to creep into our routine without notice, ask what has changed. The minor changes that go unnoticed can be associated with the four fundamental risks we previously discussed: pilot, aircraft, environment, and external pressures.

As a pilot, am I still using checklists or have I become reliant on memory to complete my checks? Do I check (Notices to Airmen) NOTAMs before every flight or only when I think it is necessary? And the aircraft: did I feel that vibration before or is it new? Was there a log book entry for it? If so,

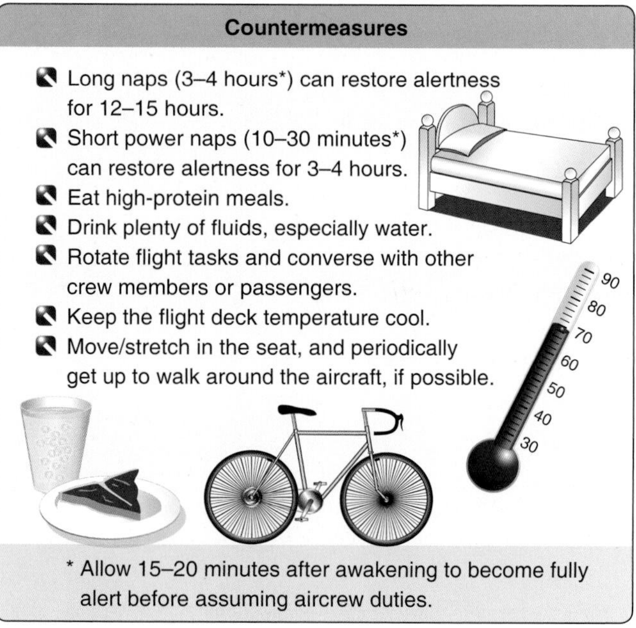

Figure 13-8. Countermeasures to fatigue according to the FAA Civil Aerospace Medical Institute (CAMI).

has it been checked?

Complacent acceptance of common weather patterns can have huge impacts on safety. The forecast was for clearing after the rain shower, but what was the dew-point spread? The winds are greater than forecast. Will this create reduced visibility in dusty, snowy areas or exceed wind limitations?

While conducting crop spraying, a new agent is used. Does that change the weight? Does that change the flight profile and, if so, what new hazards might be encountered? When things are going smoothly, it is time to heighten your awareness and become more attentive to your flight activities.

Advanced avionics have created a high degree of redundancy and dependability in modern aircraft systems, which can promote complacency and inattention. Routine flight operations may lead to a sense of complacency, which can threaten flight safety by reducing SA.

Loss of SA can be caused by a minor distraction that diverts the pilot's attention from monitoring the instruments or scanning outside the aircraft. For example, a gauge that is not reading correctly is a minor problem, but it can cause an accident if the pilot diverts attention to the perceived problem and neglects to control the aircraft properly.

Operational Pitfalls

There are numerous common behavioral traps that can ensnare the unwary pilot. Pilots, particularly those with considerable experience, try to complete a flight as planned, please passengers, and meet schedules. This basic drive to achieve can have an adverse effect on safety and can impose an unrealistic assessment of piloting skills under stressful conditions. These tendencies ultimately may bring about practices that are dangerous and sometimes illegal and may lead to a mishap. Pilots develop awareness and learn to avoid many of these operational pitfalls through effective SRM training. *[Figure 13-9]*

Controlled Flight Into Terrain (CFIT) Awareness

An emergency medical services (EMS) helicopter departed for a night flight to transport an 11-day-old infant patient from one hospital to another. No record was found indicating the pilot obtained a weather briefing before departure. The pilot had a choice of taking either a direct route that crossed a remote area of rugged mountainous terrain with maximum ground elevations of about 9,000 feet or a route that was about 10 minutes longer and followed an interstate highway with maximum ground elevations of about 6,000 feet. Radar data, which show about 4 minutes of the helicopter's flight before coverage was lost due to mountainous terrain, are consistent with the flight following the direct route.

A search was initiated about 4 hours after the helicopter did not arrive at the destination hospital, and the wreckage was located the following morning. Physical evidence observed at the accident site indicated that the helicopter was in level flight at impact and was consistent with CFIT. [Figure 13-10]

CFIT is a type of accident that continues to be a major safety concern, while at the same time difficult to explain because it involves a pilot controlling an airworthy aircraft that is flown into terrain (water or obstacles) with inadequate pilot awareness of the impending disaster.

One constant in CFIT accidents is that outside visibility is limited, or the accident occurs at night and the terrain is not seen easily until just prior to impact. Another commonality among CFIT accidents is lack of SA. This includes not only horizontal awareness, and knowing where the helicopter is over the ground, but also vertical awareness.

Training, planning, and preparation are a pilot's best defenses for avoiding CFIT accidents. For example, take some time before takeoff to become familiar with the proposed flight and the terrain. Avoidance of CFIT begins before the helicopter departs the home location. Proper planning, including applied risk mitigation must occur before the aircraft is even started. Thorough assessment of terrain, visibility, pilot experience and available contingencies must be conducted. If necessary, delay or postpone the flight while on the ground. The decision to abort the flight is much easier to make in the planning room than in the air. In case conditions deteriorate once in flight. Have contingency options available.

While many CFIT accidents and incidents occur during nonprecision approaches and landings, great measures have been taken to improve instrument training, equipment and procedures. For the qualified pilot, instrument flight should not be avoided, but rather, trained as a viable option for safely recovering the aircraft. Like any other training, frequent instrument training builds confidence and reassurance.

Good instrument procedures include studying approach charts before leaving cruise altitude. Key fixes and airport elevation must be noted and associated with terrain and obstacles along the approach path. Pilots should have a good understanding of both approach and departure design criteria to understand fully the obstacle clearance margins built into them. Some pilots have the false belief that ATC provides obstacle clearance while en route off airways. The pilot is ultimately responsible for obstacle clearance.

Operational Pitfalls
Peer Pressure It would be foolish and unsafe for a new pilot to attempt to compete with an older, more experienced pilot. The only safe competition should be completing the most safe flights with no one endangered or hurt and the aircraft returned to service. Efficiency comes with experience and on-the-job training.
Mindset A pilot should be taught to approach every day as something new.
Get-There-Itis This disposition impairs pilot judgment through a fixation on the original goal or destination, combined with a disregard for any alternative course of action.
Duck-Under Syndrome A pilot may be tempted to arrive at an airport by descending below minimums during an approach. There may be a belief that there is a built-in margin of error in every approach procedure, or the pilot may not want to admit that the landing cannot be completed and a missed approach must be initiated.
Scud Running It is difficult for a pilot to estimate the distance from indistinct forms, such as clouds or fog formation.
Continuing Visual Flight Rules (VFR) Into Instrument Conditions Spatial disorientation or collision with ground/obstacles may occur when a pilot continues VFR into instrument conditions. This can be even more dangerous if the pilot is not instrument rated or current.
Getting Behind the Aircraft This pitfall can be caused by allowing events or the situation to control pilot actions. A constant state of surprise at what happens next may be exhibited when the pilot is "getting behind" the aircraft.
Loss of Positional or Situational Awareness In extreme cases of a pilot getting behind the aircraft, a loss of positional or situational awareness may result. The pilot may not know the aircraft's geographical location, or may be unable to recognize deteriorating circumstances.
Operating Without Adequate Fuel Reserves Pilots should use the last of the known fuel to make a safe landing. Bringing fuel to an aircraft is much less inconvenient than picking up the pieces of a crashed helicopter! Pilots should land prior to whenever their watch, fuel gauge, low-fuel warning system, or flight planning indicates fuel burnout. They should always be thinking of unforecast winds, richer-than-planned mixtures, unknown leaks, mis-servicing, and errors in planning. Newer pilots need to be wary of fuselage attitudes in low-fuel situations. Some helicopters can port air into the fuel system in low-fuel states, causing the engines to quit or surge.
Descent Below the Minimum En Route Altitude The duck-under syndrome, as mentioned above, can also occur during the en route portion of an IFR flight.
Flying Outside the Envelope The pilot must understand how to check the charts, understand the results, and fly accordingly.
Neglect of Flight Planning, Preflight Inspections, and Checklists All pilots and operators must understand the complexity of the helicopter, the amazing number of parts, and why there are service times associated with certain parts. Pilots should understand material fatigue and maintenance requirements. Helicopters are unforgiving of disregarded maintenance requirements. Inspections and maintenance are in place for safety: something functioning improperly can be the first link in the error chain to an accident. In some cases, proper maintenance is a necessary condition for insurance converage.

Figure 13-9. *Operational pitfalls.*

Altitude error is another common cause of CFIT. Cases of altitude error involve disorientation with respect to the NAVAID, improper transition on approach, selecting the wrong NAVAID, or just plain lack of horizontal SA. Today's modern aircraft have sophisticated flight directors, autopilots, autothrottles, and flight management systems. These devices make significant contributions to the overall safety of flight, but they are only machines that follow instructions. They do whatever is asked of them, even if it is wrong. When commanded, they unerringly follow instructions—sometimes straight into the ground. The pilot must ensure that both vertical and horizontal modes are correct and engaged. Cross-check autopilots constantly.

Figure 13-10. *Helicopter heading straight for mountain.*

When automated flight equipment is not available, great care must be taken to prepare properly for a night flight. SRM becomes more challenging under the cover of darkness, and caution should be exercised when determining what artificial light source to use inside the aircraft. A light source that is too bright will blind the pilot from seeing outside obstacles or rising terrain. Certain colored lenses bleach out symbols and markings on a map. Conduct this planning on the ground, in a dark room if necessary, before the actual flight.

Pilots must be even more conservative with their decision-making and planning when flying at night. Flying becomes more difficult due to the degradation of our sensory perception and the lack of outside references. Beginning with preflight, looking over the helicopter with a flashlight can cause pilots to miss even the smallest discrepancy that they would easily see during the day. For example, failing to remove one or all of the tie downs and attempting to take off would probably result in a dynamic rollover accident. Whenever possible, preflight inspection should always be conducted during the day or in a lighted hangar. Depth perception is less acute; therefore, hover height should be increased to avoid contact with obstacles and hover speed should be reduced. Weather conditions can be very deceptive and difficult to detect in flight under night conditions. On a low-illumination night, it is easy to fly into clouds without realizing it before it is too late to correct.

Due to the number of recent CFIT night accidents, the NTSB issued a safety alert in 2008 about avoiding night CFIT accidents. That alert included the following information:

- Terrain familiarization is critical to safe visual operations at night. Use sectional charts or other topographic references to ensure the helicopter will safely clear terrain and obstructions all along the route.
- When planning a nighttime VFR flight, follow IFR practices, such as climbing on a known safe course until well above surrounding terrain. Choose a cruising altitude that provides terrain separation similar to IFR flights (2,000 feet above ground level in mountainous areas and 1,000 feet above the ground in other areas). Using this technique, known obstacles, such as towers, will be avoided.
- When receiving radar services, do not depend on ATC to warn of terrain hazards. Although controllers try to warn pilots if they notice a hazardous situation, they may not always recognize that a particular VFR aircraft is dangerously close to terrain.
- When ATC issues a heading with an instruction to "maintain VFR," be aware that the heading may not provide adequate terrain clearance. If any doubt exists about your ability to avoid terrain and obstacles visually, advise ATC immediately and take action to reach a safe altitude.
- For improved night vision, the FAA recommends the use of supplemental oxygen for flights above 5,000 feet.
- Obtain as much information about areas in which you will be flying, and the routes to them, by utilizing hazard maps and satellite imagery.
- Before flying at night to unfamiliar remote areas or areas with hazardous terrain, try to arrange a day flight for familiarization.
- If a pilot flies at night, especially in remote or unlit areas, consider whether a global positioning system (GPS)-based terrain awareness unit would improve the safety of the flight.

Of particular note in the 2008 safety alert is a comment regarding oxygen use above 5,000 feet. Most helicopters are neither required nor equipped for supplemental oxygen use at this altitude. Due to the physiological effect on night vision of reduced available oxygen at higher elevations, care should be taken to exercise light discipline. Interior lighting should be lowered to the lowest possible levels but must allow adequate illumination of necessary systems and instruments. This, in turn, allows greater recognition of outside obstacles and terrain features.

Limited outside visibility is one constant in CFIT accidents. In the accident cited at the beginning of this section, it appears the pilot failed to obtain a weather briefing. If the pilot had obtained one, he would probably have learned of the cloud cover and light precipitation present along his planned route of flight. The limited outside visibility probably caused the CFIT accident, since no evidence was found of any pre-impact mechanical discrepancies with the helicopter's airframe or systems that would have prevented normal operation.

Automation Management

Automation management is the control and navigation of an aircraft by means of the automated systems installed in the aircraft. One of the most important concepts of automation management is simply knowing when to use it and when not to.

Ideally, a pilot first learns to perform practical test standard (PTS) maneuvers and procedures in the aircraft manually, or hand flying. After successfully demonstrating proficiency in the basic maneuvers, the pilot is then introduced to the available automation and/or the autopilot. Obviously, in some aircraft, not all automated systems may be disengaged for basic flight. The purpose of basic flight without automation is to ensure the pilot can hand fly the maneuver when necessary.

Advanced avionics offer multiple levels of automation, from strictly manual flight to highly automated flight. No one level of automation is appropriate for all flight situations, but to avoid potentially dangerous distractions when flying with advanced avionics, the pilot must know how to manage the course indicator, the navigation source, and the autopilot. It is important for a pilot to know the peculiarities of the particular automated system in use. This ensures the pilot knows what to expect, how to monitor for proper operation, and promptly take appropriate action if the system does not perform as expected.

At the most basic level, managing the autopilot means knowing at all times which modes are engaged and which modes are armed to engage. The pilot needs to verify that armed functions (e.g., navigation tracking or altitude capture) engage at the appropriate time. Automation management is a good place to practice the callout technique, especially after arming the system to make a change in course or altitude. Callouts are verbalizations of particular flight guidance automation mode changes. In an attempt to reduce the risk for mode confusion some operators have required flight crews to callout all flight guidance automation mode changes as a means of forcing pilots to monitor the Flight Mode Annunciator (FMA).

Chapter Summary

This chapter focused on aeronautical decision-making, which includes SRM training, risk management, workload or task management, SA, CFIT awareness, and automation management. Factors affecting a helicopter pilot's ability to make safe aeronautical decisions were also discussed. The importance of learning how to be aware of potential risks in flying, how to clearly identify those risks, and how to manage them successfully were also explored.

Glossary

Absolute altitude. The actual distance an object is above the ground.

Advancing blade. The blade moving in the same direction as the helicopter. In helicopters that have counterclockwise main rotor blade rotation as viewed from above, the advancing blade is in the right half of the rotor disk area during forward movement.

Agonic Line. An isogonic line along which there is no magnetic variation.

Air density. The density of the air in terms of mass per unit volume. Dense air has more molecules per unit volume than less dense air. The density of air decreases with altitude above the surface of the earth and with increasing temperature.

Aircraft pitch. The movement of the aircraft about its lateral, or pitch, axis. Movement of the cyclic forward or aft causes the nose of the helicopter to pitch up or down.

Aircraft roll. The movement of the aircraft about its longitudinal axis. Movement of the cyclic right or left causes the helicopter to tilt in that direction.

Airfoil. Any surface designed to obtain a useful reaction of lift, or negative lift, as it moves through the air.

Airworthiness Directive. When an unsafe condition exists with an aircraft, the FAA issues an Airworthiness Directive to notify concerned parties of the condition and to describe the appropriate corrective action.

Altimeter. An instrument that indicates flight altitude by sensing pressure changes and displaying altitude in feet or meters.

Angle of attack. The angle between the airfoil's chord line and the relative wind.

Antitorque pedal. The pedal used to control the pitch of the tail rotor or air diffuser in a NOTAR® system.

Antitorque rotor. See tail rotor.

Articulated rotor. A rotor system in which each of the blades is connected to the rotor hub in such a way that it is free to change its pitch angle, and move up and down and fore and aft in its plane of rotation.

Autopilot. Those units and components that furnish a means of automatically controlling the aircraft.

Autorotation. The condition of flight during which the main rotor is driven only by aerodynamic forces with no power from the engine.

Axis of rotation. The imaginary line about which the rotor rotates. It is represented by a line drawn through the center of, and perpendicular to, the tip-path plane.

Basic empty weight. The weight of the standard helicopter, operational equipment, unusable fuel, and full operating fluids, including full engine oil.

Blade coning. An upward sweep of rotor blades as a result of lift and centrifugal force.

Blade damper. A device attached to the drag hinge to restrain the fore and aft movement of the rotor blade.

Blade feather or feathering. The rotation of the blade around the spanwise (pitch change) axis.

Blade flap. The ability of the rotor blade to move in a vertical direction. Blades may flap independently or in unison.

Blade grip. The part of the hub assembly to which the rotor blades are attached, sometimes referred to as blade forks.

Blade lead or lag. The fore and aft movement of the blade in the plane of rotation. It is sometimes called "hunting" or "dragging."

Blade loading. The load imposed on rotor blades, determined by dividing the total weight of the helicopter by the combined area of all the rotor blades.

Blade root. The part of the blade that attaches to the blade grip.

Blade span. The length of a blade from its tip to its root.

Blade stall. The condition of the rotor blade when it is operating at an angle of attack greater than the maximum angle of lift.

Blade tip. The furthermost part of the blade from the hub of the rotor.

Blade track. The relationship of the blade tips in the plane of rotation. Blades that are in track will move through the same plane of rotation.

Blade tracking. The mechanical procedure used to bring the blades of the rotor into a satisfactory relationship with each other under dynamic conditions so that all blades rotate on a common plane.

Blade twist. The variation in the angle of incidence of a blade between the root and the tip.

Blowback. The tendency of the rotor disk to tilt aft in transition to forward flight as a result of unequal airflow.

Calibrated airspeed (CAS). Indicated airspeed of an aircraft, corrected for installation and instrumentation errors.

Center of gravity. The theoretical point where the entire weight of the helicopter is considered to be concentrated.

Center of pressure. The point where the resultant of all the aerodynamic forces acting on an airfoil intersects the chord.

Centrifugal force. The apparent force that an object moving along a circular path exerts on the body constraining the object and that acts outwardly away from the center of rotation.

Centripetal force. The force that attracts a body toward its axis of rotation. It is opposite centrifugal force.

Chip detector. A warning device that alerts you to any abnormal wear in a transmission or engine. It consists of a magnetic plug located within the transmission. The magnet attracts any metal particles that have come loose from the bearings or other transmission parts. Most chip detectors have warning lights located on the instrument panel that illuminate when metal particles are picked up.

Chord. An imaginary straight line between the leading and trailing edges of an airfoil section.

Chordwise axis. For semirigid rotors, a term used to describe the flapping or teetering axis of the rotor.

Coaxial rotor. A rotor system utilizing two rotors turning in opposite directions on the same centerline. This system is used to eliminated the need for a tail rotor.

Collective pitch control. The control for changing the pitch of all the rotor blades in the main rotor system equally and simultaneously and, consequently, the amount of lift or thrust being generated.

Coning. See blade coning.

Coriolis effect. The tendency of a rotor blade to increase or decrease its velocity in its plane of rotation when the center of mass moves closer to or farther from the axis of rotation.

Cyclic feathering. The mechanical change of the angle of incidence, or pitch, of individual rotor blades, independent of other blades in the system.

Cyclic pitch control. The control for changing the pitch of each rotor blade individually as it rotates through one cycle to govern the tilt of the rotor disk and, consequently, the direction and velocity of horizontal movement.

Degraded Visual Environment (DVE). Any flight environment of reduced visibility in which situational awareness of the aircrew or control of the aircraft may be severely diminished, completely lost, or may not be maintained as comprehensively as they are during flight operations within clear or undiminished visibility. DVE conditions are further categorized into eleven different types: smoke, smog, clouds, rain, fog, snow, whiteout, night, flat light, sand, and brownout.

Delta hinge. A flapping hinge with an axis skewed so that the flapping motion introduces a component of feathering that would result in a restoring force in the flap-wise direction.

Density altitude. Pressure altitude corrected for nonstandard temperature variations.

Deviation. A compass error caused by magnetic disturbances from the electrical and metal components in the aircraft. The correction for this error is displayed on a compass correction card placed near the magnetic compass of the aircraft.

Direct control. The ability to maneuver a helicopter by tilting the rotor disk and changing the pitch of the rotor blades.

Direct shaft turbine. A single-shaft turbine engine in which the compressor and power section are mounted on a common driveshaft.

Disk area. The area swept by the blades of the rotor. It is a circle with its center at the hub and has a radius of one blade length.

Disk loading. The total helicopter weight divided by the rotor disk area.

Dissymmetry of lift. The unequal lift across the rotor disk resulting from the difference in the velocity of air over the advancing blade half and the velocity of air over the retreating blade half of the rotor disk area.

Drag. An aerodynamic force on a body acting parallel and opposite to relative wind.

Dual rotor. A rotor system utilizing two main rotors.

Dynamic rollover. The tendency of a helicopter to continue rolling when the critical angle is exceeded, if one gear is on the ground, and the helicopter is pivoting around that point.

Emergency Position Indicator Radio Beacon (ERIPB). A device used to alert search and rescue services in the event of an emergency by transmitting a coded message on the 406 MHz distress frequency, which is relayed by the Cospas-Sarsat global satellite system.

Feathering. The action that changes the pitch angle of the rotor blades by rotating them around their feathering (spanwise) axis.

Feathering axis. The axis about which the pitch angle of a rotor blade is varied. Sometimes referred to as the spanwise axis.

Feedback. The transmittal of forces, which are initiated by aerodynamic action on rotor blades, to the cockpit controls.

Flapping. The vertical movement of a blade about a flapping hinge.

Flapping hinge. The hinge that permits the rotor blade to flap and thus balance the lift generated by the advancing and retreating blades.

Flare. A maneuver accomplished prior to landing to slow a helicopter.

Free turbine. A turboshaft engine with no physical connection between the compressor and power output shaft.

Freewheeling unit. A component of the transmission or power train that automatically disconnects the main rotor from the engine when the engine stops or slows below the equivalent rotor rpm.

Fully articulated rotor system. See articulated rotor system.

Gravity. See weight.

Gross weight. The sum of the basic empty weight and useful load.

Ground effect. A usually beneficial influence on helicopter performance that occurs while flying close to the ground. It results from a reduction in upwash, downwash, and bladetip vortices, which provide a corresponding decrease in induced drag.

Ground resonance. Selfexcited vibration occurring whenever the frequency of oscillation of the blades about the lead-lag axis of an articulated rotor becomes the same as the natural frequency of the fuselage.

Gyroscopic procession. An inherent quality of rotating bodies, which causes an applied force to be manifested 90° in the direction of rotation from the point where the force is applied.

Human factors. The study of how people interact with their environment. In the case of general aviation, it is the study of how pilot performance is influenced by such issues as the design of cockpits, the function of the organs of the body, the effects of emotions, and the interaction and communication with other participants in the aviation community, such as other crew members and air traffic control personnel.

Hunting. Movement of a blade with respect to the other blades in the plane of rotation, sometimes called leading or lagging.

In ground effect (IGE) hover. Hovering close to the surface (usually less than one rotor diameter distance above the surface) under the influence of ground effect.

Induced drag. That part of the total drag that is created by the production of lift.

Induced flow. The component of air flowing vertically through the rotor system resulting from the production of lift.

Inertia. The property of matter by which it will remain at rest or in a state of uniform motion in the same direction unless acted upon by some external force.

Isogonic line. Lines on charts that connect points of equal magnetic variation.

Knot. A unit of speed equal to one nautical mile per hour.

L_{DMAX}. The maximum ratio between total lift (L) and total drag (D). This point provides the best glide speed. Any deviation from the best glide speed increases drag and reduces the distance you can glide.

Lateral vibration. A vibration in which the movement is in a lateral direction, such as imbalance of the main rotor.

Lead and lag. The fore (lead) and aft (lag) movement of the rotor blade in the plane of rotation.

Licensed empty weight. Basic empty weight not including full engine oil, just undrainable oil.

Lift. One of the four main forces acting on a helicopter. It acts perpendicular to the relative wind.

Load factor. The ratio of a specified load weight to the total weight of the aircraft.

Married needles. A term used when two hands of an instrument are superimposed over each other, as on the engine/rotor tachometer.

Mast. The component that supports the main rotor.

Mast bumping. Action of the rotor head striking the mast, occurring on underslung rotors only.

Navigational aid (NAVAID). Any visual or electronic device, airborne or on the surface, that provides point-to-point guidance information, or position data, to aircraft in flight.

Night. The time between the end of evening civil twilight and the beginning of morning civil twilight, as published in the American Air Almanac.

Normally aspirated engine. An engine that does not compensate for decreases in atmospheric pressure through turbocharging or other means.

One-to-one vibration. A low frequency vibration having one beat per revolution of the rotor. This vibration can be either lateral, vertical, or horizontal.

Out of ground effect (OGE) hover. Hovering a distance greater than one disk diameter above the surface. Because induced drag is greater while hovering out of ground effect, it takes more power to achieve a hover out of ground effect.

Parasite drag. The part of total drag created by the form or shape of helicopter parts.

Payload. The term used for the combined weight of passengers, baggage, and cargo.

Pendular action. The lateral or longitudinal oscillation of the fuselage due to its suspension from the rotor system.

Pitch angle. The angle between the chord line of the rotor blade and the reference plane of the main rotor hub or the rotor plane of rotation.

Pressure altitude. The height above the standard pressure level of 29.92 "Hg. It is obtained by setting 29.92 in the barometric pressure window and reading the altimeter.

Profile drag. Drag incurred from frictional or parasitic resistance of the blades passing through the air. It does not change significantly with the angle of attack of the airfoil section, but it increases moderately as airspeed increases.

Resultant relative wind. Airflow from rotation that is modified by induced flow.

Retreating blade. Any blade, located in a semicircular part of the rotor disk, in which the blade direction is opposite to the direction of flight.

Retreating blade stall. A stall that begins at or near the tip of a blade in a helicopter because of the high angles of attack required to compensate for dissymmetry of lift.

Rigid rotor. A rotor system permitting blades to feather, but not flap or hunt.

Rotational velocity. The component of relative wind produced by the rotation of the rotor blades.

Rotor. A complete system of rotating airfoils creating lift for a helicopter.

Rotor brake. A device used to stop the rotor blades during shutdown.

Rotor disk area. See disk area.

Rotor force. The force produced by the rotor, comprised of rotor lift and rotor drag.

Semirigid rotor. A rotor system in which the blades are fixed to the hub, but are free to flap and feather.

Shaft turbine. A turbine engine used to drive an output shaft, commonly used in helicopters.

Skid. A flight condition in which the rate of turn is too great for the angle of bank.

Skid shoes. Plates attached to the bottom of skid landing gear, protecting the skid.

Slip. A flight condition in which the rate of turn is too slow for the angle of bank.

Solidity ratio. The ratio of the total rotor blade area to total rotor disk area.

Span. The dimension of a rotor blade or airfoil from root to tip.

Split needles. A term used to describe the position of the two needles on the engine/rotor tachometer when the two needles are not superimposed.

Standard atmosphere. A hypothetical atmosphere based on averages in which the surface temperature is 59 °F (15 °C), the surface pressure is 29.92 "Hg (1013.2 Mb) at sea level, and the temperature lapse rate is approximately 3.5 °F (2 °C) per 1,000 feet.

Static stop. A device used to limit the blade flap, or rotor flap, at low rpm or when the rotor is stopped.

Steady-state flight. The type of flight experienced when a helicopter is in straight-and-level, unaccelerated flight, and all forces are in balance.

Symmetrical airfoil. An airfoil having the same shape on the top and bottom.

Tail rotor. A rotor turning in a plane perpendicular to that of the main rotor and parallel to the longitudinal axis of the fuselage. It is used to control the torque of the main rotor and to provide movement about the yaw axis of the helicopter.

Teetering hinge. A hinge that permits the rotor blades of a semirigid rotor system to flap as a unit.

Thrust. The force developed by the rotor blades acting parallel to the relative wind and opposing the forces of drag and weight.

Tip-path plane. The imaginary circular plane outlined by the rotor blade tips as they make a cycle of rotation.

Torque. In helicopters with a single, main rotor system, the tendency of the helicopter to turn in the opposite direction of the main rotor rotation.

Trailing edge. The rearmost edge of an airfoil.

Translating tendency. The tendency of the single-rotor helicopter to move laterally during hovering flight. Also called tail rotor drift.

Translational lift. The additional lift obtained when entering forward flight, due to the increased efficiency of the rotor system.

Transverse-flow effect. The condition of increased drag and decreased lift in the aft portion of the rotor disk caused by the air having a greater induced velocity and angle in the aft portion of the disk.

True altitude. The actual height of an object above mean sea level.

Turboshaft engine. A turbine engine transmitting power through a shaft as would be found in a turbine helicopter.

Twist grip. The power control on the end of the collective control.

Underslung. A rotor hub that rotates below the top of the mast, as on semirigid rotor systems.

Unloaded rotor. The state of a rotor when rotor force has been removed, or when the rotor is operating under a low or negative G condition.

Useful load. The difference between the gross weight and the basic empty weight. It includes the flight crew, usable fuel, drainable oil, if applicable, and payload.

Variation. The angular difference between true north and magnetic north; indicated on charts by isogonic lines.

Vertical vibration. A vibration in which the movement is up and down, or vertical, as in an out-of-track condition.

Vortex ring state. A transient condition of downward flight (descending through air after just previously being accelerated downward by the rotor) during which an appreciable portion of the main rotor system is being forced to operate at angles of attack above maximum. Blade stall starts near the hub and progresses outward as the rate of descent increases.

Weight. One of the four main forces acting on a helicopter. Equivalent to the actual weight of the helicopter. It acts downward toward the center of the earth.

Yaw. The movement of a helicopter about its vertical axis.

Index

F

G

H

I

K

L

M

N

O

P

R

S

T

U

V

W